Practical Chlorine Dioxide

Volume II

Applications

Greg D. Simpson, PhD

Greg D. Simpson & Associates
635 Westport Parkway
Suite 312
Grapevine, TX 76051

Greg D. Simpson & Associates
635 Westport Parkway
Suite 312
Grapevine, TX 76051

817-251-7109

ISBN: 0-9771985-1-0

This book is dedicated to

The Three Bills

Dr. Bill Averett

Dr. Bill Clements

Dr. Bill McCord

Friends and Encouragers

This book is dedicated to

[illegible]

[illegible]

[illegible]

Dr. Bill McCord

[illegible]

Table of Contents

Preface

Chlorine dioxide is a chemical whose time has come. While it has been used historically as a very dilute aqueous solution because of its superior disinfection properties, it is now seeing increased usage as a gas-phase disinfectant, sanitizer, and sterilizer. Its exceptional properties as a gas-phase disinfectant have become recognized through the cleanup of the Hart Senate Office Building in Washington, DC, as well as other buildings contaminated by anthrax spores. More recently, it has been in the news due to its use in cleaning up mold-contaminated buildings in New Orleans after Hurricane Katrina. Its gas-phase disinfection properties are also being increasingly studied for microbiological control in food processing, in medical applications, and in sterilizing equipment used in the medical and dental fields.

Although it has been around for a long time, very few people have an in-depth understanding of its properties and proper use, particularly from an application viewpoint.

To address this need in the marketplace, Volume II deals with a number of the more conventional applications. Future volumes will include other applications.

Volume I laid the foundation for those companies who desire to begin working with this versatile chemical. It dealt with the chemical properties, reaction chemistry, safety issues, environmental issues, analytical issues, generation chemistry and equipment, and materials of compatibility for those wishing to build equipment to make or use ClO_2.

It is my goal, through these books, to both promote its use in applications for which it is the best choice and to ensure that those who use this chemical will use it safely.

Acknowledgements

There are many who contributed to this work. For the review of several of the chapters, I thank Dr. C. Cayce Warf, Dr. Jennifer Miller, Dr. Anthony Pitochelli, Dave Griffith, and Roger Etherington.

For the patience to tolerate my strange hours and erratic sleep habits during the writing of this text, I thank my wife Marilyn.

For friendship and encouragement, I thank my friends Keith Hall, Howard Bourgeois, George Gooch, Dr. Tracy Broussard, and especially Dr. Bill Clements, who passed away shortly before the completion of this book.

For pointing out a glaring inconistency in Volume I, which was subsequently corrected, I thank Jim Rauh.

For teaching me the importance of good technical writing, I thank Dr. Gene O. Carlisle, Professor of Chemistry, Mathematics and Physics at West Texas A&M University, Dr. R. Linn Belford, Professor of Chemistry at The University of Illinois, and especially John Meakim, President of Fresh Creek Technologies, Inc.

For her valuable time *and willingness* to undertake the daunting task of editing this volume, I thank Patty Dumler, President of ChemCal.

Special Acknowledgement

The fact that ClO_2 has survived at all as a potential disinfectant is remarkable. It is an unfortunate fact that the political pressures brought to bear by vocal individuals or groups who neither understand nor care about the underlying science, have much more to do with regulations that are put forth than they should. As a result, many of the environmental regulations are not based on either sound logic or reason. A number of examples follow:

Example 1. As one example, the regulatory level for copper in water discharged by a once-through power plant along the Mississippi River was at one time set so low that modern analytical technology could not even measure it! In fact, the copper level that this power plant could discharge was two or three orders of magnitude lower than the naturally-occurring copper concentration in water entering the power plant! It is this kind of nonsense that American industry has to tolerate and that is to a great extent responsible for the mass exodus of American industry to other countries. The textile industry has essentially left the US. The last petroleum refinery to be built in the US was built decades ago. Absurd environmental restrictions have been largely to blame. While there is a need to protect the environment, regulations should be based on sound science and not on some agenda of special interest groups.

Example 2. One example of the conflict between special interest groups and scientists surrounds the use of ClO_2 in pulp bleaching. Some environmental groups promoted the elimination of chlorine dioxide from pulp bleaching. This "Totally Chlorine-Free" position, taken by well-meaning but poorly informed groups or individuals, claimed that if a molecule even contained chlorine, it was harmful! (As if one specific element in this universe were evil.) Since chlorine was bad by definition, and since these people had little knowledge of the reaction chemistry of chlorine dioxide vs. that of chlorine, there was a push to eliminate chlorine dioxide from bleaching pulp to make paper, because chlorine is a part of the molecule (Do these people not put salt on their food? Gasp! It contains chlorine).

The second group, the "Elemental Chlorine-Free" group, the group which included many scientists, took a typically scientific approach. In effect, they said, "Let's look at the data." When chlorine was used to bleach pulp, dioxins and furans and other chlorinated organics were formed. These compounds have not been shown to be friendly to the environment. But when chlorine dioxide was used to bleach pulp, several things happened. The paper was whiter, stronger, and the dioxins and furans were in many cases reduced to levels below that which could be detected by modern analytical methods. The amount of chlorinated organics formed was orders of magnitude less than that formed when chlorine was used; the nature of these chlorinated organics also differed from that which was formed when chlorine was used. Scientific testing showed that these organics were generally hydrophilic, which means these organics would not accumulate in the fatty tissue of aquatic organisms, i.e., the organics would not bio-accumulate. Fortunately the hard science appears to be winning this argument.

Example 3. When ClO_2 was used to mitigate infestation by zebra mussels, the chlorite levels allowed in water exiting a once-through plant which has been treated with low levels of ClO_2 were about 2 orders of magnitude lower than the allowable chlorite levels in potable water. This would make it illegal to use potable water to cool one of these plants. Make sense?

Example 4. In another area, ClO_2 has been proven to be *THE* most effective treatment to date for ballast water. Control of the transport of non-native organisms from one part of the globe to another is thought by many to be necessary, as cholera outbreaks and other problems have been linked to this transport. Still, treating the ballast water of a ship at sea (seawater is composed of up to 3.5% NaCl – that's 35,000 mg/L NaCl) with a few mg/L of ClO_2 (~ < 5 mg/L) has been shown in both laboratory and field tests to control organisms far more effectively than by any other method. However, when the ship ballast water is discharged in the Great Lakes to take on freight, regulators looked at the sub-mg/L concentration of chlorite ion remaining in the ballast water and *totally ignored* the impact that discharging a million gallons of brine would have on the water and on aquatic organisms (the LC_{50} for Cl^- for many fresh water organisms is <10,000 mg/L). It seems reasonable that the big picture should be explored, as it is impossible to separate the toxic effect of < 2 mg/L of chlorite ion, from the toxic effect of 35,000 mg/L of NaCl towards freshwater organisms. Based on these criteria, if one were to fill the ballast tanks of a freighter with their garden hose, if their potable water were disinfected with ClO_2, it would be illegal to discharge this water into the Great Lakes! As of this writing, the limit for chlorite has been set at 13 µg/L (parts per billion). That's 0.013 mg/L (parts per million). Make sense?

Final Comments. Finally, as this is being written, the price for gasoline has broken the $3.00/gallon barrier. Everyone in the news media, politicians included, are pointing their finger at the President and the Evil Oil Companies. This is not something that has happened in the last 4 or 8 years. No one seems to reflect on the impact absurd environmental regulations have had on this issue. No major refinery has been built in the United States for decades. Absurd environmental restrictions are to blame. "No drilling off of California coasts," say California legislators. "No drilling in the great Alaskan wilderness," say environmentalists. Those holding these positions have NO right to complain about the price of gasoline (They should be allowed to freeze in the dark). While the US should have been pursuing alternative fuel sources (and several of the large oil companies have spent large sums of money in this area), the political correctness of pursuing agendas to keep getting re-elected have kept many politicians from focusing on this area.

Many such examples exist, but this is the subject for another book and as I write this I am sure my blood pressure is elevated to a dangerous level, so I will get off this soap box. Still, it is frustrating for a scientist, any scientist, even those who work for regulatory agencies, to watch as "concerned" bureaucrats are swayed by the political correctness of environmental politics.

It was Richard P Feynman, Nobel Laureate, who said,

> "Reality must take precedence over public relations....Mother Nature cannot be fooled."

In other words, wanting the world to be a certain way and the world being that way are two different things.

May I take a moment here to say that I have the utmost respect for the scientists in the regulatory agencies? Just to pursue a career as a scientist requires a unique perspective in finding out the truth about the world, the way it really is. So a great many of the scientists are dedicated individuals. It is unfortunate that many of the decision-making politicians and bureaucrats in a given regulatory agency weigh politics or political correctness over the relative importance of scientific data.

Despite the pressures from public relations, there have been some dedicated scientists who have worked tirelessly to try and bring some logic and reason to environmental regulations, at least in the potable water industry, based on sound scientific principles and studies.

One such individual is Dr. Donald Gates. Don is a scientist who is highly respected equally by industrial peers and regulatory scientists because of his intellectual honesty and integrity. He has authored numerous technical papers dealing with all aspects of chlorine dioxide, and has authored The Chlorine Dioxide Handbook, an excellent monograph dealing with issues relating to ClO_2 for potable water. While unappreciated by his previous employer, those who have been involved in the chlorine dioxide area for some time are aware of the efforts he and a few others have made to help make sure that regulations for chlorine dioxide in the potable water area in particular are based on sound science.

Several years ago, regulations for oxidant in potable water treated with chlorine dioxide were 1.0 total oxidant, which included the combination of chlorine dioxide, chlorite, and chlorate. Knowing that the study on which these limits were based was flawed, Don and others began a seven-year effort to bring a more scientific perspective to this issue.

After pointing out the inadequacies of the original study, another study was commissioned. The new study was set up properly, and the data generated. Once this new data was gathered, the data actually supported an ***increase*** in limits, with new regulatory limits for chlorine dioxide being set to 0.8 mg/L and chlorite ion being set to 1.0 mg/L. This in itself was remarkable. During a time when environmental activism and political correctness was resulting in decreasing limits for any chemical species generally, there was actually an increase in the limit for ClO_2 and chlorite ion.

This did not happen in a vacuum, but was the result of highly competent and well-respected scientists, traveling to Washington on a weekly or monthly basis to meet with and discuss the research, presenting the data in a way that regulators, who readily acknowledged that they could not hope to maintain an accurate perspective of the details involved due to their many activities and time constraints, could understand.

This special acknowledgement is to those scientists involved, with special acknowledgement to my friend and former coworker, Dr. Don Gates, who has never gotten the recognition or appreciation he deserves.

Thanks, Don.

Special Acknowledgement

List of Tables

List of Tables

List of Figures

List of Figures

Treatment of Cooling Systems with ClO_2

A. Introduction

The use of ClO_2 for disinfection of cooling towers has undergone phenomenal growth, particularly in the smaller, comfort cooling market and in hospitals, due to the well-established efficacy of ClO_2 for control of both biofilm and Legionella. The limiting factor to date has been the lack of a suitable cost-effective generator for small systems.

Chapter I provides an overview of available scientific literature, along with practical sections discussing such things as frequency and duration of treatment, selection of optimum dosage, and injection point selection.

The treatment of cooling towers with ClO_2 was pioneered by Olin Water Services, a subsidiary of Olin corporation [Sussman and Ward 1977; Ward et al. 1977; Ward 1977; Ward et al. 1978; Sussman 1978; Freymark and Rauh 1978; Freund et al. 1984]. A number of studies have been reported for use of ClO_2 to treat large industrial cooling towers with various process contaminants, including methanol [Freund et al. 1984; Pickrell et al. 1999], grain alcohol [Freund et al. 1984; Pacheco et al. 1989; Downing and Simpson 2004], ammonia or amines [Laxton et al. 2001; Pickrell et al. 1999; Sussman and Ward 1977; Ward et al. 1977; Ward et al. 1978], glycol [Pacheco et al. 1989], phenol [McGuire and Dishinger 1984], and light hydrocarbons [Simpson et al. 1993b].

Cooling towers in various types of facilities have been treated successfully, including those in food plants [Rio Linda 1984; Sussman and Ward 1977], once-through systems [Kemp and Okimoto 1982; Pacheco et al. 1989; Ambrogi 1996; Belluati et al. 1996; Masullo et al. 1997; Pickrell et al. 1999], refineries [McGuire and Dishinger 1984; Sussman and Ward 1977; Simpson et al. 1993b], steel mills [Simpson 1994a; Belluati et al. 1996], electronic printed circuit plants [Simpson and Miller 2000; Sims 2000], power plant applications [Geraci et al. 1993; Belluati et al. 1996; Dunn 1999; Pacheco et al. 1989; Sussman and Ward 1977; Pickrell et al. 1999], power plants with once-through cooling using seawater [Kemp and Okimoto 1982; Geraci et al. 1993; Ambrogi 1996; Belluati et al. 1996], and small, comfort cooling applications [Simpson 1994b; Simpson 1999a; Sims 2000; Simpson 2001] such as hospitals [Bova et al. 2004].

Some reports were directed specifically toward certain areas of interest, such as high efficiency film fill [Simpson 1993b; Nickels et al. 2000; Laxton et al. 2001], compatibility with azoles [Coschigano and Go 1993; Simpson 1999c] or phosphonates [Vaska and Go 1992; Simpson 1993b], the nature of bacteria and the quantity of bacteria associated with drift [Adams et al. 1978a; Adams et al. 1978b; Adams et al. 1980], and the control of zebra mussels [Rybarik et al. 1995; Brooks et al. 1996; Smolik et al. 1995a; Rusznak et al. 1994; Smolik et al. 1995b; Downing et al. 1995a; Downing et al. 1995b; Hale et al. 1996; Tsou et al. 1995], Asian clams [Goss et al. 1977; Cameron et al. 1987; Cameron et al. 1989], other macrofouling organisms [Kemp and Okimoto 1982; Simpson et al. 2005], and Legionella [Dunn 1999; Turvey 1999]. Others have suggested use of ClO_2 as a cleanup method [Silva 2001] and as a method for reducing corrosion rates [Pacheco et al. 1987; Marckini 1997].

ClO_2 has been used in severe situations in which either the contaminant level is sufficiently high so that chlorine is ineffective [Rauh 1976; Sussman and Ward 1977; Freund et al. 1984; McGuire and Dishinger 1984; Pacheco et al. 1989; Simpson 1993; Simpson et al. 1993b; Pickrell et al. 1999; Downing and Simpson 2004], or the dosage of chlorine required to achieve control would be so high that an unacceptable corrosion rate would result.

B. Literature Review

In one early report, the use of ClO_2 as a cooling tower biocide was described [Rauh 1976]. Major performance advantages included the fact that it worked well in systems where chlorine was ineffective and that a measurable ClO_2 residual was not required for good control. One cooling system was mentioned where the typical feedrate of chlorine gas of 1200 lbs per day was replaced by 24 lbs per day of ClO_2. Major reductions in bacterial numbers were observed within the first minute of contact.

Several case histories were presented and are summarized in Table 1-1.

Table 1-1. Case History Summaries

Type of plant	Recirc rate	ΔT °F	Cl_2 feedrate	ClO_2 feedrate
Aluminum caster	4,000	10—12	-	-
Ammonia plant	80,000	25—27	1200	24
Vegetable oil refiner	12,000	-	200	-

In each of the case histories, substantial improvement in results as measured by appearance, by the control of noxious odors, by bacterial counts, and by economics was reported [Rauh 1976].

In a second early report, energy savings achieved for a number of case histories were described briefly [Sussman 1977]. These case histories are summarized in Table 1-2.

Table 1-2. Case History Summaries

Type of plant	Circ Rate gpm	Cl_2 (lb/day)	ClO_2 (lb/day)	Ratio Cl_2/ClO_2
Ammonia plant	14,500	160—300	5	32—60
Ammonia plant	80,000	1200	24	50
Chemical plant	20,000	107	4.6	23
Petrochemical plant	85,000	1000	19	53
Refinery	10,000	250	12	21
Refinery	30,000	500	13.5	37
Refinery (once-through)	170,000	4200	56	75
Power plant (once-through)	110,000	200	15	13
Vegetable oil refinery	12,000	200	22.5	9

These reports provide some insight into the relative demand of chlorine vs. ClO_2 for a contaminated system. Expressing the chlorine dosage as a function of recirculation rate (Cl_2 dosage in lb/day divided by recirculation rate in MM lb/day), the $Cl_{2\ RR}$ mg/L varies from 0.4 to 2.1. In previous work, a rule of thumb, presented for recirculating systems treated continuously with chlorine, showed that a "clean" system with few contaminations will typically have a $Cl_{2\ RR}$ mg/L value from 0.14—0.20 mg/L [Simpson 1993].

In general, as system contamination increases, the required dosage of chlorine increases correspondingly, while the increase in ClO_2 required for that same system increases at a much slower rate. For example, Table 1-2 shows that for one contaminated system there was a 60-fold reduction in the lbs of ClO_2 vs. the lbs of Cl_2 required for treatment.

In another report the advantages of ClO_2 as a biocide in ammonia plants were described in several case histories. Most of the data had either been reported in an earlier report or information needed to extract actual feedrate data was not given [Ward et al. 1978].

Several studies were implemented to investigate the numbers of bacteria found in cooling tower drift or aerosols [Adams et al. 1978a; Adams et al. 1978b; Adams et al. 1980]. One study was discontinued because bacterial reductions found when chlorine was replaced by ClO_2 were so great that the study could not be completed [Adams et al. 1978a].

In 1982, results from another study were reported [Kemp and Okimoto 1982]. In this once-through seawater power plant cooling system, various aquatic organisms resulted in biofouling of the main condenser. Although the dosage of ClO_2 was quite low, 0.16—0.21 mg/L, feed of ClO_2 for 15 minutes every 8 hours resulted in an environment that encouraged higher life forms that could relocate to do so. A cleaner system resulted.

In a case history reported in 1984, use of ClO_2 in a grain alcohol plant was described in which the intermittent application was highlighted [Freund et al. 1984]. The plant had several towers, each of which treated different parts of the system. Some of the systems had to be taken off-line weekly for cleaning of biofilm; others required cleaning twice per year. After optimization of ClO_2 treatment (exact dosages were not specified), the massive amounts of biofilm and required cleaning were virtually eliminated.

A novel case history was reported in 1984 [McGuire and Dishinger 1984]. In this unusual case history, concerns about how to handle waste water containing fairly high levels of phenol led to the use of the wastewater as makeup to a cooling tower to let the bacteria destroy the phenol. Microbial growth was uncontrolled, and fouling of a critical exchanger resulted. To address this fouling problem, chlorine was applied to the tower with the goal of preventing fouling in the one critical exchanger but without inactivating bulk water bacteria that were consuming the phenol. The chlorine was sufficiently slow at removing the fouling material that by the time the exchanger had been cleaned, disinfection of the phenol bacteria resulted, and a higher than acceptable phenol level was found in the plant effluent. ClO_2 was then applied intermittently and directly before that particular exchanger. Heat transfer dramatically increased, the biofouled exchanger was cleaned rapidly without killing the phenol bacteria in the bulk water, and control of phenol residuals in the plant effluent remained at an acceptable level.

The use of ClO_2 in a cooling system contaminated with ethylene glycol was described in another report [Pacheco et al. 1989]. The ethylene glycol caused massive microbiological growth, fouled slat fill, plugged screens, and high corrosion rates. Low alkalinity coupled with a high chlorine feedrate resulted in a significant pH depression so that caustic had to be added daily to maintain pH in an acceptable range. ClO_2 was optimized and fed at 1 mg/L for 1 hour, 6 times per day. Carbon steel corrosion rates were reduced from a high of 24 to a low of 0.3 mpy. Copper corrosion rates also saw a reduction from a high of 3 mpy to a low of 0.2 mpy.

To investigate the efficacy of ClO_2 as an antifoulant in once-through power plant cooling systems, asbestos panels were submerged in the cooling system of two power plant cooling systems in one study [Geraci et al. 1993]. Even at the lowest treatment used, 0.7 mg/L, fouling organisms did not attach to the panels.

In the same year, a list of several case histories, summarized in Table 1-3, was reported [Simpson et al. 1993].

Table 1-3. Case History Results					
Plant	CorrRate CS/Adm Before	CorrRate CS/Adm After	Biocide Type	Before Amount	ClO_2 Amount
Ammonia Plant	-	-	Chlorine	900 lbs/day	20 lbs/day
Natural Gas Plant	3.0/0.0	4.0/0.0	2 Biocides	200 lbs/week	15 lbs/ 2-days
Steel Mill	8/1.0	4/0.1	Bleach/NaBr	100 gal/day	7 lbs/ 2-3-days
Refinery	8/1.0	0.25/0.0	Chlorine	20 lbs/day	7 lbs/ 3-days
Refinery	4/1.0	1.6/0.0	HTH	50 lbs/week	30 lbs/ 2-days
Natural Gas Plant	4/1.0	3/0.1	2 Biocides HTH	200 lbss/week 20 lbs/week	5 lbs/ 3-4-days
Natural Gas Plant	4/1.0	3/0.1	Chlorine 1 Biocide	25 lbs/day 100 lbs/week	15 lbs/day
Natural Gas Plant	2/0.3	2/0.0	Chlorine	25 lbs/day	5 lbs/ 2-3-days

These results were similar to those reported by earlier researchers in that a substantial improvement in corrosion rate was observed when chlorine either alone or in combination with non-oxidizing biocides was replaced with ClO_2.

In 1993, results were reported of a field trial using ClO_2 to treat high efficiency film fill [Simpson 1993a]. These results, described in more detail in Chapter 4 of this text, showed that ClO_2 could not only prevent fouling but could clean fouled fill, as long as some water was flowing through the system.

In 1995, a report was made of 4 case histories which deal specifically with the biofilm-removing capabilities of ClO_2 and the positive benefits that resulted from the use of ClO_2 [Simpson 1995]. In a batch mini steel mill, the number of heats per mold was raised from 8—12 to over 300 by improved control of biofilm. In a Southwestern natural gas processing plant, the increase in production was estimated by plant personnel to be 15%, due solely to removal of biofilm by ClO_2. In another gas plant, cleanup of fouled high efficiency film fill resulted in a 2 °F drop in basin water temperature because the fill was able to provide more adequate cooling. In a Southwestern petroleum refinery, cleanup of a critical biofouled exchanger resulted in a decrease in process temperature from 107 °F to 53 °F within about 48 hours. No changes to production rates were made. This drop in process temperature resulted solely from cleanup of biofilm from the biofouled exchanger.

A 1997 report described the effectiveness of ClO_2 in a variety of industrial and oil field systems [Laxton and Simpson 1997] for removal of biofilm, cleanup of iron and iron sulfide, disinfection, control of algae, and reduction in corrosion.

A later report described the impact that use of ClO_2 had on production [Pickrell et al. 1999]. The control and removal of biofilm in three case histories had a dramatic impact on production in chemical plants and power generation in power plant utilities. In one Gulf Coast petrochemical plant, carbon steel corrosion rates were reduced from over 10 mpy to below 4 mpy at the time the paper was presented and was continuing to trend downward. Odors and foaming, which had been problematic, were reduced. The cooling tower basin water temperature, which had been ~ 100 °F, had dropped to ~ 90 °F within 48 hours after implementation of feed of ClO_2,

The second case history described the use of ClO_2 to control biofouling of a once-through power plant which was located downstream of a food processing facility whose wastewater treatment plant frequently discharged organic material that was easily metabolized by bacteria. The result was the rapid formation of biofilm in the downstream power plant. The continuous feed of chlorine, with dechlorination, worked initially. However, when the frequency and duration of treatment was curtailed by regulation, chlorine became ineffective. Use of ClO_2, at 3 X per day for 40 minutes or 4 X per day for 30 minutes, effectively controlled biofouling. In addition to biofouling control, the intermittent approach prevented zebra mussel settlement which had plagued nearby plants that also used this water for once-through cooling. The cost for microbiological control was cut by more than 50%, from about $350,000 to less than $150,000 annually.

In the third case history, the frequent ammonia leaks of a Western ammonia plant produced biofouling that was unable to be controlled by a variety of disinfectants and biocides. Only with use of ClO_2 was the plant able to achieve control. The cost of microbiological control was reduced from over $800,000 to less than $300,000 annually, with better biofilm control, lower corrosion, and increased production. The increase in the production rate was conservatively estimated by plant personnel to be greater than $800,000 per year.

Use of ClO_2 in Italy was described in another report [Ambrogi 1999]. ClO_2 was applied at very low dosages to once-through cooling systems on a low level, continuous use basis for the successful prevention of biofouling. As use of ClO_2 on a continuous basis was considered to be too expensive for full scale use, the author suggested that lower concentrations or intermittent use should be explored.

At that same conference, another report described the successful intermittent use of ClO_2 for biofouling control at low levels of 0.05—0.25 mg/L [Belluati et al. 1999]. Case histories were given for two steel plants, two power plants, and a paper mill utilizing seawater. In addition, a case history for a petrochemical plant using freshwater was provided.

Another report discussed the use of ClO_2 in a Rocky Mountain power plant cooling system for control of Legionella [Dunn 1999]. An economic and performance comparison was made between ClO_2 and the previous biocide, which was chlorine gas combined with various non-oxidizing biocides. ClO_2 was the most cost-effective biocide evaluated. In addition, only by use of ClO_2 could Legionella be controlled.

Further, the use of ClO_2 for control of biofouling in a once-through nuclear power generation station was described [Nalpas and Partridge 1999]. A dosage of 0.2 mg/L for 2 hours per day effectively controlled biofouling.

In another paper, workers presented the results of a four-year study using secondary sewage effluent as makeup to a power plant cooling tower with film fill. Good results were obtained with intermittent treatment of ClO_2 from 1—2.5 hours per day, with a residual of 0.05—0.10 mg/L [DeBenedetto et al. 1999].

A report was made of a novel method for producing ClO_2 [Nickel et al. 2000]. The most striking observation about this report was the 4-year case history. During this period, intermittent use of an exceptionally low level of ClO_2 (0.01—0.05 mg/L) achieved good control of both bacteria and biofilm, in a power plant with high efficiency film fill and gray water makeup.

In another report that same year, two case histories were described, both of which utilized exceptionally low feedrates of ClO_2 for control of bacteria, biofilm, and algae [Simpson and Miller

2000]. In one case history, ClO_2 was used in an electronic chip manufacturing plant at a dose of 0.1 mg/L, intermittently, with the frequency of treatment ultimately being reduced to once per week with good results. In the second case history, ClO_2 was produced by ultraviolet activation of chlorite [Simpson 2001a], and ClO_2 was dosed at 0.006 mg/L to control algae. More details of these case histories were reported in 2001 [Simpson 2001b].

The use of ClO_2 for control of biofouling of the high efficiency film fill in a grain alcohol plant cooling water system was reported later [Downing and Simpson 2004]. In a cooling system that had historically been treated with chlorine, a plant expansion about 6 years prior to implementation of ClO_2 had created a situation where chlorine did not work. Even though numerous non-oxidizing biocides were incorporated into the program along with chlorine, the fill collapsed into the basin on an annual basis for 5 consecutive years. Plant production was so critical that the system could not be shut down to maintain the cooling system, so much of the fill remained in the basin. ClO_2 was applied and optimized, and the results were exceptional. ClO_2 is still being used.

ClO_2 has been shown to work when chlorine cannot. It effectively removes biofilm, thus improving heat transfer and, many times, resulting in improved production. As Rauh mentioned in 1976, ClO_2 can be very effective even, when fed at dosages too low to measure.

C. Identifying Treatment Objective

In order to design a program to treat a cooling system with ClO_2, one must clearly identify the treatment objective, because the mode of treatment (continuous vs. intermittent, frequency and duration of treatment, applied dosage, and injection point) will be dictated by the treatment objective.

Treatment objectives might be to keep main condensers clean in a once-through power plant system, to control biofouling in high efficiency plastic fill, or to dissolve the sludge in a cooling tower basin. Such sludge might contain sufficient residue of the copper-chromate-arsenic wood preservative to result in the sludge's being classified as a hazardous waste. Objectives might also include control of pathogenic organisms such as Legionella.

In essence there are a variety of ways to use ClO_2 to treat a tower effectively. The primary differences are those of economics and objectives.

D. Treatment Philosophy – Continuous vs. Intermittent

There is some disagreement as to whether continuous or intermittent treatment should be done. Early work showed that control of biofilm with chlorine was more effective with continuous use, while control with ClO_2 was more effective with intermittent use [Mayack et al. 1984]. In practical applications, treatment type appears to be a function of the raw water quality. In systems where raw water is not well disinfected and has a high degree of bacterial contamination, either continuous use or frequent intermittent use is required [Simpson 1999d; Simpson 1999e].

In general, when the makeup water is of good quality, i.e., the water has few contaminants and low bacterial counts, such as well-disinfected potable water or well water, intermittent application of ClO_2 is very effective [Simpson et al. 1993b]. For systems which have a high level of bacteria coupled with a high level of bacterial nutrient, a frequent intermittent application [Pickrell et al. 1999; Simpson et al. 1993b] or frequent intermittent application at very low levels [Nickels et al. 2000; Simpson and Miller 2000] has been found to be effective. If the makeup water is highly contaminated, however, treatment may have to be more frequent.

E. Selecting the Right Dosage

Historically, many have adopted a rule of thumb to treat at 5 mg/L based on the system volume. That this approach worked in many cases is a tribute to the versatility of ClO_2. Over time, the application methodology continues to be refined, with subsequent decreases in frequency, duration, and application dosage. The actual dosage for optimal treatment is tied to the ClO_2 requirements of the particular system being treated.

Basically there are two types of ClO_2 demand in a system. First, there is the "*Bulk Water Demand*". This demand includes anything in the bulk water, including contaminants, metal ions, bacteria, algae or other components that might consume ClO_2. A demand study, discussed in more detail in Volume I in Chapter 8, gives a measure of the dosage of ClO_2 required to meet the *bulk water* requirements of this material.

The second type of demand, i.e., "*System Demand*," includes any ClO_2-consuming material not found in the bulk water, i.e., material attached to system walls or piping. This generally includes attached bacteria and associated biofilm, corrosion products, organics, and demand created by the materials of construction of the water-contact surfaces. In most systems, the system demand is substantially greater than the bulk water demand. As ClO_2 reacts rapidly with this material, much more rapidly than other oxidants, it must be included in demand considerations, at least until cleanup has been accomplished.

As a result of these demands, just feeding 5 mg/L based on system volume does not adequately address the needs of many systems.

The approach of some has been to apply ClO_2 at about 1 mg/L, based on recirculation rate, until a residual approaching that applied is measured in the hot return [Simpson et al. 1993b]. In general this only happens after the system demand as well as the bulk water demand has been met.

Some have successfully applied ClO_2 intermittently at very low dosages. The approach of using very low dosage intermittently only works in systems in which the makeup water is very well disinfected [Nickels et al. 2000; Simpson and Miller 2000].

Therefore, the right dosage is very system dependent and must be optimized based on the results of monitoring tests.

The type of cooling system will also dictate the dosage to some extent. For example, the cooling water to the primary condensers of once-through power plant cooling water systems is normally split, so that half of the condenser can be shut down for cleaning without completely shutting down the entire system. In those situations where there are concerns over the concentration of the residual chlorite ion in the receiving water, it is not uncommon to treat one half of a primary condenser at a time, just prior to the primary condenser, at low dosages and in short application periods, at some to-be-determined frequency. In this way, the minimum amount of ClO_2 can be applied to the most critical heat transfer surface, thus keeping the primary condenser free from biofilm without impacting the aquatic life in the receiving water.

F. Selecting the Right Duration and Frequency of Intermittent Treatment

In some situations, such as in systems that have well water or well disinfected makeup, frequency of treatment can be every day to every 3 days [Simpson et al. 1993b; Simpson et al. 1994; Laxton and Simpson 1997]. In other systems where influent water quality is highly contaminated from a nutrient and bacteriological standpoint, the frequency of treatment may have to be several times per day [Simpson et al. 1993b; Laxton et al. 2001].

For a once-through power plant application where biofilm control is critical to good heat transfer, one to three or four times per day treatment (depending upon the nature of the raw water at the intake) at a level just above demand is done to keep the main condenser operating at peak efficiency [Pickrell et al. 1999].

In systems where Legionella has been a problem, ClO_2 has been applied in frequent, intermittent doses, low-level continuous use, or even in conjunction with certain non-oxidizing biocides [Dunn 1999].

The optimum duration and frequency of treatment must be based on treatment goals, and the times and frequencies must be determined from routine testing. One additional point should be made. The effect of even intermittent treatment appears to be cumulative, e.g., over time a lower frequency and duration of treatment is possible [Simpson et al. 1993b].

G. Selecting the Right Injection Point

In conventional systems where the goal is to control biofilm in cooling water exchangers, ClO_2 is normally injected into the pump suction to minimize its use [Simpson et al. 1993]. For towers with high efficiency film fill, keeping the fill clean becomes the primary goal of treatment, and thus feed might be split between the pump suction and the return header just before the water sees the top of the fill [Simpson 1993a]. On rare occasions, tower basin sludge might become a hazardous waste because of the leaching out of wood preservatives which commonly contain copper, chromate, and arsenic. In such cases, injection of ClO_2 through a chlorine-like header to "sweep" the basin might be desired, so that the basin sludge is removed over time during normal treatment.

H. References

[Adams et al. 1978] Adams, A., Garbett, M., Rees, H., and Lewis, B., "Bacterial Aerosols from Cooling Towers," *Journal of the Water Pollution Control Federation*, 2362(October 1978).

[Adams et al. 1978] Adams, A. and Lewis, B., "Bacterial Aerosols Generated by Cooling Towers of Electrical Generating Plants," TP-191A, Cooling Tower Institute Annual Meeting, Houston, TX, January 23-25, 1978.

[Adams et al. 1980] Adams, A., Garbett, M., Rees, H., and Lewis, B., "Bacterial Aerosols from a Cooling Tower Using Waste Water Effluent as Makeup Water," *Journal of the Water Pollution Control Federation*, 52(3), 498(March 1980).

[Allegheny County Health Department 1997] Allegheny County Health Department, Approaches to Prevention and Control of Legionella Infection in Allegheny County Health Care Facilities (2nd ed.) ACHD, Pittsburgh, PA, 1997.

[Ambrogi 1993] Ambrogi, R., "Fouling Control Systems for Water Treatment," *Oebalia*, XIX, 355(1993).

[Ambrogi 1996] Ambrogi, R., "Environmental Impact of Biocidal Antifouling Alternative Treatments of Sea Water Once-Through Cooling Systems," First European Symposium on

Chlorine Dioxide and Disinfection, Rome, November 7-8, 1996.

[Belluati et al. 1996] Belluati, M., Bartole, L., and Bressan, G., "Once-Through Cooling Systems Antifouling Treatment by Chlorine Dioxide," First European Symposium on Chlorine Dioxide and Disinfection, Rome, November 7-8, 1996.

[Bova et al. 2004] Bova, G., Sharpe, P., and Keane, T., "Evaluation of Chlorine Dioxide in Potable Water Systems for Legionella Control in an Acute Care Hospital Environment," International Water Conference, IWC-04-11, October 2004.

[Brooks et al. 1996] Brooks, G., Matisoff, G., and Bourland, B., "Toxicity of Chlorine Dioxide to Adult Zebra Mussels," *Journal of the American Water Works Association*, 88, 93(August, 1996).

[Burton and Liden 1978] Burton, D. T., and Liden, L. H., "Biofouling Control Alternatives to Chlorine for Power Plant Cooling Water Systems: An Overview," Water Chlorination: Environmental Impact and Health Effects, (Jolley, R. L, et. al, eds.), Vol. 2, Ann Arbor Sci. Publ., Ann Arbor, Mi., 1978.

[Caffaro 1995] Caffaro Technical Bulletin, "Micro and Macrofouling Control Using Chlorine Dioxide of Sea Water Used in Large Industrial Systems," 1995.

[Cameron et al. 1987] Cameron, G. N., Symons, J. M., Spencer, S. R., and Ma, J. Y., "Comparison of Various Disinfectants for Control of the Asiatic Clam (*Corbicula fluminea*)," Proceedings, American Water Works Association Annual Conference, 1987.

[Cameron et al. 1989] Cameron, G. N., Symons, J. M., Spencer, S. R., Ma, J. Y., "Minimizing THM Formation During Control of the Asiatic Clam: A Comparison of Biocides," *Journal of the American Water Works Association*, 81(10), 53(1989).

[Coschigano and Go 1993] Coschigano, M., and Go, W., "The Stability of Azoles to Chlorine and Chlorine Alternatives for Microbiological Control," International Water Conference, Paper No. IWC-93-63, 375(1993).

[CTI 1994] Cooling Tower Institute, "Application of Oxidizing Biocides," June, 1994.

[DeBenedetto et al. 1999] DeBenedetto, R., Freedman, A., Boschetti, D., and Hays, G., "Cooling Towers & Secondary WasteWater Makeup: Four Years' Operating Experience," International Water Conference, Technical Paper, IWC-999-42, Pittsburgh, PA, 1999.

[Downing and Simpson 2004] Downing, S. and Simpson, G., "Control of Biofouling in Grain Alcohol Plant Cooling Systems," unpublished report, 2004.

[Downing et al. 1995a] Downing, S., Rusznak, L., and Smolik, N., "Chlorine Dioxide-An Oxidizing Agent for Water Disinfection and Zebra Mussel Control," Third Annual Symposium on ClO_2, New Orleans, LA, September 14-15, 1995.

[Downing et al. 1995b] Downing, S., Smolik, N., Rusznak, L., Anderson, J., and Hale, L., "The Use of Chlorine Dioxide for Zebra Mussel Control – A Perspective of Treatment Histories," Third Annual Symposium on ClO_2, New Orleans, LA, September 14-15, 1995.

[Dubost 1996] Dubost, E., "Chlorine Dioxide as an Alternative to Traditional Biocides in a Sea Water Cooling System," MS Thesis, University of San Francisco, 1996.

[Dunn 1999] Dunn, Steven A., "Alternative Oxidizing Biocide Experience at Hayden Station," Southwest Chemistry Conference, 1999.

[Freund et al.1984] Freund, J., Booker, J. and Ward, W., "Biocontrol-Key to Heat Exchanger Efficiency in Grain Alcohol Plant," Proceedings International Water Conference, 1984.

[Freymark and Rauh 1978] Freymark, S. and Rauh, J., "Selective Oxidation of Industrial Wastewater Contaminants by Chlorine Dioxide," Proceedings Mid. Atlantic Industrial Waste Conference, University of Delaware, Newark, DE, June 19-20, 1978.

[Garey 1981] Garey, J., "A Review and Update of Possible Alternatives to Chlorination for Controlling Biofouling in Cooling Water Systems of Steam Generating Stations," Water Chlorination, Environmental Impact and Health Effects, Jolley, et. al., eds., 4(42), 453(1981).

[Geraci et al.1993] Geraci, S., Ambrogi, R., Festa, V., and Piraino, S., "Field and Laboratory Efficacy of Chlorine Dioxide as Antifouling in Cooling Systems of Power Plants," *Oebalia*,

XIX, 383(1993).

[Goss et al.1977] Goss, L., Jackson, J., Flora, H., Isom, B., Gooch, C., Murray, S. Burton, C. and Bain, W., "Control Studies on Corbicula for Steam-Electric Generating Plants," Proceedings, First International Corbicula Symposium, 139, 1977.

[Hale et al.1996] Hale, L., Freymark, S., Mallen, E., and Lippincott, B., "Use of Chlorine Dioxide to Prevent Zebra Mussel Settlement," Sixth International Zebra Mussel and Other Aquatic Nuisance Species Conference, Dearborn, MI, March, 1996.

[Hamilton et al.1996] Hamilton E., Seal D., Hay, J., "Comparison of Chlorine Dioxide and Chlorine for Control of Legionella in a Hospital Potable Water Supply," *Journal of Hospital Infection,* 32, 156(1996).

[Hood et al. 2000] Hood, J., Cheape, G, Mead, A., Curran, E., "Six Years' Experience with Chlorine Dioxide in the Control of *Legionella pneumophila* in the Potable Water Supply of Glasgow Royal Infirmary," 4th Decennial Conference in Nosocomial and Healthcare Associated Infections, Atlanta, GA, March 7, 2000.

[Ito and Seeger 1980] Ito, K. and Seeger, M., "Effects of Germicides on Microorganisms in Can Cooling Waters," *Journal of Food Protection*, 43(6), 484(1980).

[Kemp and Okimoto 1982] Kemp, P. and Okimoto, N., "The Use of Subtoxic Levels of Chlorine Dioxide to Prevent Surface Condenser Fouling in Once-Through Seawater Cooling Exchangers," International Water Conference, 1982.

[Kim et al. 2002] Kim, B., Anderson, J., Mueller, S., Gaines, W., and Kendall, A., "Literature Review – Efficacy of Various Disinfectants Against Legionella in Water Systems," *Water Research,* 36, 4433(2002).

[Laxton and Simpson 1997] Laxton, G. and Simpson, G., "Chlorine Dioxide: The Versatile Oxidizer," *Water Management International*, (The Annual Review of the Water and Waste Water Industry) 35, 1997.

[Laxton et al.2001] Laxton, G., Sylvester, C., and Cornelius, S., "Chlorine Dioxide Can Provide Improved Performance and Decreased Cost in Ammonia Plant Cooling Water Systems," Gulf Coast Ammonia Producers Association, Baton Rouge, LA, April 9-11, 2001.

[Marckini 1997] Marckini, S., "Chlorine Dioxide Reduces Corrosion Rates," Water Technology, 20(12), 56(1997).

[Masullo et al. 1997] Masullo, A., Malphurs, J., Berkstresser, L., and Eltomi, S., "Chlorine Dioxide as a Biocide in Once-Through Cooling Water and Auxiliary Cooling Water Systems at Florida Power and Light's Cape Canaveral Plant," Proceedings of the 59th Annual American Power Conference, Illinois Institute of Technology, 59-2, 968(1997).

[Mayack et al. 1984] Mayack, L., Soracco, R., Wilde, E., and Pope, D., "Comparative Effectiveness of Chlorine and Chlorine Dioxide Regimes for Biofouling Control," *Water Research*, 18(5), 593(1984).

[McGuire and Dishinger 1984] McGuire, L. and Dishinger, T., "Chlorine Dioxide Solves Biofouling Problems in a Refinery Cooling Tower Used for Phenol Destruction," TP 84-06, CTI, Houston, TX, February, 1984.

[MDHMH 2000] Maryland Department of Health and Mental Hygiene, "Report of the Maryland Scientific Working Group to Study Legionella in Water Systems in Healthcare Institutions," MDHMH, Baltimore, MD., 2000.

[Miller and Simpson 1999] Miller, J. and Simpson, G., "Chemical Control of Legionella," Association of Water Technologies Annual Meeting, Palm Springs, CA, October 26-30, 1999.

[Nalpas and Partridge 1999] Nalpas, D., and Partridge, J., "Applications of ClO_2 in Recirculated Cooling Systems," Second European Symposium on Chlorine Dioxide and Disinfection, Paris, 155, June 24-25, 1999.

[Nickels et al. 2000] Nickels, A., Harriman, C., Nickels, A., Freedman, A., and Harris, A., "A Novel Method of Applying Chlorine Dioxide," AWT Annual Conference, Nov 4, 2000.

[Pacheco et al. 1987] Pacheco, A., Dishinger, T., and Tomlin, J., "Experiences in Controlling

Microbially-Induced Corrosion," Corrosion 87, National Association of Corrosion Engineers Annual Meeting, San Francisco, CA, March 9, 1987.

[Pacheco et al. 1989] Pacheco, A., Durham, H., Dhillon, R., and Edward, C., "The Use of Chlorine Dioxide to Control Microbiological Growth in an Ethylene Glycol Contaminated Cooling Tower.. A Case History," CTI 1989 Annual Mtg, TP89-14.

[Pickrell et al. 1999] Pickrell, W., Germer, M., Miller, J., and Simpson, G., "Plants Increase Production through Use of Chlorine Dioxide," First International Conference: Practical Industrial Water Treatment Technology for the New Millennium, Houston, Texas, January 27-29, 1999.

[Rauh 1976] Rauh, J., "Chlorine Dioxide: A Cooling Water Microbiocide," Proceedings of the Fourth Annual Industrial Pollution Conference, Water Wastewater Equipment Manufacturers Association, McLean, VA, LIII, 20, 1976.

[Rio Linda 1984] Rio Linda, "Heinz-Stockton 30-Day Progress Report of ClO_2 Treatment of Evaporator Cooling Towers," August 21, 1984.

[Rittenhouse 1991] Rittenhouse, R. C., "Industry Weapons Grow in Biofouling Battles," *Power Engineering*, 95, 17(Oct., 1991).

[Rusznak et al. 1994] Rusznak, L., Mincar, G., Smolik, N., "Chlorine Dioxide - A Molluscicidal Agent for Adult Zebra Mussel Eradication," Proceedings Fourth International Zebra Mussel Conference, Madison, WI, March 1994.

[Rybarik et al. 1995] Rybarik, D., Byron, J., Germer, M., "Chlorine Dioxide Treatment for Zebra Mussel Control," Fifth International Zebra Mussel and Other Aquatic Nuisance Organisms Conference, Toronto, CA, February 1995.

[Schlesinger 1977] Schlesinger, H., "Economics of Alternatives for OTEC Biofouling Protection," Ocean Thermal Energy Conversion Biofouling and Corrosion Symposium, Seattle, WA, Oct. 10-12, 1977.

[Silva 2001] Silva, A., "Controlling Biofilms: Cooling Tower Operators Can Use Chlorine Dioxide as a Cleanup Method," Water Technology, 24(3), 96(2001).

[Simpson 1993a] Simpson, G., "Use of Chlorine Dioxide to Clean Small Diameter High Efficiency PVC Film Fill," Cooling Tower Institute, New Orleans, LA, February, 1993.

[Simpson 1993b] Simpson, G., unpublished results, 1993.

[Simpson 1994b] Simpson, G., "Chlorine Dioxide Offers a Cooling Tower Option," *Water Technology*, 80(April 1994).

[Simpson 1995] Simpson, G., "Biofilm: Removal and Prevention with Chlorine Dioxide," Third Annual Symposium on ClO_2, New Orleans, LA, September 14-15, 1995.

[Simpson 1999a] Simpson, G., "New ClO_2 Feed System for Comfort Cooling Towers - Field Trial Report," October 13, 1999.

[Simpson 1999b] Simpson, G., "The Reaction of ClO_2 with Phosphonates," unpublished results, 1999.

[Simpson 1999c] Simpson, G., "The Reaction of ClO_2 with Azoles," unpublished results, 1999.

[Simpson 1999d] Simpson, G., "Field Trial Report – Refinery Cooling Tower Treatment with ClO_2," - unpublished results, 1999.

[Simpson 1999e] Simpson, G., "Field Trial Report – Power Plant Cooling Tower Treatment with ClO_2," unpublished results, 1999.

[Simpson 2001a] Simpson, G., "Chlorine Dioxide Generator," US Patent 6,171,558, January 9, 2001.

[Simpson 2001b] Simpson, G., "Control of Biofilm in Cooling Towers with ClO_2," Fourth International Symposium on Chlorine Dioxide, Las Vegas, NV, February 15-16, 2001.

[Simpson and Miller 2000] Simpson, G. and Miller, J., "Control of Biofilm with ClO_2," Association of Water Technologies Annual Meeting, Honolulu, HA, October 28 – November 5, 2000.

[Simpson et al. 1993a] Simpson, G., Miller, R., Laxton, G., and Clements, W., "A Focus on Chlorine Dioxide: The 'Ideal' Biocide," Paper No. 472, Corrosion 93, New Orleans, LA, March

8-12, 1993.

[Simpson et al. 1993b] Simpson, G., Laxton, G., Miller, R., and Clements, W., "A Focus on Chlorine Dioxide: Biocide of Choice for 'Stressed' Cooling Water Systems," WaterTech 93, Houston, TX, 1993.

[Simpson et al. 1994] Simpson, G., Laxton, G., Ferriola, J., and Hogg, R., "A Focus on Chlorine Dioxide: Increasing Mold Life in Electric Arc Furnace Mini Steel Mills," WaterTech '94, Houston, TX, November 9-11, 1994.

[Simpson et al. 2005] Simpson, G., Alarid, V., Matchim, D., and Hale, E., "From Seawater to Potable Water: Challenges and the Role of Chlorine Dioxide," unpublished report, 2005.

[Sims 2000] Sims, J., Northstar Consulting Co., "ClO_2 for Microbiological Control in a Comfort Cooling Tower - Trial Report," June 28, 2000.

[Smolik et al. 1995a] Smolik, N., Rusznak, L., Anderson, J., and Hale, L., "The Use of Chlorine Dioxide for Zebra Mussel Control - A Perspective of Treatment Histories," Proceedings Fifth International Zebra Mussel and Other Aquatic Nuisance Organisms Conference, Toronto, Ontario, Canada, February 1995.

[Smolik et al. 1995b] Rusznak, L., Smolik, N., Hale, L., and Freymark, S., "Examination of the Potential of Chlorine Dioxide for use in Zebra Mussel Veliger Control," Proceedings Fifth International Zebra Mussel Conference, Toronto, Ontario, Canada, 1995.

[Sussman 1978] Sussman, S., "Problems and Opportunities in Cooling Water Treatment," Chemical Marketing Research Association Meeting, NY, May 10, 1978.

[Sussman and Ward 1977] Sussman, S., and Ward, W. J., "Microbiological Control with Chlorine Dioxide Helps Save Energy," *Materials Performance*, 16(7), 24(1977).

[Tsou et al. 1995] Tsou, J., Coyle, J., Pall, S., Mussalli, Y., and Martin, P., "Evaluation of the Use of Chlorine Dioxide to Control Zebra Mussels," Proceedings Fifth International Zebra Mussel and Other Aquatic Nuisance Organisms Conference, Toronto, Canada, February 1995.

[Turvey 1999] Turvey, M., "The Use of Chlorine Dioxide for Controlling Legionnaires Disease," Second International Conference on Chlorine Dioxide, Paris, June 24, 1999.

[Vaska and Go 1992] Vaska, M., and Go, W., "Evaluation of Alternatives to Gaseous Chlorine for Cooling Water Microbiological Control," Cooling Tower Institute Annual Meeting, Technical Paper No TP92-14, 1992.

[Waite et al. 1978] Waite, T., Jorden, R., and Kawaratani, R., "Evaluation of Alternative Chemical Treatments for Biofouling Control in Electric Power Facilities," Water Chlorination: Environmental Impact and Health Effects, (Jolley, R. L, et. al, eds.), Vol. 2, Ann Arbor Sci. Publ., Ann Arbor, Mi., 1978.

[Ward 1977] Ward, W., "Chlorine Dioxide, a 'New' Selective Oxidant/Disinfectant for Wastewaters," Proceedings of the Forum on Ozone Disinfection, International Ozone Institute, Syracuse, NY, 382(1977).

[Ward et al. 1977] Ward, W., Lee, J., and Freymark, S., "Chlorine vs. Chlorine Dioxide for Ammonia Plant Bio Control," Symposium on Safety in Ammonia Plants, AICHE, August 28-31, 1977.

[Ward et al. 1978] Ward, W., Lee, J., and Freymark, S., "Advantages of Chlorine Dioxide as Biocide," Ammonia Plant Safety, Vol 20, A CEP Technical Manual, prepared by editors of Chemical Engineering Progress, AIChE, 1978.

[Yu et al. 1977] Yu, H., Richardson, G., and Hedley, W., "Alternatives to Chlorination for Control of Condenser Tube Bio-Fouling," EPA-600/7-77-030, March 1977.

Slime Control in Alkaline Whitewater Systems With Chlorine Dioxide

A. Introduction

Bacteria cause numerous problems in paper mills. Such problems include the production of explosive gases, e.g., hydrogen, hydrogen sulfide, and methane. Bacteria also produce bacterial biofilm, which causes additional problems.

Biofilm forms readily in alkaline white water systems due to the temperature and nutrient levels found in such systems. Bacterial biofilm, which can become quite thick on walls, can dislodge under a variety of conditions and result in either defects in the finished paper or lost production due to tears in the paper.

Conventional methods of control include use of expensive non-oxidizing biocides (antimicrobials). A biocide with good disinfection properties for the bulk whitewater may not be effective on the sessile bacteria and may be ineffective at preventing the formation of the biofilm. Many non-oxidizing biocides lack the ability to penetrate biofilm and control the biological growth within. Most non-oxidizing biocides lack the ability to remove an established biofilm.

The use of chlorine dioxide in biofilm control in alkaline white water systems is described in this chapter. A brief overview of the problems associated with bacterial growth in alkaline whitewater, the properties of biofilm and bacteriology of alkaline white water, the chemistry of and equipment for ClO_2 generation, and feed and control strategies are given. In addition, an economic comparison between the use of ClO_2 vs. non-oxidizing biocides for one mill is discussed.

B. The Problem with Bacterial Biofilm

A number of problems arise from a healthy bacterial biofilm. Within a biofilm, a diverse biological community or consortium is established. Metabolic by-products produced by one bacterium are often utilized as a food source by another. This interdependence results in the formation of a highly complex and diverse biological community. Within the biofilm anaerobic conditions are created which are conductive to growth of the sulfate-reducing bacteria, *Desulfovibrio desulfuricans*. These sulfate-reducing bacteria, referred to as SRB, produce sulfide ion as a metabolic by-product [Postgate 1984]. Left unchallenged, healthy communities of SRB can produce sufficient sulfide to create hazardous breathing conditions [Wahlstrom 1969].

Bacteria and its associated biofilm can also cause corrosion. Corrosion can be caused not only by direct bacterial action [Costerton and Geesey 1986], in particular by the anaerobic and highly corrosive SRB [McKoy 1980], but corrosion can also result from action of bacterial enzymes [Boivin et al. 1990], which may survive cell death [Ford et al. 1990]. It is commonly assumed that destruction or inactivation of bacteria within a biofilm is sufficient to control corrosion resulting from bacterial activity. Some researchers have suggested that corrosion resulting from bacterial activity may not be controlled until the biofilm has been completely removed [Hamilton and Maxwell 1986; Characklis et al. 1991].

Biofilm tends to be patchy and non-uniform over the surface [Characklis and Marshall 1990]. The difference in the oxygen concentration at the metal surface which results from this non-uniformity can also cause corrosion [Updegraff 1955].

Microbiological growth in pulp and paper mill alkaline white water systems also causes other problems including the production of explosive gases. In some cases explosions have been attributed to bacteria in closed systems [Rowbottom 1993; Eidsaa et al. 1979; Johnsen et al. 1974] King 1990; Robichau 1991; Rowbottom 1989; Wahlberg and Brink 1988].

In addition to causing corrosion and breathing hazards, biofilm can cause production problems. Under certain conditions massive slime deposits may form that can be as thick as an inch or more. This biofilm can slough off walls under certain conditions of thickness and turbulence [Wiatr 1994]. When these slime-based deposits break off, imperfections in the sheet of paper can result. If the sheet of paper tears, a shutdown of the paper machine may be required while the surfaces of the machine are cleaned mechanically. Sometimes stock chests are boiled out at high pH and high temperature to dissolve these slime masses. In the case of sheet breaks, the machine must be rethreaded and downtime and lost production results [Robertson 1994].

C. Biofilm Development and the Bacteriology of Alkaline Whitewater

Many bacteria prefer to reside on surfaces (sessile) rather than the bulk water (planktonic). In fact, in a natural aquatic environment, for every planktonic bacterium found in the bulk water, there are from 1000 to 50,000 bacteria living on surfaces [Watkins and Costerton 1987]. It is now widely recognized that the attached, sessile bacterial population is more numerous and important than the planktonic one, and that misleading results can be obtained if this fact is overlooked [Videla et al. 1990; Costerton and Lashen 1984].

Many of the sessile bacteria produce a material referred to as extracellular polymeric substances (EPS). This EPS layer, comprised primarily of polysaccharides, functions by protecting the bacterium from any toxic chemicals which it may encounter. This sticky EPS layer is referred to as "slime" or bacterial biofilm and provides a place for other bacteria, debris, dirt, wood fibers and paper making additives to lodge [Characklis and Marshall 1990].

Initially sessile aerobic bacteria grow, using carbon sources adsorbed to the metal surface or dissolved in the water, often producing slime. As this population increases, localized areas of low oxygen arise, allowing the growth of anaerobic bacteria in microenvironments within the biofilm. These bacteria carry out fermentation reactions, which liberate low molecular weight organic compounds as by products. SRB can use these compounds as carbon and or energy sources. Their growth can produce sulfide and potentially stimulate corrosion [Sanders and Hamilton 1986].

Figure 2-1. Three Phases of Biofilm Growth

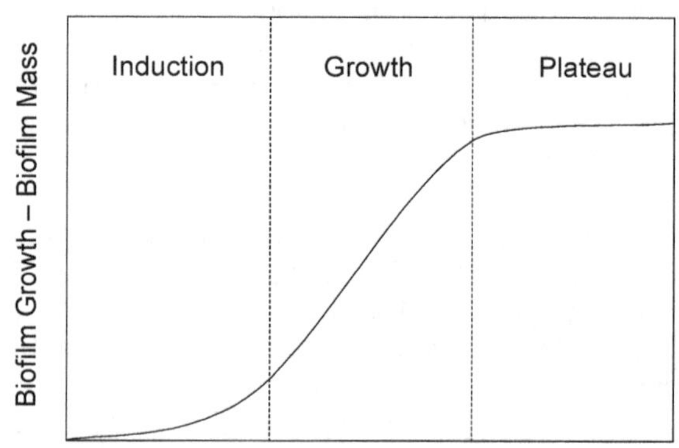

Biofilm develops in 3 stages, induction, growth, and plateau [Bryers and Characklis 1979; Characklis 1979]. This development is shown schematically in Figure 2-1. In the induction phase, organics in the system adsorb onto a surface. These organics in some way promote bacterial attachment to the surface. The duration of this period is a function of the attachment surface roughness and composition. The length of the induction period increases with decreasing nutrient loading rates.

The growth phase is that phase where the biofilm thickness is increasing. The logarithmic rate of biofilm development in this phase can be dependent on nutrient loading rate, wall surface temperature, bulk water temperature, and mechanical shear forces created by the velocity of the water.

The plateau phase is that phase where the biofilm reaches its maximum thickness. The maximum biofilm thickness is controlled by the interplay of shear stress and nutrient loading, i.e., the biofilm growth is encouraged by increasing nutrient loading and discouraged by shear stress forces created by flowing water.

As the biofilm thickness increases, conditions beneath and within the biofilm change. Within a healthy biofilm established in an aerated system, research has shown that a biofilm with a thickness of 12 μ is sufficient to create anaerobic conditions sufficient to promote the growth and development of SRB [Costerton and Geesey 1986; Watkins and Costerton 1987; Sanders and Hamilton 1986].

The nature of biofilm that grows in a system that is nutrient rich, such as an alkaline white water system, is different from a biofilm that would be found in a nutrient limited system. In nutrient limited systems, the biofilm is tenacious and difficult to remove. Biofilm in nutrient rich systems, such as in alkaline white water systems, grows rapidly and is relatively easy to dislodge [Wiatr 1994]. Thus it is imperative that biofilm in an alkaline white water system be minimized or prevented to control or reduce breaks and defects and the lost production that results.

A previous report demonstrated that, although the relative distribution of organisms of an alkaline fine paper system are nearly identical to those found in acid systems, the absolute number of viable bacteria is about an order of magnitude higher in alkaline systems than in acid systems. The predominant bacteria found in alkaline white water include Flavobacterium and Pseudomonas, both notorious slime-formers [Goldstein 1984].

D. Monitoring of Bacteria and Bacterial Biofilm

Numerous tests have been used to monitor the bacteria in the bulk water. These tests provide some information, although it has been well established that a biocide that may be effective on planktonic bacteria may in fact not be very effective at all on sessile bacteria. Thus relying solely on the results of bulk water testing to confirm the effectiveness of the microbiological treatment program is questionable. Equipment inspections along with other biological indicators should be used in conjunction with bulk water bacterial testing to establish the effectiveness of the treatment program.

There are several types of bulk water bacterial tests that are available and used with varying degrees of success. Each type is based on monitoring actual colony growth or the ability of the bacteria to produce or metabolize certain materials. A brief summary of these methods is described below:

1. Standard Plate Count

The standard plate count [American Public Health Association 1976] (SPC) is the standard method of monitoring planktonic or bulk water bacterial activity and it is this test to which all other tests are compared. The test involves collecting the sample, making serial dilutions, placing the various doses into a specific medium, incubating the plates for a period of time, and counting the colonies which grow.

While the SPC method is the industry standard, it suffers from three drawbacks. First, the method is manpower intensive and requires substantial laboratory and analytical technique to perform. Second, the time for the test to produce usable results is 48—72 hours. Third, the method will only allow quantification of those bacteria that will utilize the media as a food source. These three limitations have led to the development of numerous alternatives that overcome one or more of these limitations.

2. Dip Slides

A number of companies market quick microbiological test kits utilizing easy to use dip slide technology [Carter and McGraw 1979; Kajdasz et al. 1984]. The dip slide, commonly used in some applications, is a tab coated with a biological medium. The slide is dipped into the water to be tested, shaken to remove any drops of water, and inserted into the holder. It is incubated for 24—72 hours, and the colonies grow on the slide media. Rather than counting individual colonies, the order of magnitude of colonies is determined by reference to a comparison chart.

This test minimizes the manpower time and technique required, yet does little to reduce the time for incubation. And, like the SPC, it only quantifies bacteria that utilize the media as a food source.

3. ATP Luminescence

All living organisms contain adenosine triphosphate, ATP. ATP is a material which is generally involved in respiration of a given organism. When the organism dies, the ATP is released to the water where it is degraded by enzymes. ATP reacts with the luciferin:luciferase enzyme system found in fireflies. Light is given off during this reaction. Quantitation of the light and assuming an average amount of ATP per bacterium allows an estimate of bacterial population [Carter and McGraw 1979; Holmsen et al. 1966; Stanley and Williams 1969; Engstrom and Tully 1994; Van Dyke et al. 1969; Young-Bandala and Kajdasz 1983]. The conventional ATP-based method eliminates the time of incubation but other problems make it unlikely for this method to see much use until much more work is done. ATP is not organism specific, in that anything that lives contains ATP. In addition, ATP can survive cell death for some period of time depending upon the organism and conditions and thus viable bacterial levels reported from this test could be substantially higher than actual.

A number of companies have introduced very user friendly luminescence type devices. Several of these tests incorporate all the necessary reagents in a single test swab. Such systems have found use in the food processing industry to provide rapid indication of cleanliness of a surface [Russell 1995]. However, it is difficult if not impossible, presently, to provide any meaningful correlation between the SPC and the results obtained by this method. While the ATP method may provide a much more accurate picture of the numbers of bacteria in a given system, some correlation between SPC results and ATP results needs to be developed to bridge the gap which currently exists between these methods. Or, a correlation between the ATP results and the condition of the system needs to be developed. Still, these tests show much promise in that they do not require much user sophistication, results are obtained very rapidly, and they provide a measure of total bacterial activity in the bulk water.

4. Dye-Based Tests

A number of tests have been devised based on oxidation/reduction of colored dyes, such as resazurin and 2,3,5-triphenyltetrazolium chloride (TTC). Such tests may be as accurate if not more so than other tests and thus hold much promise in replacing or augmenting the standard pour plate method [Heck and Hollis 1984; Farkas et al. 1987; Putko 1997]. A dye-based test has been shown to have a high degree of correlation to SPC results in some systems. Thus this test may overcome all three limitations of the SPC method. It is rapid, i.e., the higher the bacterial loading, the more rapidly the color develops; it is user friendly; and it appears to correlate well with the SPC method.

5. Measuring Biofilm

Testing of the bulk water is important, as it is the bacteria present in the bulk water, which will establish sessile populations and ultimately form biofilm. Testing of the surface is also important, however. One commonly used method for monitoring the sessile population involves the use of a device referred to as the "Modified Robbins Device" [Costerton and Lashen 1984; Characklis 1990; Eycott 1990; Kajdasz 1996; Ruseska et al. 1982].

The Modified Robbins Device, MRD, is essentially a device that allows small fixed reproducible areas in contact with the water to become populated with bacteria. These small areas are cleaned of their biofilm, the biofilm is dissolved in water and the water is tested for bacteria.

Efficacy of a biocidal material to the planktonic population is generally quite different than to the sessile population [Kajdasz et al. 1984; Eycott 1990; Kajdasz 1996]. For this reason, testing of the bulk water alone may not provide adequate information regarding the optimal biocide to use on the sessile population.

E. Bacterial Control Strategies

Various biocides are used to control the growth of biofilm producing bacteria. Such biocides includes the following:

Alkyl dimethyl benzyl ammonium chloride (ADBAC)
2,2-dibromo-3-nitrilopropionamide (DBNPA)
Dodecyl guanidine hydrochloride (DDGH)
Methylene bisthiocyanate
Glutaraldehyde
Carbamates
Isothiazoline
Sulfones

Generally, antimicrobials are bacterial poisons. Each of the antimicrobials listed has its strong and weak points. Like any bacterially active agent, each antimicrobial is more effective on some bacteria and less effective on others. Each antimicrobial has a place, and for some applications the choice as to which biocide is best for that application is unequivocal.

Bacterially contaminated systems are dynamic. The biocide that works best today may not be the best tomorrow. Due to natural selection, bacterial strains become predominant which have a higher resistance to a given non-oxidizing biocide. For this reason, periodic testing of biocidal efficacy on both planktonic and sessile bacteria should be done. To avoid this problem it is not uncommon for two different biocides to be fed alternately [Purvis and Tomlin 1991].

F. Properties of ClO_2 and Control of Biofilm

Chlorine dioxide, a gas at ambient temperatures that is very soluble in water, is produced at the point of use as a dilute aqueous solution. It has unique performance characteristics that make it the ideal choice for disinfection or control of biofilm in many cases. Some of these performance characteristics, described in detail elsewhere [Simpson et al. 1993], are listed below:

- Very fast rate of disinfection
- Effective at low dosages
- Non-reactive with most organics
- Non-reactive with ammonia
- Disinfection independent of pH

Chlorine dioxide is also very effective on control and removal of biofilm. A substantial amount of work has been done on the use of chlorine dioxide and its effectiveness on biofilm [Simpson 1995; Simpson et al. 1996; Simpson et al. 1993].

One of the major observations regarding the use of chlorine dioxide for control of biofilm is that when chlorine dioxide reacts, most of the time it accepts one electron and reverts to chlorite ion. The role of the chlorite ion in disinfection and biofilm control has been relatively ignored until the past few years, although it is known that chlorite is a relatively poor disinfectant [Masschelein 1991]. However, the chlorite ion can play a very substantial role in bacteriological and biofilm control in water systems.

Many of the bacteria in a healthy biofilm produce organic acids. Many of these organic acids can react with chlorite ion to form chlorine dioxide directly, or the acids lower the pH in some parts of the biofilm so that chlorous acid is formed. The chlorous acid reverts to chlorine dioxide, which becomes active at retarding aftergrowth in a system and controlling biofilm.

This phenomenon was observed by Clark and Langley [Clark and Langley 1990], who in 1990 received a patent for use of chlorite ion in biofilm control. Their patent involves feed of aqueous chlorite to a system and then following this with an addition of bacterial nutrient. The nutrient results in a high level of activity within the biofilm, specifically the production of organic acids and hence pH reduction. The acids produced react with the chlorite to produce ClO_2 and biofilm removal results. This was indirectly confirmed by others [Hoehn et al. 1990], who noted that in some potable water systems treated with chlorine dioxide, the chlorine dioxide disappears midway through the distribution system and reappears at the end. This reappearance may, in fact, result from bacterial biofilm regeneration of ClO_2.

Even in the pulp and paper industry, the advantages of chlorite in an acid system [Buonanno and Shore 1962] and of chlorine dioxide in alkaline whitewater systems [Cravens 1996; Baker 1980; Baker 1981a; Baker 1981b; Nelson 1982a; Nelson 1982b; Sanford 1986; Latshaw 1994; Cowley 1995] have been well known for some time. Buonanno and Shore observed that use of chlorite alone controlled biofilm very effectively in acid systems. In fact, they observed that paper made in systems using chlorite as a slime control agent imparted some shelf life to bread or milk over and above paper in the same system that did not receive treatment with chlorite. Clearly it is not the chlorite that is providing bacterial disinfection, but at the pH's used in acid systems, the chlorite is converted to chlorine dioxide. So it is clear that although chlorite ion alone is a poor disinfectant, it can play a very substantial role in limiting bacterial regrowth and biofilm formation.

G. ClO_2 – Chemistry and Generation

The generation of chlorine dioxide has been described elsewhere in more detail [Simpson 2005]. ClO_2 is generated in two ways:

- Oxidation of chlorite
- Reduction of chlorate

In pulp bleaching, or other large applications where several tons per day of ClO_2 are required, chlorate reduction is the method of choice. For smaller such as alkaline white water slime control, cooling water microbiological control, poultry chill water disinfection, and vegetable washing and disinfection, oxidation of chlorite is the method of choice.

Smaller applications most commonly utilize chlorine, either molecular chlorine or acid plus bleach to oxidize chlorite as shown in the following equation:

$$Cl_2 + 2ClO_2^- \Rightarrow 2ClO_2 + 2Cl^- \qquad (2\text{-}1)$$

How efficiently this reaction proceeds is a function of generation chemistry and generator design. A discussion of both these topics has been dealt with in significant detail elsewhere [Masschelein 1991; Emmenegger and Gordon 1967; Sussman and Rauh 1976; Aieta and Roberts 1981; Aieta 1984; Tang and Gordon 1984; Aieta and Roberts 1986a; Aieta and Roberts 1986b; Gates 1989; Pitochelli and Mason 1994; Pitochelli 1995, Simpson 2005].

From a safety standpoint, generation chemistry and generator design are very important. From a reliability and performance standpoint, generator design is critical. The selection of the best generator for a given application depends upon the needs of the end user. Automation, efficiency of ClO_2 production, safety, and other such considerations are factors which will lead to the selection of the best system.

H. Feed and Control Strategies

Selection of feedpoints can be critical to the efficacy of chlorine dioxide, such as injecting ClO_2 into an open chest, or into the suction side of a fan pump. Providing proper contact time and mixing to eliminate the potential for release of gaseous ClO_2 are extremely important. Control parameters depends on the plants historical operating data, e.g., frequency of holes or breaks in the sheet given standard plate count quantification. It is difficult to identify the best feed points without operating data from the mill.

Chlorine dioxide is normally added to multiple points in the system. Injection points include the saveall, the whitewater chest, the broke chest, and the stock chests as shown in Figure 2-2. In certain cases, ClO_2 can be added to machine chests, wall boxes, or head boxes. But controlling bacterial activity in the earlier parts of the system helps to eliminate biofilm development in systems closer to the paper machine. The best feed point for each potential application point will be dictated by the constraints of the equipment. For most applications, feeding to pump suction is desirable and to a level several feet below the liquid level. In some cases, due to piping or other limitations, this may not be possible and feed directly into a tank or chest is required.

Feedrate is normally controlled to achieve a residual in the pump discharge of ≤ 0.5 mg/L. Since chlorine dioxide is a very rapid disinfectant, and since it is largely non-reactive with the vast majority of organics, this residual will generally be sufficient to provide the necessary microbiological control. There may be certain organic additives that will create a demand for ClO_2, which may require alterations in normal feedpoints. In other systems, use of certain dyes may eliminate use of chlorine dioxide in certain parts of the system, unless the mill has an automatic color additives system, which will automatically compensate for any variation in dye present in the water. Certain sizing materials may also have an affinity for ClO_2. Sulfited stock with residual sulfite also has a natural affinity for oxidizers.

Figure 2-2. Typical Feedpoints for ClO_2

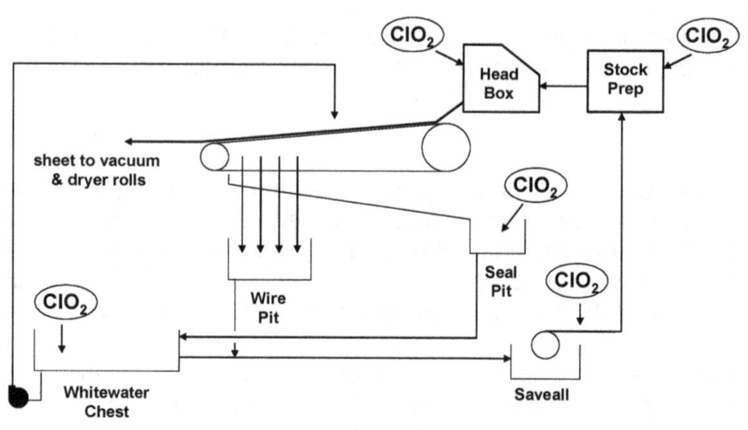

Frequency of feed varies substantially, depending upon the constraints of a given mill. Some mills opt for manual continuous feed, some opt for automated or DCS controlled continuous feed and others choose automated or DCS intermittent feed. While continuous feed provides the security of having good microbiological control, intermittent feed provides a balance between good microbiological control and economics. For those opting for intermittent feed it is absolutely critical to stay on top of microbiological counts as well as hole counts from the paper machine. When one paper break can results in up to 45 minutes or more in lost production time, it is important to minimize any breaks due to microbiological activity. When compared to the cost of a paper break, the cost of an effective microbiological control program is easily justified.

Most mills have determined the bacterial levels that can be tolerated in a given system. The choice of the correct control limits for bulk water bacteria is typically something that is done over the course of time and will be dictated by that level which insures minimum problems in the system due to microbiological activity. The proper limits provide a good balance between economics of biocide usage and good control. For example, if a system will allow a bulk water bacterial level of 10^5 CFU/ml (colony-forming units/ml) without resulting in formation of a sufficiently thick slime layer to cause problems in the system, unacceptably high corrosion rates, or accumulation of explosive gasses, it does not make good economic sense to control bulk water bacteria at 10^2 CFU/ml.

I. Economic Evaluation

In one fine paper mill gaseous chlorine was used as the microbiological control agent. A plant-wide decision to eliminate gaseous chlorine in the plant occurred about the same time as a conversion from acid to alkaline paper. This resulted in an evaluation of chlorine dioxide as an alternative microbiocide. The benefits, both from an environmental and process safety management perspective, resulted in chlorine dioxide being used as the primary whitewater biocide. Since chlorine was used in the acid system, and chlorine dioxide was used in the alkaline system, any direct comparison is not meaningful.

In this mill non-oxidizing biocides are periodically evaluated. Costs for microbiological control in alkaline whitewater using non-oxidizing biocides have varied from \$2.00—\$5.00 per ton of finished product. Chlorine dioxide in this particular mill runs substantially less than \$2.00 per ton of finished product with better results.

In one evaluation, a quat/guanidine blend was tested in an application down stream of the chlorine dioxide application. Despite the benefit of receiving whitewater pretreated with chlorine dioxide, costs to achieve an acceptable level of microbiological control was approximately 85% higher than costs of chlorine dioxide.

J. ClO_2 for Food Grade Paper Applications

Since chlorine dioxide is used for potable water disinfection, it is appropriate to use this versatile disinfectant in food grade paper applications.

Because Dairyman's Standards (the most stringent standard for microbiological contamination of paper products) are imposed on food grade paper production for direct food contact, costs of microbiological control are considerably greater than for fine paper. This is because it is difficult to inactivate bacterial spores, particularly the genus Bacillus, that survive the extreme temperatures of the dryers in the papermaking process. Chlorine dioxide has been found to be a very effective sporicide in food grade paper applications [Bendt 1985; Conkey 1981], in potable water applications [Ridenour et al. 1949; Sokolova et al. 1969], and in some food processing applications [Foegeding et al. 1986; Ito and Seeger 1980]. Unlike chlorine, chlorine dioxide is relatively non-reactive with most of the organics found in alkaline whitewater. As a result, a large part of the chlorine dioxide fed will be available for disinfection. Thus the bacterial activity can be effectively reduced to almost any desired level by controlling the chlorine dioxide feedrate.

K. Summary

The use of chlorine dioxide for alkaline whitewater microbiological control is not new. Only recently have some of the benefits been understood from a mechanistic perspective. A summary of the important benefits of chlorine dioxide is given below:

- Chlorine dioxide is a very effective biofilm control agent.
- Chlorine dioxide reacts very rapidly. This means it can be applied immediately before the problem area, unlike many conventional antimicrobials, which are generally slow acting.
- Because chlorine dioxide remains relatively non-reactive with the vast majority of organics, the use dosage of chlorine dioxide is generally a few mg/L. Thus, corrosion rates resulting from use of chlorine dioxide is typically very low, i.e., much less than other oxidizers. In addition, minimization of the biofilm reduces microbiologically influenced corrosion on equipment
- Chlorite ion keeps working as both a bacteriostat and biofilm control agent, even after the chlorine dioxide has reacted.

- By controlling the biofilm growth, the frequency of boilouts can be reduced and the potential for unscheduled downtime due to paper breaks can be minimized.
- Effectively controlling the biofilm growth minimizes the hole count thus maintaining the quality of the finished sheet.
- Odors resulting from bacterial fermentation, phenols, sulfides, or mercaptans are virtually eliminated by use of chlorine dioxide.

L. References

[Aieta 1984] Aieta, E., "The Kinetics of Chlorine Dioxide Generation for Water Treatment," Doctoral Dissertation, Stanford University, (1984).

[Aieta and Roberts 1981] Aieta, E. and Roberts, P., "Chlorine Dioxide Chemistry: Generation and Residual Analysis," Chemistry in Water Reuse, Cooper, W., ed., Chapter 20, 429 (1981).

[Aieta and Roberts 1986a] Aieta, E. and Roberts, P., "Kinetics of the Reaction between Molecular Chlorine and Chlorite in Aqueous Solution," *Environmental Science and Technology*, 20, 50(1986).

[Aieta and Roberts 1986b] Aieta, E. and Roberts, P., "Application of Mass-Transfer Theory to the Kinetics of a Fast Gas-Liquid Reaction: Chlorine Hydrolysis," *Environmental Science and Technology* 20(1), 44(1986).

[American Public Health Association 1976] American Public Health Association, Standard Methods for the Examination of Water and Wastewater, 14th Edition, New York, 1976.

[Baker 1980] Baker, E., "Chlorine Dioxide Proves an Effective Biocide in Alkaline Whitewater Systems," *Pulp and Paper*, 54(8), 140(1980).

[Baker 1981a] Baker, E., "Slime Control in Alkaline Paper Machine Systems - Chlorine Dioxide," Proceedings Papermakers Conference, 49(1981).

[Baker 1981b] Baker, E., "Using Chlorine Dioxide for Slime Control in Alkaline Paper Machine Systems," *TAPPI Journal*, 64(8), 91(1981).

[Bendt 1985] Bendt, H., "How to Control Bacteria Growth in Food Grade Paperboard Production," *Pulp and Paper*, 119(November 1985).

[Boivin et al. 1990] Boivin, J., Laishley, E., Bryant, R., and Costerton, J., "The Influence of Enzyme Systems on MIC," Paper No. 128, Corrosion 90, Las Vegas, NV, April 23-27, 1990.

[Bryers and Characklis 1979] Bryers, J. and Characklis, W., "Measurement of Primary Biofilm Formation," EPRI Condenser Biofouling Control Symposium, Atlanta, GA, March 25-29, 1979.

[Buonanno and Shore 1962] Buonanno, J. and Shore, J., "Slime Control Agents and Disinfectants in Paper Mill White Water," US Patent 3,046,185 (July 24, 1962).

[Carter and McGraw 1979] Carter, D. and McGraw, P., "The Monitoring and Control of Microbicides," *Water Technology*, September/October 1979.

[Characklis 1979] Characklis, W., "Biofilm Development and Destruction in Turbulent Flow," Cooling Tower Institute Technical Paper # TP 204A, January 22-24, 1979.

[Characklis 1990] Characklis, W., "Biofouling: Effects and Control," Biofouling and Biocorrosion in Industrial Water Systems, Flemming, H.-C. and Geesey, G., eds, Proceedings of the International Workshop on Industrial Biofouling and Biocorrosion, Stuttgart, Germany, September 13-14, 1990.

[Characklis and Marshall 1990] Characklis, W. and Marshall, K., Biofilms, John Wiley & Sons, Inc., New York, NY, 504(1990).

[Characklis et al. 1991] Characklis, W., Little, B., Stoodley, P., and McCaughey, M., "Microbial Fouling and Corrosion in Nuclear Power Plant Service Water Systems," Paper No. 281, Corrosion 91, Cincinnati, OH, March 11-15, 1991.

[Clark and Langley 1990] Clark, J. and Langley, D., "Biofilm Control," US Patent 4,929,365 (May 29, 1990).

[Conkey 1981] Conkey, J., "Sporicidal Activities of Chlorine, Chlorine Dioxide, and Peracetic Acid in a Simulated Papermaking Furnish," *TAPPI Journal*, 64(4), 101(April, 1981).

[Costerton and Geesey 1986] Costerton, J. and Geesey, G., "The Microbial Ecology of Surface Colonization and of Consequent Corrosion," Biologically Induced Corrosion, NACE, Houston, TX, 223(1986).

[Costerton and Lashen 1984] Costerton, J. and Lashen, E., "Influence of Biofilm on Efficacy of Biocides on Corrosion-Causing Bacteria," *Materials Performance*, 13(1984).

[Cowley 1995] Cowley, G., "Chlorine Dioxide as a Biocide in Paper Applications," Third International Chlorine Dioxide Symposium, New Orleans, LA, September 15, 1995.

[Cravens 1996] Cravens, B., "Stabilized Chlorine Dioxide for Microorganism Control," *TAPPI Journal*, 49(8), 53A(August 1966).

[Eidsaa et al. 1979] Eidsaa, G., Johnsen, K., Tangen, R., "An Explosion Caused by Microbial Hydrogen Formation," *Biotechnology Letters*, 1(1), 31(1979).

[Emmenegger and Gordon 1967] Emmenegger, F. and Gordon, G., "The Rapid Interaction between Sodium Chlorite and Dissolved Chlorine," *Inorganic Chemistry*, 6(3), 633(March 1967).

[Engstrom and Tully 1994] Engstrom, G. and Tully, J., "Monitoring Biological Control in Cooling Systems," Cooling Tower Institute Annual Meeting, Paper No. TP94-04, Houston, TX, February 13-16, 1994.

[Eycott 1990] Eycott, R., "Controlling Biofilms in Cooling Water Systems, Industrial and Process Water Treatment," *Water and Wastewater Treatment*, 40(1990).

[Farkas et al. 1987] Farkas, J., Jones, E., Ormerod, D., "A Simple, Rapid Means for Detecting Excessive Biological Activity in Pulp and Paper Mill Systems," *TAPPI Journal*, 165 (1987).

[Foegeding et al. 1986] Foegeding, P., Hemstapat, V., and Giesbrecht, F., "Chlorine Dioxide Inactivtion of Bacillus and Clostridium Spores," *Journal of Food Science*, 51(1), 197 (1986).

[Ford et al. 1990] Ford, T., Black, J., and Mitchell, R., "Relationship Between Bacterial Exopolymers and Corroding Metal Surfaces," Paper No 110, Corrosion 90, Las Vegas, NV, April 23-27, 1990.

[Gates 1989] Gates, D., "Chlorine Dioxide Generation Technology and Mythology," American Water Works Association Water Quality Technology Conference, Philadelphia, PA, Nov 12-16, 1989.

[Goldstein 1984] Goldstein, S., "Microbiology in Alkaline Fine Paper Systems," Proceedings 1984 TAPPI, Research and Development Conference, Appleton, WI, 125(1984).

[Hamilton and Maxwell 1986] Hamilton, W. and Maxwell, S., "Biological and Corrosion Activities of Sulfate Reducing Bacteria Within Natural Biofilms," Biologically Induced Corrosion, NACE, Houston, TX, 131(1986).

[Heck and Hollis 1984] Heck, N. and Hollis, C., "Rapid Biocide Selection Test for Aqueous Systems," TAPPI Research and Development Conference, 121(1984).

[Hoehn et al. 1990] Hoehn, R., Dietrich, A., Farmer, W., Orr, M., Lee, R., Aieta, E., Wood, D., and Gordon, G., "Household Odors Associated with the Use of Chlorine Dioxide," *Journal American Water Works Association*, Research and Technology, 166(April, 1990).

[Holmsen et al. 1966] Holmsen, H., Holmsen, I, and Bernhardsen, A., "Microdetermination of Adenosine Diphosphate and Adenosine Triphosphate in Plasma with the Firefly Luciferase System," *Analytical Biochemistry*, 17, 456(1966).

[Ito and Seeger 1980] Ito, K. and Seeger, M., "Effects of Germicides on Microorganisms in Can Cooling Waters," *Journal of Food Protecttion*, 43(6), 484(1980).

[Johnsen et al. 1974] Johnsen, K., Tangen, R., and Eidsaa, G., "Explosion in Storing Tower with HSSC-Pulp," *Norsk Skogindustri*, 28(3), 60(1974).

[Kajdasz 1996] Kajdasz, R., "Measuring the Biocidal Efficacy against Sessile Sulfate-Reducing Bacteria," *Industrial Water Treatment*, 39(March/April 1996).

[Kajdasz et al. 1984] Kajdasz, R., Young-Bandala, L., and Einstman, R., "Biocidal Efficacy with Respect to Sessile and Planktonic Organisms," Paper No. IWC-84-60, International Water Conference, Pittsburgh, PA, October 22-24, 1984.

[King 1990] King, V., "Microbial Production of Explosive/Toxic Gases in the Paper Industry," TAPPI Proceedings, TAPPI Press, Atlanta, GA, 185 (1990).

[Latshaw 1994] Latshaw, C., "Chlorine Dioxide: Effective, Broad-Spectrum Biocide for White-Water Systems," *TAPPI Journal*, 78(4), 163(April, 1994).

[Masschelein 1984] Masschelein, W., "Experience with Chlorine Dioxide in Brussels: Generation of Chlorine Dioxide," *Journal of the American Water Works Association*, 76(1), 70(1984).

[Masschelein 1991] Masschelein, W., "Chlorine Dioxide," Chemical Oxidation, Technology for the Nineties, First International Symposium, Vanderbilt University, Nashville, TN, February 20-22, 1991.

[McKoy 1980] McKoy, J., Microbiology of Cooling Water, Chemical Publishing Company, New York, NY, 1980.

[Nelson 1982a] Nelson, T., "Appleton Papers finds Chlorine Dioxide to be an Alternative to Conventional Biocides in Alkaline Systems," *TAPPI Journal*, 65(6), 69-73(1982).

[Nelson 1982b] Nelson, T., "Chlorine Dioxide: An Alternative to Conventional Biocides in Alkaline Systems," Annual Meeting - TAPPI, 261(1982).

[Pitochelli 1995] Pitochelli, T., "Chlorine Dioxide Generation Chemistry," Third International Symposium: Chlorine Dioxide: Drinking Water, Process Water, and Wastewater Issues, New Orleans, LA, September 14-15, 1995.

[Pitochelli and Mason 1994] Pitochelli, T. and Mason, J., "Large-Scale Field Applications of Chlorine Dioxide," Oxidative Treatment of Pollutants in Wastewater, Intertech Conferences, Philadelphia, PA, March 21-24, 1994.

[Postgate 1984] Postgate, J., The Sulfate-Reducing Bacteria, Second Edition, Cambridge University Press, 1984.

[Purvis and Tomlin 1991] Purvis, M. and Tomlin, J., "Microbiological Growth and Control in the Papermaking Process," TAPPI Short Course Notes, Chemical Processing Aids, 69(1991).

[Putko 1997] Putko, J., Industrial Mechanical Equipment Company, private communication, March, 1997.

[Ridenour et al. 1949] Ridenour, G., Ingols, R., and Armbruster, E., "Sporicidal Properties of Chlorine Dioxide", *Water Works and Sewage*, 96, 276(1949).

[Robertson 1994] Robertson, L., "Prevention of Microbial Adhesion," 1994 Biological Sciences Symposium, 225(1994).

[Robichau 1991] Robichau, W., "Controlling Anaerobic Bacteria to Improve Product Quality and Mill Safety," *TAPPI Journal*, 74(2), 149(1991).

[Rowbottom 1989] Rowbottom, R., "Bacteria Cause Fatal Explosion at Corrugating Medium Mill," *Pulp and Paper - Canada*, 90(4), 75(1989).

[Rowbottom 1993] Rowbottom, R., "Risks of Bacterial Hydrogen Generation in White Water Systems," *TAPPI Journal*, 76(1), 97(1993).

[Ruseska et al. 1982] Ruseska, I., Robbins, J., Costerton, J., and Lashen, E., "Biocide Testing Against Corrosion Causing Oil-Field Bacteria Helps Control Plugging," *Oil and Gas Journal*, 253(March 8, 1982).

[Russell 1995] Russell, S., "ATP Bioluminescence: It's Portable and Real-Time," *Broiler Industry*, 18(1995).

[Sanders and Hamilton 1986] Sanders, P. and Hamilton, W., "Biological and Corrosion Activities of Sulfate Reducing Bacteria in Industrial Process Plant," Biologically Induced Corrosion,

NACE, Houston, TX, 47(1986).
[Sanford 1986] Sanford, T., "Chlorine Dioxide Controls Deposits in Alkaline System," *Southern Pulp and Paper*, 20(July 1986).
[Simpson 1995] Simpson, G., "Biofilm: Removal and Prevention with Chlorine Dioxide," Third International Symposium on Chlorine Dioxide: Process Water, Drinking Water and Water Waste Issues, New Orleans, LA, September 14-15, 1995.
[Simpson et al. 1993] Simpson, G., Laxton, G., Miller, R., and Clements, W., "A Focus on Chlorine Dioxide: The 'Ideal' Biocide," Paper No. 472, Corrosion 93, New Orleans, LA, March 8-12, 1993.
[Simpson et al. 1993] Simpson, G., Laxton, G., Miller, R., and Clements, W., "A Focus on Chlorine Dioxide: Biocide of Choice for 'Stressed' Cooling Water Systems," Industrial Water Treatment, WaterTech 93, Houston, TX, 1993.
[Simpson et al. 1996] Simpson, G., Ferriola, J., and Hogg, R., "Increasing Mold Life in Electric Arc Furnace Mini Steel Mills," *Industrial Water Treatment*, 28(1), 36(1996).
[Simpson 2005] Simpson, G., Practical Chlorine Dioxide: Volume I – Foundations, Greg D. Simpson & Associates, publishers, 2005.
[Sokolova et al. 1969] Sokolova, N., Troitskii, I., and Zhuchlsova, N., "Effectiveness of Chlorine Dioxide as a Disinfectant Against Spore Forming Bacteria in Water," *Tr. Tsent. Nauch-Issled Dzinfek. Inst.*, 20, 179(1969).
[Stanley and Williams 1969] Stanley, P. and Williams, S., "Use of the Liquid Scintillation Spectrometer for Determining Adenosine Triphosphate by the Luciferase Enzyme," *Analytical Biochemistry*, 29, 381(1969).
[Sussman and Rauh 1976] Sussman, S. and Rauh, J., "Use of Chlorine Dioxide in Water and Wastewater Treatment," Proceedings of IOI Meeting, Cincinnati, OH, 344(1976).
[Tang and Gordon 1984] Tang, T-F. and Gordon, G., "Stoichiometry of the Reaction between Chlorite Ion and Hypochlorous Acid at pH 5," *Environmental Science and Technology*, 18(3), 212(1984).
[Updegraff 1955] Updegraff, D., "Microbiological Corrosion of Iron and Steel," University of Oklahoma Short Course on Corrosion, Norman, OK, April 7, 1955.
[Van Dyke et al. 1969] Van Dyke, U., Stitzel, R., McClellan, T., and Szustkiewicz, C., "An Automated Procedure for the Sensitive and Specific Determination of ATP," *Clinical Chemistry*, 45(4), 3(January 1969).
[Videla et al. 1990] Videla, H., Mele, M., Silva, R., Bianchi, F., and Canales, C., "A Practical Approach to the Study of the Interaction Between Biofouling and Passive Layers on Mild Steel and Stainless Steel in Cooling Water," Paper No. 124, Corrosion 90, Las Vegas, NV, April 23-27, 1990.
[Wahlberg and Brink 1988] Wahlberg, T. and Brink, G., "Bacteria Caused Explosion," *Svensk Papperstidning – Nordisk Cellulosa*, 91(11), 6(1988).
[Wahlstrom 1969] Wahlstrom, J., "Beware of Hydrogen Sulfide - How Protective Measures Save Employees," *Pulp and Paper*, 43(5): 139(1969).
[Watkins and Costerton 1987] Watkins, L. and Costerton, J., "Growth and Biocide Resistance of Bacterial Biofilms in Industrial Systems," *Chemical Times and Trends* 1987.
[Wiatr 1994] Wiatr, C., "Development of Biofilms," 1994 Biological Sciences Symposium, 203(1994).
[Young-Bandala and Kajdasz 1983] Young-Bandala, L., and Kajdasz, R., "A Rapid Method for Monitoring microbial Fouling in Industrial Cooling Water Systems," Paper No IWC-83-45, International Water Conference, Pittsburgh, PA, October 1983.

Ballast Water Disinfection with ClO_2

A. Ballasting of Water – What Is It and Why Is It Done?

About 80% of the world's trade occurs via sale or exchange of goods which are moved by global shipping. In order to maintain proper propeller depth, to compensate for wind or water currents, to increase the stability of the ship, or to maintain stress loads, water is taken on board. This water is called ballast. Maintaining optimum ballast reduces stresses on the ship and provides greater stability and maneuverability of the ship during transit.

Tankers usually carry cargo in one direction and return empty. The ballast carried by a ship is pumped from the port where cargo is unloaded, or en route. Because organisms range in size from viruses to fish located in the water taken as ballast, there is the potential for organisms native to one port to be introduced to the destination port, where they are not native and are considered alien species. Over the years, ballasting and discharge of ballast water as cargo is unloaded have destabilized aquatic ecosystems around the world. Even fully loaded ships which can claim to carry no ballast on board (NOBOB) contribute to the problem, since such ships retain residual volumes of unpumpable water and sediment in their ballast tanks. This residual contains a wide variety of aquatic organisms which can become resuspended when new ballast water is taken on.

Today, an estimated 10,000 million tons of ballast water transport an estimated 15,000 potentially invasive species in about 75,000 ships annually. One estimate of the cost of invasive species in the US alone is about $137 billion /year (Pimental et al. 2000; Simberloff et al. 2000). It has been suggested that non-indigenous invasions are the most dangerous threat to our planet today.

As there is no standard vessel, no standard voyage, and no standard port, it will be challenging for regulators to develop any kind of standard protocol. The lack of an international standard has resulted in the passing of regional laws, enacted to promote ballast water exchange as a means of controlling invasive species. These laws have been made in response to the complaint that the International Maritime Organization (IMO) Ballast Water Working Group has been too slow in developing standards for process evaluation.

B. Background—The Problem with Ballast Water

Ballast water discharge has become a concern for two major reasons. Firstly, ballast water normally contains aquatic nuisance organisms, many of which are not indigenous to the area of ballast water discharge [Carlton 1996; Ruiz et al. 1997]. Currently, at least 162 non-indigenous aquatic species which have colonized the Great Lakes have been identified. Australia is host to more than 175 marine pests that have been introduced through ballast water.

These alien organisms create a potential threat to the local ecosystem, in that some of the organisms may find a suitable habitat but with few natural predators. The zebra mussel is an excellent example of this type of organism. Thought to have been introduced to the US from ballast water discharge of a Russian freighter in Lake St. Claire in the mid 1980s, the zebra mussel has multiplied at a prodigious rate. The extremely high fecundity (a mature female zebra mussel can produce as many as 1.5 million veligers in a single season) coupled with the lack of a natural predator has served to allow the zebra mussel to multiply at extremely high rates. These mollusks have been found as far south as New Orleans and across the Great Lakes and have spread to a large part of the surface water of the Northeastern US.

Live zebra mussels have appeared as far west as Los Angeles, California, attached to a boat on a trailer [Mackie and Schloesser 1996]. The zebra mussel has changed entire ecosystems, driving out less competitive filter feeders. The zebra mussel grows in communities that can clog intake pipes and restrict water flow and has cost industry an estimated 5 billion dollars.

In other parts of the world, in particular Asia and Australia, coastal fisheries have been devastated by the introduction of dinoflagellate species such as Alexandrium. In Australia, the Northern Pacific Seastar (*Asterias amurensis*) has invaded the Derwent River estuary threatening yields in nearby scallop fisheries. The paralytic shellfish-poisoning species of dinoflagellate alga has been responsible for the closure of shell fisheries in southern Tasmania for periods of up to six months to protect public health.

A second reason for concern about ballast water is that it contains pathogenic organisms and viruses. Ballast water discharged from vessels from South America has been responsible for the dissemination of cholera in the US. In 1991, shellfish in Mobile Bay, Alabama, were found contaminated with the epidemic strain of cholera [McCarthy and Khambaty 1994]. In addition, others have also suggested that ballast water might also be the transport mechanism for intermediate hosts for pathogens such as cholera [Colwell and Spira 1992; Epstein 1993].

Introduction of toxic dinoflagellates from the discharge of sediments in Australia, New Zeeland, and Japan have produced red tides that have killed fish and invertebrates. Some have produced human neurotoxins that accumulate in mussels and clams, causing toxic effects in humans who eat these organisms. The American Atlantic coast comb jellyfish (*Mnemiopsis leidyi*) introduced into the Black and Azov Seas in the early 1980s consumed much of the seas' crustacean zooplankton, contributing to the decline of the fishery industry in six nations.

These problems have in recent years resulted in a closer scrutiny of ballast water handling practices. A number of states and international organizations are proposing legislation to mandate more effective management practices. Some have suggested that ballast water discharges should be regulated under the Clean Water Act. The state of California has developed a "total maximum daily load" for exotic species under § 303(d) of the Clean Water Act [Johnston 1999].

The following section presents several options for ballast water treatment, including treatment with chlorine dioxide (ClO_2). The pros and cons of the use of ClO_2 for this purpose as well as its efficacy towards the various organisms likely to be encountered are discussed. The environmental impact, safety, and economics of its use are presented. The results of work done at the University of Rhode Island School of Oceanography are described.

C. Conventional Approaches for Ballast Water Disinfection

A number of possible approaches are being considered for mitigation of this problem. These include the following:

- ❑ Ballast water exchange – This is currently the most common method employed for ballast water treatment. Concerns over safety of exchange during transit have been expressed. This approach will not remove any significant amount of biofilm attached to the walls and piping of the ballast water system.

- ❑ Ballast water heating and flushing – Heating has been used effectively [Rigby et al. 1999]. However, while sufficient heat may be generated to kill most of the phytoplankton in the water, insufficient heat is generated to kill some of the hardier bacterial pathogens such as cholera. Significant modification would be required to the ballast system of a vessel to achieve the temperatures required to kill pathogenic bacterial spores [Rigby and Hallegraeff 1994]. Some have suggested that potentially important pathogens, specifically *Aeromonas salmonicida* and *Vibrio cholerae,* will not be controlled by this method [Oemcke 1999].

- ❑ Filtration and hydrocyclones – Filtration has been used with varying success to remove most of the bacteria and larger organisms. Problems have occurred in various filtration operations due to plugging of filter media with material produced by plankton [Edyvean and Snedden 1985].

- ❑ Ultraviolet radiation – This is thought to be the most environmentally innocuous approach to disinfection. The low-pressure mercury vapor lamp puts out UV at the highest intensity at 253.6 nm, with the wavelength for best inactivation of bacteria being near that. Since almost all organic and many inorganic compounds absorb strongly at this wavelength, questions arise regarding the efficacy of UV in water with a suspended solids level of more than 20 mg/L and ideally 10 mg/L [White 1986b].

  If medium- or high-pressure mercury vapor lamps are to be used, the high heat generation of the lamps create potential maintenance difficulties concerning fouling of the quartz sleeves surrounding the lamps and the inorganic scale formation.

  The efficacy of UV inactivation of some higher life forms has been questioned [Montani et al. 1995].

  Initial capital costs are projected to be very high, and the periodic replacement of both the lamps and the quartz sleeves, which can become solarized by the UV light rendering the quartz sleeve non-transmissive to UV, must also be considered.

  Finally, with UV, unlike with chemical disinfectants, there is no measurable residual, which means that the sole method of verifying effectiveness is through monitoring bacterial counts.

- ❑ Chemical approaches – Various chemicals, including chlorine, ozone, hydrogen peroxide, and various non-oxidizing biocides have been suggested. Each has its benefits and drawbacks.

The ClO_2 chemical approach detailed below offers the benefit of very low dosage requirements, biofilm removal potential, effectiveness towards a wide range of life forms, and, with recent developments, low toxicity.

D. Chlorine Dioxide—A Potential Chemical for Ballast Water Control

Chlorine dioxide is an oxidant that fits a range of applications [Laxton and Simpson 1997]. It is either relatively non-reactive or very slow to react with the vast majority of organics likely to be encountered at mg/L concentrations, and thus most of the applied oxidant goes toward bacterial disinfection, > 99% of which occurs in the first minute of contact.

The disinfection characteristics of ClO_2 have been recognized for many years. Regarding its disinfection kinetics, it is ranked among oxidants only to ozone as a disinfectant [Hoff and Geldrich 1978], and it is ranked first regarding the dosages required to achieve a kill in contaminated systems [Hoff and Geldrich 1981; Simpson et. al 1993].

The resistance organisms have to inactivation by oxidants follows the general guideline shown below [Hoehn 1976]:

Heterotropic bacteria < viruses < cysts < spores

ClO_2 does not react to any significant degree with the water and thus maintains its disinfection characteristics over a wide pH range. In fact, the disinfection efficiency of ClO_2 increases somewhat with pH [Bernarde et al. 1967].

ClO_2 has been shown to be effective at inactivating bacteria responsible for numerous diseases. These bacteria include *Klebsiella pneumonae* [Harakeh et al. 1985], Legionella [Miller and Simpson 1999], *Listeria monocytogenes* [Lin et al. 1996], Salmonella [Thiessen et al. 1984], Shigella [Ridenour and Armbruster 1949], Staphylococcus [Baran et al. 1973], Streptococcus [Oliver et al. 1989], and many other pathogenic bacteria.

In numerous studies, ClO_2 has been shown to be a superior virucide [Aieta et al. 1980; Berg et al. 1979]. For example, in one study Human Immunodeficiency Virus (HIV) was reduced with ClO_2 by about 5—6 orders of magnitude in medical-waste contaminated water [Farr and Walton 1993]. In other studies, ClO_2 was found to be more effective towards poliovirus than either ozone or chlorine [Harakeh and Butler 1980; Taylor and Butler 1982]. In another study, ClO_2 was found to be effective at inactivating polioviruses 1, 2, and 3, and herpes simplex 1 and 2 [Smith and McVey 1973].

ClO_2 has shown to be second only to ozone for inactivation of the very difficult-to-control organisms such as the Cryptosporidium oocyst. The Ct factor describes the concentration (C) of a given disinfectant applied for a specific time (t) to insure adequate inactivation of a specific pathogen. In Table 3-1, the Ct values for inactivation of Cryptosporidium are shown [Korich et al. 1990]. For example, to control Cryptosporidium sufficiently to meet potable water standards, 5—10 mg/L of ozone must be applied for one minute, or 0.5—1.0 mg/L must be applied for 10 minutes. It is clear from this table that neither chlorine nor monochloramine are adequate to control this pathogen.

Table 3-1. Ct Estimates for *Cryptosporidium parvum* Oocyst

Disinfectant	Ozone	ClO_2	Cl_2	Monochloramine
Ct value	5—10	78	7200	7200

In one plant test, ozone and ClO_2 were tested for inactivation of Giardia cysts. Results showed that ClO_2 worked as well as ozone, and that even the 1 mg/L limit for ClO_2 is sufficient over a contact time of a few minutes to effectively control this pathogen [Wallis et al., 1990].

Bacterial spores are among the most difficult organisms to inactivate. Bacillus spores [Ridenour et al. 1949], and Clostridium spores [Ito and Seeger 1980] can be controlled by use of ClO_2 [Foegeding et al. 1986]. In fact, ClO_2 has been shown to be among the very few biocides which can achieve Dairyman Standard (< 200 organisms per gram of finished paper), the bacteriological standard required for paper produced for direct food contact [Conkey 1981; Bendt 1985].

The control of taste and odors in potable water due to growth of algae was the first commercial use of ClO_2 [Aston 1947]. ClO_2 was found to be much more successful and economical than use of copper sulfate for this purpose [Ringer 1955]. Today, much of the potable water use of ClO_2 is to control taste and odors caused by algae.

ClO_2 has been shown to be effective at controlling *Dreissena polymorpha* (zebra mussel) [Matisoff et al. 1996] and *Corbicula fluminea* (Asian clam) [Cameron et al. 1989]. In another report, very low levels of ClO_2 were sufficiently effective to control various organisms in a power plant water intake system [Kemp and Okimoto 1982].

In summary, the perceived benefits of ClO_2 as a superior disinfect at very low dosages (mg/L and sub-mg/L levels) led to an investigation into its use as a ballast water disinfectant [McClennen and Simpson 2001; Perlich and Goodsill 2004].

E. Experimental—Use of ClO_2 as a Potential Ballast Water Treatment

To investigate the feasibility of using ClO_2 as a treatment for control of organisms found in ballast water, a study was undertaken at the University of Rhode Island, Graduate School of Oceanography. Bench scale tests were done using 40-liter containers in a dark, temperature-controlled environment. Concentrations of ClO_2 tested were 0, 1, 5, and 10 mg/L.

The tanks were mixed continuously with paddles operating at 30 rpm. Various organisms, including zooplankton, phytoplankton, *Vibrio cholerae,* and a virus, MS2, were added to separate tanks.

The decrease in the concentration of ClO_2 was tracked for the various doses added and is shown in Figure 3-1.

Figure 3-1. ClO_2 – Persistence

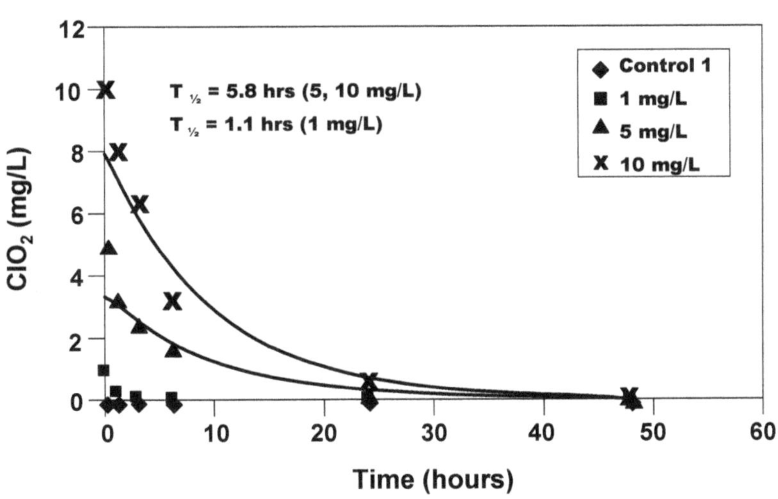

Typical of applications of ClO_2 in other systems, it appears that the ClO_2 does its work of inactivating organisms rapidly, within the first minute or two, with any remaining ClO_2 oxidizing other components of the seawater at a much slower rate.

The phytoplankton as measured by both total chlorophyll and most-probable number analysis, were reduced at all ClO_2 dosages, as shown in Figure 3-2.

Figure 3-2. Chlorophyll Study

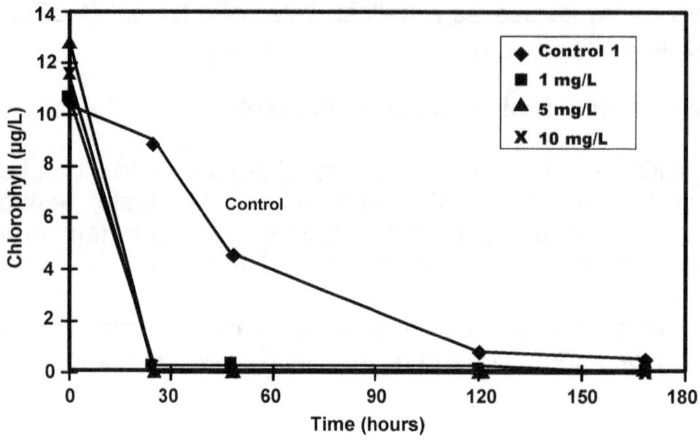

These reductions were expected because of the previous work done on algae in fresh water. ClO_2 has been shown to inactivate algae at low dosages effectively.

The zooplankton investigated included bivalves, ctenophores, polychaetes, calanoids and non-calanoid copepods. Survival of the zooplankton is shown in Figure 3-3.

Concentrations of ClO_2 above 5 mg/L were sufficient to kill 98.4% of a diverse array of commonly occurring zooplankton species.

A concentration of < 5 mg/L ClO_2 might be sufficient to inactivate an acceptable percentage of organisms, but further work is required to determine this with certainty.

The initial results were puzzling because the common practice is to count live organisms, i.e., those which showed movement. In this study, no live organisms were found for 10 mg/L ClO_2, but no dead organisms were found either. Only what appeared to be parts resulting from some cataclysmic affect on the organism remained.

Figure 3-3. Zooplankton Study

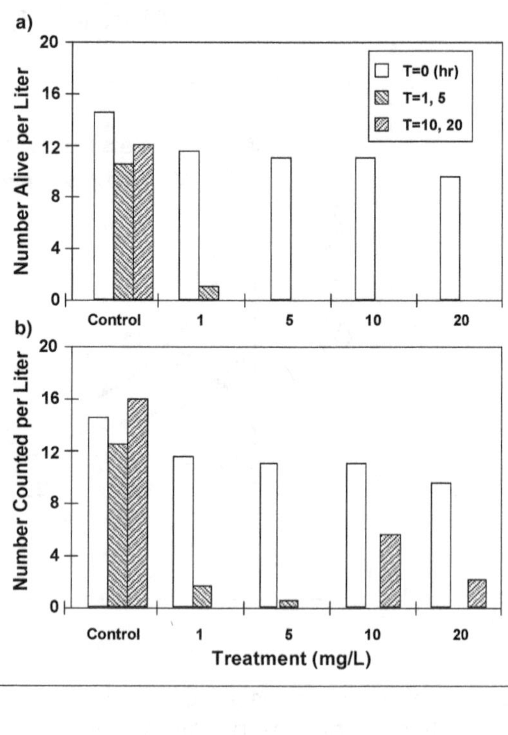

For bacteria, the 1 mg/L treatment with ClO_2 reduced bacterial numbers significantly, although the rate of bacterial regrowth was significant. This suggests that treatment with ClO_2 at 1 mg/L would not result in substantial reduction over the duration of time that ballast water would be held prior to discharge. Treatments at 5 and 10 mg/L showed no bacterial recovery within this time frame. These findings are illustrated in Figure 3-4.

Figure 3-4. Bacterial Reduction Study

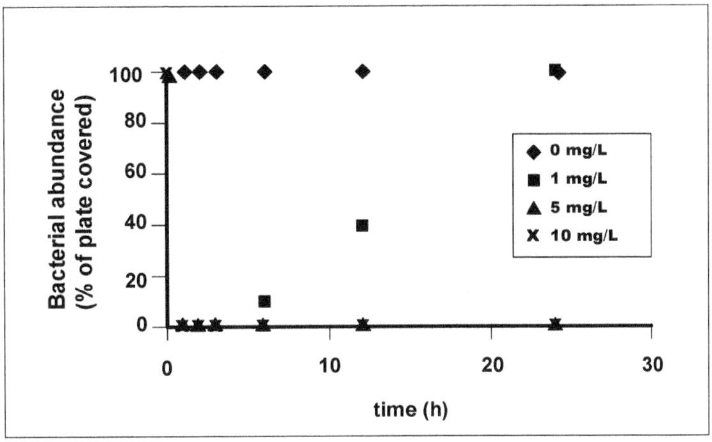

Vibrio chlolerae, the organism responsible for cholera, was totally inactivated at all concentrations of ClO_2 tested.

The RNA bacteriophage MS2 exhibited limited survival in the 1 mg/L ClO_2 treatment, with no survival in the 5 and 10 mg/L treatments.

A more detailed report of these results has been made [McClennen and Simpson 2001; McClennen et al. 2003; Oviatt et al. 2003].

F. Current Status

Ecochlor [Perlich and Goodsill 2004], has been issued a patent for a method and design for a special generator/precursor feed system appropriate for installation onboard a ship. The system was designed with safety in mind and is shown in Figure 3-5. This system has been on board one ship since November 2004. By October 2005 a second installation was completed, with a number of others pending.

Figure 3-5. Ecochlor ClO_2 Generating System

G. References

[Aieta and Roberts 1984] Aieta, E. and Roberts, P., "The Chemistry of Oxo-Chlorine Compounds Relevant to Chlorine Dioxide Generation," Fifth Conference on Water Chlorination:

Environmental Impact and Health Effects, June 3-8, 1984.

[Aieta et al. 1980] Aieta, E., Berg, J., Roberts, P., and Cooper, R., "Comparison of Chlorine Dioxide and Chlorine in Wastewater Disinfection," *Journal of the Water Pollution Control Federation*, April 1980.

[Aston 1947] Aston, R., "Chlorine Dioxide Use in Plants on the Niagra," *Journal of the American Water Works Association*, 39(7), 687(1947).

[Baran et al. 1973] Baran, W., Dawson, L., and Lechowich, R., "Influence of Chlorine Dioxide Water Treatment on Numbers of Bacteria Associated with Processed Turkey," *Poultry Science*, 52, 1053(1973).

[Bendt 1985] Bendt, H., "How to Control Bacteria Growth in Food-Grade Paperboard Production," *Pulp and Paper*, 119(November, 1985).

[Berg et al. 1979] Berg, J., Aieta, E., and Roberts, P., "Effectiveness of Chlorine Dioxide as a Wastewater Disinfectant," Progress in Wastewater Disinfection Technology, A. Venosa, Ed., EPA-600/9-79-018, EPA, Cincinnati, OH, 1979.

[Bernarde et al. 1967] Bernarde, M., Snow, W., and Olivieri, V., "Chlorine Dioxide Disinfection Temperature Effects," *Journal of Applied Bacteriology*, 30, 159(April 1967).

[Bissing et al. 1997] Bissing, A., Coogan, J., Lips, P., Morgan, G., and Ressler, B., "Advanced Monochromatic UV System for the Treatment and Control of Microorganisms in Metal Removal Fluids," Proceedings of the AAMA Metalworking Fluids Symposium II, Detroit, MI, September 15-18, 1997.

[Cameron et al. 1989] Cameron, G. N., Symons, J. M., Spencer, S. R., Ma, J. Y., "Minimizing THM Formation During Control of the Asiatic Clam: A Comparison of Biocides," *Journal of the American Water Works Association*, 81(10), 53-62(1989).

[Carleton 1996] Carleton, J., "Marine Bioinvasions: The Alteration of Marine Ecosystems by Non-Indigenous Species," *Oceanography*, 9, 39(1996).

[Clark and Langley 1990] Clark, J. and Langley, D., "Biofilm Control," US Patent 4,929,365, May 29, 1990.

[Colwell and Spira 1992] Colwell, R. and Spira, W., "The Ecology of *Vibrio Cholera*," Cholera, Barua, D., and Greenough, W., (eds.), Plenum Medical Book Co., 107(1992).

[Conkey 1981] Conkey, J., "Sporicidal Activities of Chlorine, Chlorine Dioxide, and Peracetic Acid in a Simulated Papermaking Furnish," *TAPPI Journal*, 64(4), 101(April 1981).

[Costilow et al. 1984] Costilow, R., Uebersax, M., and Ward, P., "Use of Chlorine Dioxide for Controlling Microorganisms During the Handling and Storage of Fresh Cucumbers," *Journal of Food Science*, 49, 396(1984).

[Cowley and Znajewski 1998] Cowley, G. and Znajewski, T., "ECF Technology: The ClO_2 Generation System of the Future," Western Canada Water and Wastewater Association Conference, 50th Annual Conference, October 25-28, 1998.

[Edyvean and Snedden 1985] Edyvean, R. and Snedden, A., "The Filtration of Plankton from Seawater," *Filtration and Separation*, 184(May/June 1985).

[Epstein 1993] Epstein, P., "Algal Blooms in the Spread and Persistence of Cholera," *Biosystems* 31, 209(1993).

[Farr and Walton 1993] Farr, R. and Walton, C., "Inactivation of Human Immunodefficiency Virus (HIV) by a Medical Waste Disposal Process Using Chlorine Dioxide," *Infection Control and Hospital Epidemiology*, 14(9), 527(Sept. 1993).

[Fisher and Burton 1993a] Fisher, D. and Burton, D., "The Acute Effects of Continuous and Intermittent Application of Chlorine Dioxide and Chlorite on *Daphnia magna*, *Pimphales promelas*, and *Oncorhynchus mykiss*," Report No. WREC-93-B4, University of Maryland System, Maryland Institute of Agriculture and Natural Resources, Wye Research and Education Center, Queenstown, MD, 21658, August 1993.

[Fisher and Burton 1993b] Fisher, D. and Burton, D., "The Acute and Chronic Effects of Intermittent Application of Chlorine Dioxide on *Mysidopsis bahia* and *Menidia beryllina*,"

Report No. WREC-93-B5, University of Maryland System, Maryland Institute of Agriculture and Natural Resources, Wye Research and Education Center, Queenstown, MD, September 1993.

[Fisher and Burton 1995] Fisher, D. and Burton, D., "Determination of Acute Water Quality Criteria for Continuous and Intermittent Exposure of Chlorite for Freshwater Organisms," Report No. WREC-95-O3, University of Maryland System, Maryland Institute of Agriculture and Natural Resources, Wye Research and Education Center, Queenstown, MD, 21658, June 1995.

[Foegeding et al. 1986] Foegeding, P., Hemstapat, V., and Giesbrecht, F., "Chlorine Dioxide Inactivation of Bacillus and Clostridium Spores," *Journal of Food Science*, 51(1), 197 (1986).

[Freymark and Rauh 1978] Freymark, S. and Rauh, J., "Selective Oxidation of Industrial Wastewater Contaminants by Chlorine Dioxide," Proceedings Middle Atlantic Industrial Waste Conference, University of Delaware, Newark, DE, June 19-20, 1978.

[Gates 1989] Gates, D., "Chlorine Dioxide Generation Technology and Mythology," American Water Works Association Water Quality Technology Conference, Philadelphia, PA, Nov 12-16, 1989.

[Gates 1998] Gates, D., The Chlorine Dioxide Handbook, American Water Works Association Water Disinfection Series, 1998.

[Harakeh and Butler 1980] Harakeh, M. and Butler, M., "Inactivation of Viruses in Municipal Effluent by Chlorine Dioxide," Proceedings of the International Symposium on Viruses and Wastewater Treatment, University of Surrey, Guildford, September 15-17, 1980.

[Harakeh et al. 1985] Harakeh, M., Berg, J., Hoff, J., and Matin, A., "Susceptibility of Chemostat-Grown *Yersinia enterocolitica* and *Klebsiella pneumoniae* to Chlorine Dioxide," *Applied and Environmental Microbiology*, 49(1), 69 (1985).

[Hicks 1981] Hicks, B., "Chlorine Dioxide Generation Apparatus and Process," US Patent 4,247,531, Jan 27, 1981.

[Hoehn 1976] Hoehn, Robert C., "Comparative Disinfection Methods," *Journal of the American Water Works Association*, Water Technology/Quality, 302(1976).

[Hoff and Geldreich 1978] Hoff, J. and Geldreich, E., "Alternative Disinfectants for Drinking water - Ozone, Chlorine Dioxide, and Chloramines," Proceedings of the Annual Public Water Supply Engineers Conference, 20, 41(1978).

[Hoff and Geldreich 1981] Hoff, J. and Geldreich, E., "Comparison of the Biocidal Efficiency of Alternative Disinfectants," *Journal of the American Water Works Association*, 73(1), 40(1981).

[Iatrou 1991] Iatrou, A., "Removal of Chlorite by Reaction with Ferrous Iron," MS Thesis, Virginia Polytechnic Institute and State University, Blacksburg, VA, July, 1991.

[Ito and Seeger 1980] Ito, K. and Seeger, M., "Effects of Germicides on Microorganisms in Can Cooling Waters," *Journal of Food Protection*, 43(6), 484(1980).

[Johnson et al. 1993] Johnson, D., Concle, J., Hashimoto, S., and Minday, M., "Simpson Tacoma Kraft Operates Dioxin-Free with High Percentage ClO_2 Substitution," *TAPPI Journal*, 76(3), 89(March 1993).

[Johnston 1999] Johnston, C., "Why Ballast Water Discharges Should be Regulated Under the Clean Water Act," First National Conference on Marine Bioinvasions, Massachusetts Institute of Technology, January 24-27, 1999.

[Kemp and Okimoto 1982] Kemp, P. and Okimoto, N., "The Use of Subtoxic Levels of Chlorine Dioxide to Prevent Surface Condenser Fouling in Once-Through Seawater Cooling Exchangers," International Water Conference, 1982.

[Korich et al. 1990] Korich, D., Mead, J., Madore, M., Sinclair, N., and Sterling, C., "Effects of Ozone, Chlorine Dioxide, Chlorine, and Monochloramine on *Cryptosporidium parvum* Oocyst Viability," *Applied and Environmental Microbiology*, 56, 1423(May 1990).

[Lalezary et al. 1986] Lalezary, S., Pirbazari, M., and McGuire, M., "Oxidation of Five Earthy-Musty Taste and Odor Compounds," *Journal of the American Water Works Association*,

62(March 1986).

[Laxton and Simpson 1997] Laxton, G. and Simpson, G., "Chlorine Dioxide: The Versatile Oxidizer," *Water Management International*, The Annual Review of the Water and Wastewater Industry, 35(1997).

[Lin et al. 1996] Lin, W., Huang, T., Cornell, J., Lin, C., and Wei, C., "Bactericidal Activity of Aqueous Chlorine and Chlorine Dioxide Solutions in a Fish Model System," *Journal of Food Science*, 61(5), 1030(1996).

[Lykins and Griese 1986] Lykins, B. and Griese, M., "Using Chlorine Dioxide for Trihalomethane Control," *Journal of the American Water Works Association*, 78(6), 88(June 1986).

[Mackie and Schloesser 1996] Mackie, G. and Schloesser, D., "Comparative Biology of Zebra Mussels in Europe and North America: An Overview," *American Zoologist*, 36(3), 244(June 1996).

[Masschelein 1980] Masschelein, W., "The State of Art in the Use of Chlorine Dioxide and Ozone in the Treatment of Water," *Water SA*, 6(3), 116(1980).

[Masschelein 1991] Masschelein, W., "Chlorine Dioxide," Chemical Oxidation: Technology for the Nineties, First International Symposium, Vanderbilt University, Nashville, TN, February 20-22, 1991.

[Matisoff et al. 1996] Matisoff, G., Brooks, G., Bourland, B., "Toxicity of Chlorine Dioxide to Adult Zebra Mussels," *Journal of the American Water Works Association*, 93(August 1996).

[McCarthy and Khambaty 1994] McCarthy, S. and Khambaty, F., "International Dissemination of Epidemic *Vibrio cholerae* by Cargo Ship Ballast and Other Nonpotable Waters," *Applied and Environmental Microbiology*, 60, 2597(1994).

[McClennen and Simpson 2001] McClennen, C. and Simpson, G., "Ballast Water Disinfection with Chlorine Dioxide," Second International Conference on Marine Bioinvasions, New Orleans, LA, April 9-11, 2001.

[McClennen et al. 2003] McClennen, C, Oviatt, C., Simpson, G., Hargraves, P., Kelly, R., Kirs, M., Maranda, L., Moran, S., Outram, D., Smith, D., Sullivan, B., Whitman, K., "Bench Scale Assays of Toxicity Using Dilute Aqueous Solutions of Chlorine Dioxide Demonstrate Near-Total Mortality of Marine Flora and Fauna of Coastal Marine Waters," International Association for Great Lakes Research Conference Program and Abstracts, International Association of Great Lakes Research, 2205 Commonwealth Blvd., Ann Arbor, MI, 48105, USA; (46th): 112(2003).

[Miller and Simpson 1999] Miller, J. and Simpson, G., "Chemical Control of Legionella," Association of Water Technologies Annual Meeting, Palm Springs, CA, October 26-30, 1999.

[Montani et al. 1995] Montani, S., Meksumpun, S., and Ichimi, K., "Chemical and Physical Treatments for Destruction of Phytoflagellate Cysts," *Journal of Marine Biotechnology*, 2, 179(1995).

[Oemcke 1999] Oemcke, D., "Future Research on Ballast Water Treatment – A Technologist's View," First National Conference on Marine Bioinvasions, Massachusetts Institute of Technology, January 24-27, 1999.

[Oliver et al. 1989] Oliver, S., King, S., Torre, P., Shull, E., Dowlen, H., Lewis, M., and Sordillo, L., "Prevention of Bovine Mastitis by a Postmilking Teat Disinfectant Containing Chlorous Acid and Chlorine Dioxide in a Soluble Polymer Gel," *Journal of Dairy Science*, 72, 3091(1989).

[Oviatt et al. 2003] Oviatt, C., McClennen, C., Simpson, G., and Perlich, T., "Chlorine Dioxide as an Effective Ballast Water Treatment: Results of Bench-Scale Assays on Its Toxicity to Marine Organisms," 9th International Conference on Marine Engineering Systems, Helsinki, Finland, March 2003.

[Palin 1957] Palin, A., "Determination of Free Chlorine and Combined Chlorine in Water by the Use of Diethyl-*p*-phenylene Diamine," *Journal of the American Water Works Association*. 49, 873(1957).

[Parsons et al. 1997] Parsons, M., Cangelosi, A., Harkins, R., Mackey, T., and Munro, D., "Design of the Great Lakes Ballast Technology Demonstration Project," Annual Meeting of The Society of Naval Architects and Marine Engineers, Ottawa, October 16-18, 1997.

[Perlich and Goodsill 2004] Perlich, T. and Goodsill, C., "Methods, Apparatus, and Compositions for Controlling Organisms in Ballast Water," US Patent 6,773,611, August 10, 2004.

[Pimental et al. 2000] Pimental, D., Lach, L., Zuniga, R., and Morrison, D., "Environmental and Economic Costs of Nonindigenous Species in the United States," *BioScience*, 50, 53(2000).

[Pitochelli 1995] Pitochelli, T., "Chlorine Dioxide Generation Chemistry," Third International Symposium, Chlorine Dioxide: Drinking Water, Process Water and Wastewater Issues, New Orleans, LA, September 14-15, 1995.

[Pitochelli 2001] Pitochelli, A., "Composition for Generating Chlorine Dioxide," US Patent 6,197,215, March 6, 2001.

[Pitochelli and Mason 1994] Pitochelli, T. and Mason, J., "Large-Scale Field Applications of Chlorine Dioxide," Oxidative Treatment of Pollutants in Wastewater, Intertech Conferences, Philadelphia, PA, March 21-24, 1994.

[Pryke 1994] Pryke, D., "ClO_2 – Strong Performer, Bright Future," *Paper Age*, 18(April 1994).

[Ridenour and Armbruster 1949] Ridenour, G. and Armbruster, E., "Bactericidal Effects of Chlorine Dioxide," *Journal of the American Water Works Association*, 41, 537(1949).

[Ridenour et al. 1949] Ridenour, G., Ingols, R., and Armbruster, E., "Sporicidal Properties of Chlorine Dioxide," *Water Works and Sewage*, 96, 276(1949).

[Rigby and Hallegraeff 1994] Rigby, G. and Hallegraeff, G., "The Transfer and Control of Harmful Marine Organisms in Shipping Ballast Water: Behavior of Marine Plankton and Ballast Water Exchange on the MV Iron Whyalla," *Journal of Marine Environmental Engineering*, 1, 91(1994)

[Rigby and Taylor 1999] Rigby, G. and Taylor, A., "Progress in the Management and Treatment of Ship's Ballast Water to Minimize the Risks of Translocating Harmful Nonindigenous Aquatic Organisms," First National Conference on Marine Bioinvasions, Massachusetts Institute of Technology, January 24-27, 1999.

[Rigby et al. 1999] Rigby, G., Hallegraeff, G., and Sutton, C., "Novel Ballast Water Heating Technique Offers Cost-Effective Treatment to Reduce the Risk of Global Transport of Harmful Marine Organisms," *Marine Ecology Progress Series* 191, 289(1999).

[Ringer and Campbell 1955] Ringer, W. and Campbell, S., "Use of Chlorine Dioxide for Algae Control at Philadelphia," *Journal of the American Water Works Association*, 47, 740(1955).

[Ruiz et al. 1997] Ruiz, G., Carlton, J., Grosholz, D., and Hines, A., "Global Invasions of Marine and Estuarine Habitats by Non-Indigenous Species: Mechanisms, Extent and Consequences," *American Zoologist*, 37, 621(1997).

[Simberloff et al. 2000] Simberloff, D., Parker, I., and Windle, P., "Introduced Species Policy, Management, and Future Research Needs," *Frontiers in Ecology and the Environment*, 3(1), 12(2000).

[Simpson 1995] Simpson, G., "Biofilm: Removal and Prevention with Chlorine Dioxide," The Third International Symposium on Chlorine Dioxide: Process Water, Drinking Water and Water Waste Issues, New Orleans, LA, September 14 -15, 1995.

[Simpson 2001a] Simpson, G., "Control of Biofilm in Cooling Towers with Chlorine Dioxide," Fourth International Symposium on Chlorine Dioxide, Las Vegas, NV, February 15-16, 2001.

[Simpson 2001b] Simpson, G., "The Reduction of the Chlorite Ion," Fourth International Symposium on Chlorine Dioxide, Las Vegas, NV, February 15-16, 2001.

[Simpson 2001c] Simpson, G., "ClO_2 for Odor Control in Closed Recycled Paper Mills," unpublished results, 2001.

[Simpson and Miller 2000] Simpson, G. and Miller, J., "Control of Biofilm with ClO_2," Association of Water Technologies Annual Meeting, Honolulu, HA, October 28 – November 5, 2000.

[Simpson and Phillips 1998] Simpson, G. and Phillips, E., "The Environmental Impact of Chlorine Dioxide," Chemical Oxidation: Technologies for the Nineties, Eighth International Symposium, Nashville, TN, April 1-3, 1998.

[Simpson et al. 1993] Simpson, G., Laxton, G., Miller, R., and Clements, W., "A Focus on Chlorine Dioxide: The 'Ideal' Biocide," Paper No. 472, Corrosion 93, New Orleans, LA, March 8-12, 1993.

[Simpson et al. 1994] Simpson, G., Kuykendall, C., Miller, J., and Averett, B., "The Safe and Effective Use of Chlorine Dioxide," WaterTech '94, Houston, TX, November 9-11, 1994.

[Smith and McVey 1973] Smith, J. and McVey, J., "Virus Inactivation by Chlorine Dioxide and Its Application to Storm Water Overflow," Division of Environmental Chemistry, American Chemical Society, Chicago, IL, August 26-31, 1973.

[Solomon et al. 1994] Solomon, K., Bergman, H., Huggett, R., Mackay, D., and McKague, B., "A Review and Assessment of the Ecological Risks Associated with the Use of Chlorine Dioxide for the Bleaching of Pulp," 1994 International Pulp Bleaching Conference, 145(1994).

[Sproul et al. 1983] Sproul, O., Chen, Y., Engel, J., and Rubin, A., "Comparison of Chlorine and Chlorine Dioxide for the Inactivation of an Amoebic Cyst," *Environmental Technology Letters*, 4, 335(July 1983).

[Stevens 1982] Stevens, A., "Reaction Products of Chlorine Dioxide," *Environmental and Health Perspective*, 46, 101(1982).

[Taylor and Butler 1982] Taylor, G. and Butler, M., "A Comparison of the Virucidal Properties of Chlorine, Chlorine Dioxide, Bromine Chloride, and Iodine," *Journal of Hygiene (Cambridge)*, 89, 321(1982).

[Thiessen et al. 1984] Thiessen, G., Usborne, W., and Orr, H., "The Efficacy of Chlorine Dioxide in Controlling Salmonella Contamination and Its Effect on Product Quality of Chicken Broiler Carcasses," *Poultry Science*, 63, 647 (1984).

[Wadsworth and Simpson 1997] Wadsworth, J. and Simpson, G., "Control of Biofilm in Alkaline Whitewater Systems with Chlorine Dioxide," TAPPI Papermakers Conference, Nashville, TN, October 6-9, 1997.

[Wallis et al. 1990] Wallis, P., Roodselaar, A., Neuwirth, M., Roach, P., Buchanan-Mappin, J., and Mack, H., "Inactivation of Giardia Cysts in a Pilot Plant Using Chlorine Dioxide and Ozone," Proceedings of the Water Quality Technology Conference, 17, 695(1990).

[White 1986a] White, G., Handbook of Chlorination, Second Edition, Van Nostrand Reinhold, 197(1986).

[White 1986b] White, G., Handbook of Chlorination, Second Edition, Van Nostrand Reinhold, 996(1986).

Biofouling Control of Cooling Tower Fill with ClO_2

A. Introduction

The introduction of high efficiency film fill, two examples of which are shown in Figure 4-1, has resulted in some significant reductions in both the size and associated expense of building a cooling tower.

Figure 4-1. Examples of Film Fill

As a result of the lower capital cost, many of the new cooling towers have been designed with this fill. In addition, there has been a trend toward replacement of existing fill with this new material, as replacement of old fill was required or improvements in cooling efficiency were sought.

Cooling tower fill design has evolved over the past several decades. As the conventional slat fill (Figure 4-2) has been replaced by the PVC film fill (Figure 4-1), the cost of cooling tower construction has declined considerably, because a smaller tower can be built to achieve the same results.

Figure 4-2. Schematic of Slat Fill

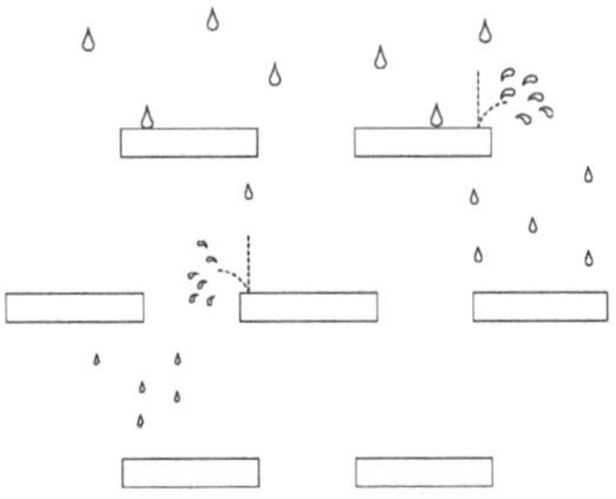

The reduced construction cost of a cooling tower which results from use of this fill has been somewhat offset by the costs associated with the loss in efficiency caused by the increased fouling tendency of this type of fill.

The problems associated with the use of this fill have been described by others [Monjoie et al. 1993; Robinette and Tvedt 1999]. Although a numerous types of film fill construction materials have been evaluated, PVC has been the material of choice [Meek 1967].

As the conditions within a fill are ideal for growth of some types of bacteria, it has been demonstrated that some fill designs are inherently more prone to biofoul, and hence more difficult to treat than others. One early problem which surfaced with PVC fill was the tendency of certain bacteria, specifically Pseudomonas, a notorious slime-former, to attach and grow on the

PVC. Many of the published reports made note that the bacterial exopolymers provides the "glue" that holds other deposits in place in the fill [Monjoie et al. 1993; Hobson and Fitchett 1992; Gill et al. 1994; Robinette and Tvedt 1999]. The combination of bacterial slime and the inorganic deposits which it traps and holds in place can become quite heavy, and a number of these cooling tower fills have collapsed.

B. Case History 1

A West Texas natural gas processing plant had both of the fill types shown in Figure 4-1. Their fill had become fouled to the extent that the cooling tower was not cooling sufficiently, and production was suffering.

The plant's cooling tower is shown in Figure 4-3. The small cell shown in this figure was equipped with the small fill shown in Figure 4-1 and had a very high fouling potential.

The larger cells were equipped with the larger fill, and the fouling potential was much lower.

To investigate the feasibility of use of ClO_2 for cleaning the tower's fill, a field trial was undertaken.

The tower was entered prior to the trial to investigate the fill condition. Each piece of fill in this cell was 12" X 12" X 48," weighed 42 lbs when clean, and was placed into the tower in a crosshatch manner, four sections high.

Figure 4-3. Cooling Tower

The top section appeared to be relatively clean, but when subsequent sections were pulled, the fouling became apparent, reaching a maximum at the third section from the top. A piece of this fill was removed and photographed. It took four men to lift the piece of fill.

The fill was replaced, and ClO_2 was fed to the return riser of this smaller cell at a rate of 2—5 mg/L for 24 hours.

During this period of time, a very large amount of biological material was washed out of the fill from below with a water hose. So much material was washed out that the dual fine stainless mesh screens had to be pulled and cleaned manually (Figure 4-4). Use of the block and tackle normally used to pull the screens was too slow to prevent the starvation of the recirculation pumps.

Figure 4-4. Screen Fouling

Over the next 36 hours, the composition of fouling material caught on the screen changed from mostly organic material to inorganic material (Figure 4-5). A sample of this deposit was taken and analyzed and found to be composed primarily of silica.

Figure 4-5. Deposits on Screen after 36 Hours

As the water became cooler toward the bottom of the fill, just prior to its exiting the fill, the solubility of silica (silica becomes less soluble as the temperature decreases) was exceeded and the deposition of silicates resulted.

At the end of this treatment, the small tower cell was again entered; the same fill section was removed once again; and this section was again photographed. The section of fill treated with ClO_2 is shown alongside the piece of fill before treatment in Figure 4-6.

One important conclusion was made during this field trial as well as in subsequent field trials: ClO_2 has the ability to unstick the glue that holds biofilm/deposits in place. Best results are obtained when there is some turbulence or change in flow after the ClO_2 treatment to help dislodge the material loosened by the ClO_2.

Figure 4-6. Treatment of Fill with ClO_2

Before

After

C. Case History 2

Background: A mid-western grain alcohol plant produced about 62 million gallons of ethanol per year from fermentation of grain. An overview of the plant is shown in Figure 4-7. At the heart of the process was the "A" Cooling Tower which had a nominal recirculation rate of about 6000 gpm. This recirculation rate can vary significantly during the day, with flows being pinched back on some of the spiral coolers to achieve the optimum process temperature.

Historically, adequate control of microbiological growth was achieved with chlorine gas. After a plant expansion where three additional spiral coolers were installed, and after a move away from gaseous chlorine, microbiological problems began. Although plant capacity increased from 38 to 62 MM gal /yr, no expansion to the cooling system was done.

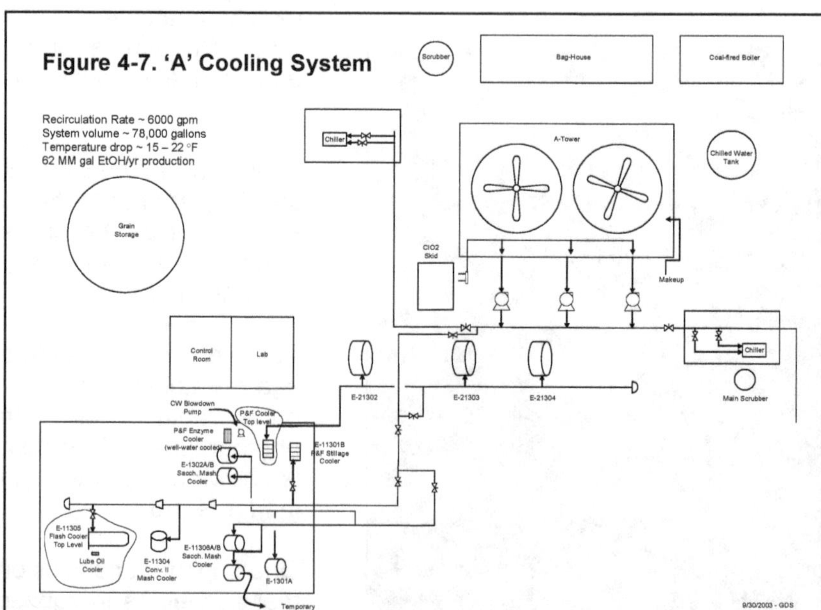

Figure 4-7. 'A' Cooling System

This small two-cell tower had a system volume thought to be around 78,000 gallons, with a ΔT of ~ 15—22 °F. The tower had high efficiency packed film fill. The fill material was composed of long plastic corrugated sheets, about 4 feet by about 10—12 feet. The fill was hung in the tower vertically and the fill lay perpendicular to the cooling tower. The fill was arranged in two rows, one toward the inside of the tower and the other toward the outside of the tower. Only the outside fill could be inspected during operation. The design was such that during hot summer months when more water flowed, the outside of the tower saw significantly more water. During the colder winter months, less water flowed outside. This design had the advantage of protecting the fill from freezing during cold weather, but it had the disadvantage of the outside fill seeing intermittent water flow, which made keeping the fill free of microbiological growth difficult.

For five years, every year since the plant expansion and prior to this field evaluation of ClO_2, the fill had collapsed due to the increase in fill weight due to biogrowth. As the cooling system was essential to maintain maximum production rates, the cooling system was not taken out of service to remove the portion of fill that had collapsed into the basin. That portion of collapsed fill was replaced, and as much of the collapsed fill in the basin that could be reached was removed after each event. Thus much of this collapsed fill was left in the tower basin. While replacement fill was relatively expensive, the cost of lost production which would result from downtime to remove the entire collapsed fill far outweighed any other economic consideration. The fill in the basin provided a place for bacteria to grow.

Numerous microbiological control methodologies had been investigated and used, including bleach, bromine, various organic biocides, and combinations of these. These chemistries had either been ineffective, very expensive, or both. Therefore, in a continuing effort to obtain a cost-effective solution to this problem, ClO_2 was considered. This case history documents the results obtained during the initial field trial and the conclusions reached.

Protocol: The ClO_2 generator used was a manual, 3-chemical, eductor-driven 250 lb/day unit [Simpson 2005]. A booster pump was included with the skid and makeup water was piped to the booster pump. A PVC header was constructed to feed a balanced flow of generated aqueous ClO_2 (~ 2000 mg/L) to the basin, just before the suction side of each of the three recirculation pumps.

The protocol normally followed was to feed ClO_2 at the rate of about 1.0 mg/L into pump suction, based on recirculation rate. Thus at a recirculation rate of ~ 6000 gpm, or 72 million pounds per day, the feedrate of ClO_2 required to achieve a 1.0 mg/L dosage was 72 lb/day.

After the generator was turned on, the ClO_2 residual was measured on a water sample collected from pump discharge and compared to the applied dosage. The difference was a measure of the ClO_2 demand of the bulk water. That is, as the measured residual approached the applied dosage, the demand for ClO_2 was being met: ClO_2 was reacting with those bulk water contaminants which created a demand for ClO_2. As the bulk water demand was met, the ClO_2 residual in the pump discharge increased and leveled off at a concentration near the applied dosage.

The residual on the hot return was also measured. This residual, when compared to the applied ClO_2 dosage, was a measure of system demand for ClO_2. (System demand is the demand of materials on surfaces for ClO_2 [Simpson 2006].

Although very low levels of ClO_2, i.e., 0.05—0.10 mg/L, have been shown by others [Nickels et al. 2000; Simpson and Miller 2000; Simpson 2001] to control biofilm effectively, plant personnel decided to adopt a more conservative approach and use higher dosages.

Results: Baseline data was taken for about 2 weeks prior to the startup of this field trial (Table 4-1). At the start of the field trial, bleach was being fed to the back of the tower to "sweep" the basin. A chlorine residual of at least 2.2 mg/L (the upper limit of the Hach DPD free chlorine test) was present just prior to the implementation of ClO2 feed. The bleach feed system was turned off prior to the implementation of the ClO_2 feed. After about 15—20 minutes, the chlorine residual dropped to below detectable levels.

Generator startup was uneventful. Given the condition of the fill and the possibility of an unknown organic contamination in the tower, an initial dosage of 2.0 mg/L was targeted. The flow of ClO_2 solution to each of the 3 injection stingers was balanced. Several times per day, ClO_2 residuals were measured on the supply and hot return. As there was no convenient sample point on the hot return, plant personnel removed part of the cover on the deck and left if off so that a sample of water returning to the tower could be taken during the trial.

Initially, at an applied dosage of about 2.0 mg/L, the dosage measured in pump discharge immediately increased to about 2.2 mg/L. This difference may have indicated that our estimated flow was high by about 10%, or the difference may have been due to testing interferences with DPD (the most common test for chlorine). This increased dosage indicated that the bulk water was well disinfected and that there were few contaminants in the bulk water which would consume ClO_2. The return residual was slightly less.

Table 4-1. Background Data for Field Trial

	Cl_2	Turb	T.H.	Alk	pH	Cu	Fe	ΔT
9/15	0.03—022	-	535—350	280	7.8—7.9	-	-	20—23
9/16	0.14—0.41	-	420—660	270	7.7—8.4	-	-	21—22
9/17	0.23—0.53	-	750—900	208—350	7.6—8.2	-	-	16—21
9/18	0.15—0.60	-	640—745	210—220	7.7—7.8	-	-	23—23
9/19	0.43—2.02	33	600—765	200—240	7.8—8.0	0.0	.19	23—25
9/20	0.41—0.85	-	700—750	200—280	7.8—8.0	-	-	21—23
9/21	0.13—0.30	-	650—720	200—230	7.8—8.1	-	-	20—23
9/22	0.04—0.30	50	485—680	200—280	7.7—7.9	.007	.22	23—25
9/23	0.02—0.14	44	570—710	200—220	7.7—8.1	.005	.15	20—24
9/24	0.09—1.79	91	830—1000	220—250	7.6—7.8	.042	.60	23—24
9/25	0.21—0.43	29	730—775	200—220	7.4—8.0	0.0	.12	21—26
9/26	0.0—0.11	50	515—690	200—230	7.8—8.0	.005	.14	20—20
9/27	0.14—0.22	-	895—930	200—240	7.8—8.0	-	-	19—24
9/28	0.17—2.08	-	640—750	200—210	7.2—7.9	-	-	22—25
9/29	-	30	-	-	-	0.0	.06	-
9/30-a	-	31	-	-	-	0.0	.07	-
9/30-p	-	28	-	-	-	-	-	-
10/1-a	-	19	-	-	-	0.0	0	-
10/1-p	-	12	-	-	-	0.0	.06	-
10/2-a	-	14	-	-	-	0.0	.23	-

Then, with time, the residual of ClO_2 in the supply water began to decrease. By the end of the day, the residual had decreased to about 25% of what it had been initially, as shown in Figure 4-8.

An ethanol leak could explain the decline in the residual, either directly as a result of some component of the ethanol stream reacting with ClO_2, or indirectly as a result of increased bacterial activity. To investigate the possibility of a leak, the generator was turned off at the end of day 1, and no additional microbiological control chemicals were added. If a leak existed, and if an increase in ClO_2 demand occurred, the generator would have to run longer on startup of day 2 to achieve the same residual.

Figure 4-8. ClO_2 Residual Supply and Return – Day 1 – 9/30

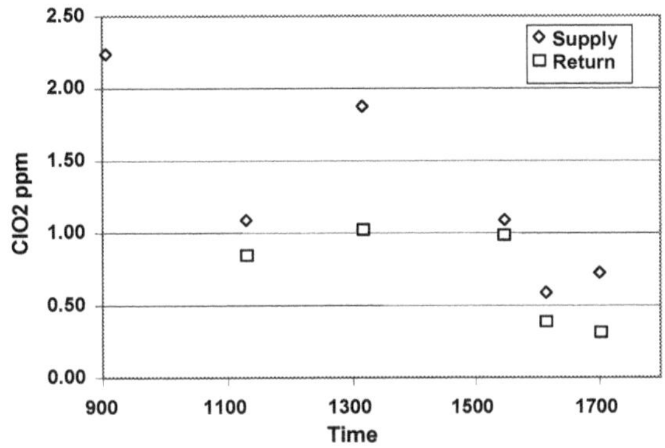

On day 2, the generator was turned on, and a good residual was measured immediately in pump discharge and also in the hot return, suggesting that a leak was not likely. Initially, the residual was almost as high as it had been on the previous day. Then, as the day progressed, the residual began to decline. About half way through the day, the residual hit a minimum and began to increase. By the end of day 2, the residual had increased to near the applied ClO_2 dosage. This trend is shown in Figure 4-9.

Figure 4-9. ClO_2 Residual Supply and Return - Day 2 – 10/1

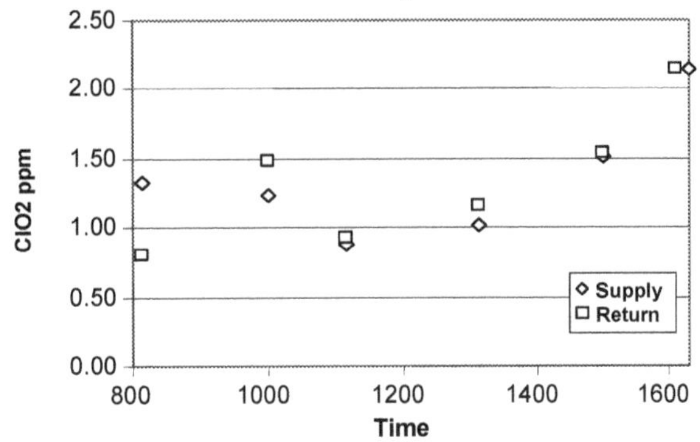

The generator was left on during the night of day 2 and during day 3 with the dosage unchanged to see if the variability in residual was a phenomenon of plant operation or if it was a phenomenon of the "cleanliness" of the system. The trend observed in day 3 is shown in Figure 4-10.

Figure 4-10. ClO_2 Residual Supply and Return - Day 3 – 10/2

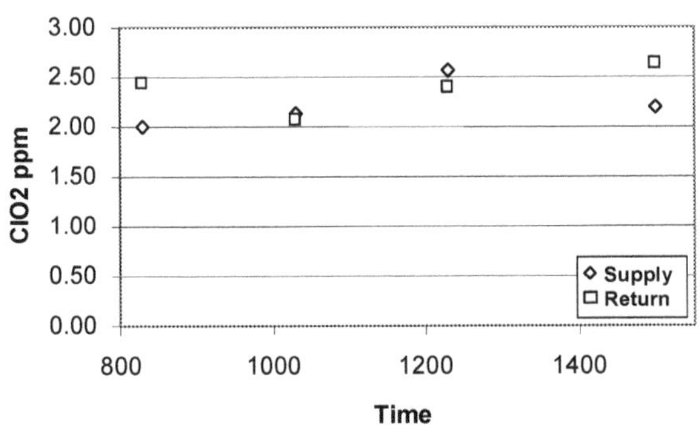

At the end of day 3, it seemed that the residual variability observed during days 1 and 2 was a phenomenon of system cleanliness, not of cooling water flow variability. However, there was a plant outage at this time, so this conclusion must be viewed with caution. The supply/return residual of ClO_2 remained close to the applied dosage over day 3, indicating that the system had become relatively clean, in that everything that reacts rapidly with ClO_2 had reacted. Then, the applied dosage was reduced to about 0.6 mg/L at the end of day 3. The goal was to allow the ClO_2 to be applied continuously as long as possible during this evaluation phase. Reducing the feedrate to this level was expected to produce results comparable to that at 2.0 mg/L while minimizing the amount of ClO_2 being lost over the cooling tower.

Inspection of the fill before ClO_2 showed a substantial buildup of biofilm in the fill. Some sections of fill were easy to move apart, and some were very compressed. Periodic inspection of the fill during days 1—3 showed increasing improvement. At the end of day 3, when the fill was pulled apart for inspection, the biofilm fell in sheets from the fill. The ClO_2 had apparently reduced or inhibited the ability of the biofilm and associated bacteria to attach to the surfaces. As a result, the biofilm was readily removed. Several plant personnel noted the improvement.

Iron and Copper Levels. Figure 4-11 shows the iron and copper levels from the background through the first few days of ClO_2 treatment. In general, while the iron and copper levels were acceptable prior to ClO_2, the data indicated an improvement with the implementation of ClO_2.

Figure 4-11. Cu- Fe Before-After ClO_2

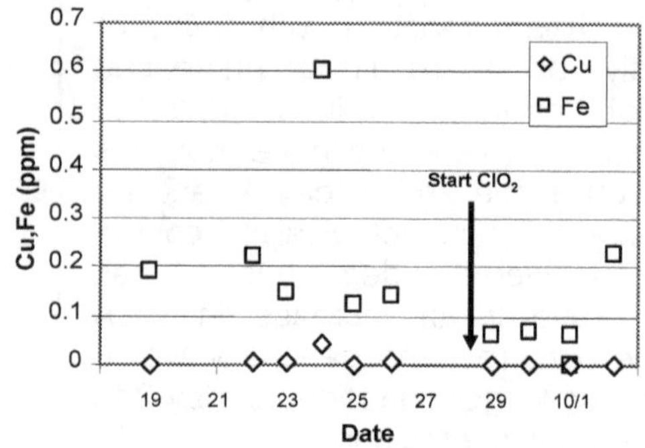

Figure 4-12. Turbidity Before – After ClO_2

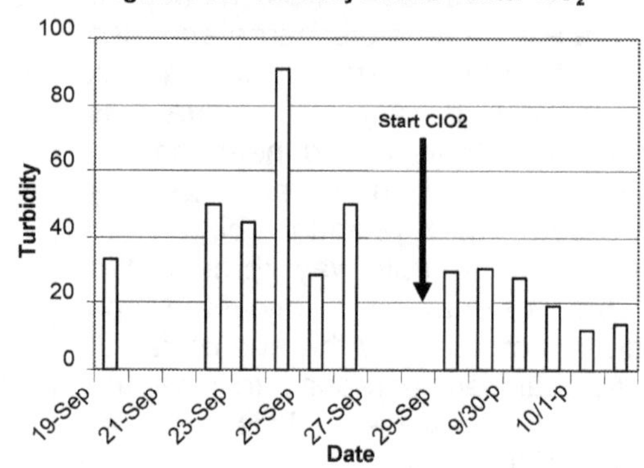

Turbidity. Figure 4-12 shows turbidity as the tower transitioned from chlorine bleach to ClO_2. The clarity of the water improved throughout the trial.

Figure 4-13. Cl_2 Residual Before ClO_2

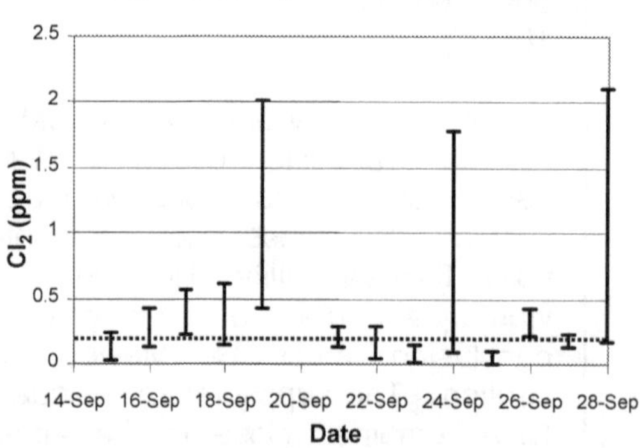

Chlorine Residual. Figure 4-13 shows the highs and lows of the chlorine as measured on a daily basis by plant personnel during the pre-trial data-gathering period. Much of the time, the measured Cl_2 residual fell below the dotted line at 0.2 mg/L Cl_2. Given the interferences, a good assumption is that if the Cl_2 concentration is measured to be < ~0.2 mg/L with DPD, the Cl_2 residual can be assumed to be zero. This rule is particularly true for systems heavily contaminated by anaerobic bacterial biofilm. The low chlorine level coupled with the pH of operation reduced the effectiveness of the microbiological program. Effective use of chlorine in such a system as this requires that a residual of chlorine of > 0.2 mg/L be present at all times. Unfortunately, it is difficult to insure that a free chlorine residual is present at all times. Once the residual is lost, bacteria begin to grow biofilm. It is difficult to recover with chlorine. For example, one water treatment expert has said, "It is not possible to clean up a biofouled system simply by resuming a discontinued microbiological control program (speaking of chlorine). A fully developed biofilm can be removed only through intensified on-line or off-line treatment..." [Kozelski 1983]. Fortunately a microbiological program based on chlorine dioxide can recover. This fact allows intermittent treatment with chlorine dioxide, even in fairly contaminated systems and with low ClO_2 dosages.

Bacteriology. Figure 4-14 shows the results of using serial dilution bottles (one common method for measuring bacterial activity [Dixie 2000]) for aerobic bacteria. With this method, sequential dilutions of water to be tested are inoculated into the bottles, which contained bacterial nutrient and a color indicator. A color change after 3—4 days provided a positive indication of bacterial activity. For example, if five bottles turned, the bacterial growth was in the 10^5 colony forming units per mL (CFU/mL).

Figure 4-14. Aerobic Bacteria

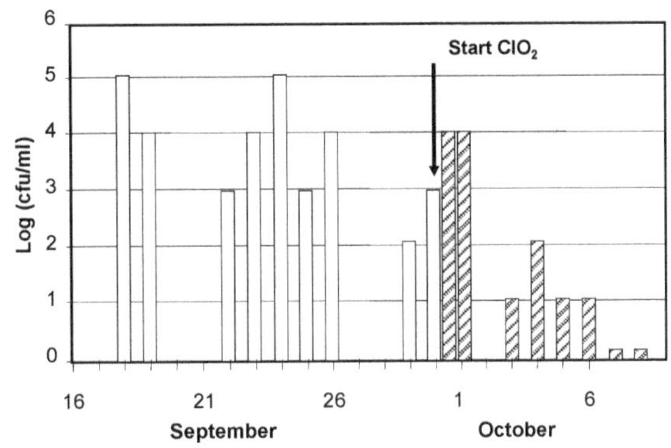

As only five bottles were inoculated (typically this is sufficient), when all five bottles changed color, bacterial levels were *at least* 10^5. Bacterial results improved throughout the trial.

Figure 4-15. Sulfate Reducing Bacteria

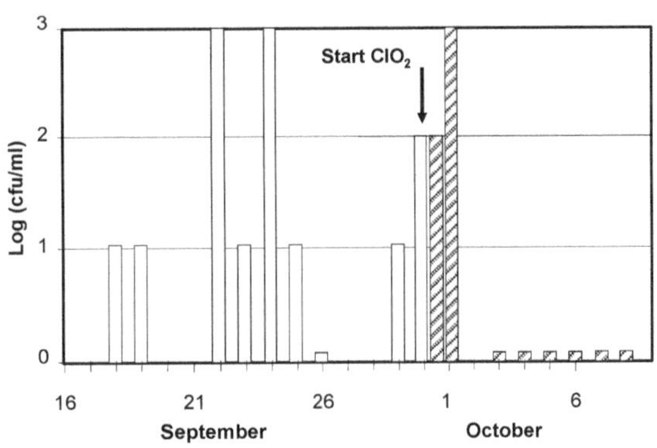

Figure 4-15 shows the results of the anaerobic sulfate-reducing bacteria (SRB) vials. SRB are a serious potential problem because they are predominantly sessile, i.e., they reside on surfaces. They produce H_2S and are particularly troublesome on stainless 304 metallurgy. Their mode of attack can be to turn this particular metal into swiss cheese, with the SRB forming tunnels which run throughout the metal. In general, the data shows a marked improvement with implementation of ClO_2. It is not unusual to see one or two high readings after initiation of feed of ClO_2 because the ClO_2 strips biofilm and releases the anaerobic bacteria lying underneath.

Today, the plant continues to use ClO_2 and has optimized usage to insure optimum cost-effectiveness. Results have continued to improve, the water turbidity is better than ever, and the fill, which had been a major maintenance problem, has continued to clean up.

Discussion: A number of questions that relate to the trial and ClO_2 related have been posed.

1. **Why did we see an initial ClO_2 residual near the applied dosage, and why did the dosage decrease with time?** With the amount of bleach being fed, and with the high

residual that was present, the bulk water was fairly well disinfected. Thus, the concentration of contaminants in the bulk water required little ClO_2. Two possibilities exist which explain the decrease in ClO_2 residual in the supply water. The first possibility is that there was a significant increase in recirculation rate during days 1 and part of day 2. In order for this to be true, the recirculation rate would have to have increased from about 6,000 gpm to about 18,000—24,000 gpm, which would be unreasonable given the design of the plant. The second possibility involves the ability of ClO_2 to cause sloughing of biofilm. As ClO_2 passes over the fill, biofilm is sloughed and the biofilm disperses into the bulk water, causing additional demand for ClO_2. When this phenomenon occurs, something is dispersed which reacts with the ClO_2. This "something" might be bacterial biofilm, or, more probably, given the speed of reaction which would be required, might be anaerobic biofilm which is loaded with SRB and consequently sulfides. The reaction of ClO_2 with reduced sulfur compounds, and in particular sulfide, is very rapid. The second phenomenon has been observed in other cooling towers and explains why SRB, though not measured initially, are commonly found at very low levels during a field trial.

2. **After 3 days, the applied, supply, and hot return ClO_2 residuals are pretty close. Does this mean that the fill is clean?** Inspection of the fill showed this not to be the case. It actually means that the ClO_2 had reached and dislodged the easy-to-remove material. The phenomenon is much like on-line cleaning of boiler scale. Although removal of boiler scale can be done, it is a very slow, time-consuming process. In general, it is much easier to prevent biofilm (or boiler scale) formation than it is to remove it once it has formed. In this case, a highly compacted mass of biofilm existed inside the cooling tower fill. In some of the slices of fill, a great deal of cleaning had taken place. Those slices of fill not reached by ClO_2 due to very little water flow remained fouled. In the Southeast corner of this particular tower, biofilm appeared to be very concentrated, and it was in this area that the fill appeared to be most highly compacted, judging from the difficulty of separating the fill for inspection.

3. **Why is ClO_2 promoted as being so much better than chlorine for biofilm control?** Biofilm penetration and effect of chlorite ion are two primary reasons for the promotion of ClO_2.

   Chlorine penetrates a bacterial biofilm very slowly. It reacts with biofilm fragments at the surface of the biofilm and is inactivated well before it can penetrate and disinfect the bacteria inside [Chen and Stewart 1996; de Beer et al. 1994; Xu et al. 1996]. In a cooling tower, the effectiveness of chlorine is a function of the rate at which chlorine penetrates the biofilm vs. the rate of production of biofilm by bacteria. The outcome of this competition depends upon the type and amount of available nutrient. In cooling towers with high nutrient loading (i.e., systems where the rate of biofilm formation > the rate of biofilm removal), oxidants such as chlorine and bromine are ineffective [Trulear and Wiatr 1988].

   The effectiveness of ClO_2 is directly related to its reaction chemistry. Although ClO_2 has the capacity to accept up to 5 electrons in an oxidation-reduction reaction, under normal circumstances it accepts only one electron and forms chlorite ion, ClO_2^-. Although it is not very effective as a disinfectant, in general the chlorite ion plays a significant role in biofilm removal. The role of chlorite ion in controlling or removing biofilm has been described in more detail in another report [Simpson 1995].

One group of researchers has shown that adding a chlorite solution to a system laden with biofilm resulted in removal of the biofilm once a bacterial nutrient has been added to the water [Clark and Langley 1990]. The nutrient caused a rapid increase in metabolism of bacteria present in the biofilm, producing acids. These acids caused an activation of the chlorite solution to form ClO_2, and the ClO_2 formed removed the biofilm.

In other reports [Masschelein 1991], similar observations as to the slow bacterial recovery of systems treated with ClO_2 have been described. Thus, the chlorite ion formed when ClO_2 reacts is available to be reactivated to ClO_2 when it encounters a part of the biofilm that is metabolically active.

4. **What are the factors involved in the growth and development of biofilm and why can't chlorine control it in some cases?** Several well known factors are involved in biofilm development. Of these, the amount and nature of the nutrients available to the bacteria are two of the most important. In a water system, the first thing bacteria do is to attach to surfaces and produce a polymeric substance (biofilm) that protects them from external toxins. Once a biofilm is established in a system, keeping the system free from biofilm is a race between how fast chlorine can burn the biofilm off and how fast the bacteria can produce the biofilm. In systems highly contaminated by materials such as ethanol that are easily metabolized by bacteria, the bacteria may win the race. Thus, maintaining microbiological control in such systems can be challenging when chlorine is used.

5. **A common question asked is, "I am currently using X lbs of chlorine gas. How much ClO_2 will it take to replace it?"** This is a difficult if not impossible question to answer. The reason lies in the differences between the reaction chemistries of the various oxidants. Although chlorine and chlorine dioxide are both classified as oxidants and have chlorine in the molecule, they exhibit considerable differences with regard to their reaction characteristics. The pH dependence, reaction with organics, and reaction with ammonia nitrogen or organically bound nitrogen (such as protein) vary with each oxidant.

   - **Chlorine:** Chlorine is a strong oxidant. When chlorine gas is added to water, the chlorine reacts with the water to form hypochlorous acid [HOCl: the oxidation potential = 1.49V)] and hydrochloric acid. Hypochlorous acid is a powerful disinfectant.

     - The pH dependence of disinfection with chlorine is well known. Depending upon pH, hypochlorous acid can dissociate to form hypochlorite ion [OCl^-]. At a pH of about 7.5, for example, about half the chlorine will exist as hypochlorous acid and half as hypochlorite ion.
     - Hypochlorite ion is a notoriously poor disinfectant; testing has shown the biocidal efficacy of hypochlorite to be from 1/20 to 1/300 as effective as HOCl, depending upon the organism tested [Robeck 1981].
     - Chloramines are stable byproducts of reactions of chlorine with various nitrogen containing compounds. They are generally much less biocidally active than HOCl [White 1986].

   - **Chlorine Dioxide:** Chlorine dioxide (aqueous ClO_2 oxidation potential = 0.95V) reacts exclusively by oxidation. The primary reaction product of chlorine dioxide is chlorite ion, ClO_2^-. Although chlorite ion is known to have poor disinfectant properties, it has been shown to retard regrowth in a disinfected system [Pickrell et al. 1999].

- ClO_2 remains a gas when dissolved in water and, unlike chlorine or bromine, does not react significantly with the water.
- As a result, chlorine dioxide disinfection remains relatively constant over a fairly wide pH range, i.e., from about 6—10.
- ClO_2 at concentrations commonly used in cooling water applications (i.e., ~ 0.1— ~ 1.0 mg/L) reacts not at all with ammonia or primary amines and only very slowly with secondary amines.

In summary, a pound-to-pound comparison of oxidants cannot generally be made because with operating pH, chlorine can exist as hypochlorous acid and hypochlorite ion. Each has different biocidal properties. In addition, differing system conditions such as the nature and amount of organic contaminants, as well as the presence or absence of ammonia or amines, creates a significantly different demand for chlorine vs. ClO_2. Because of these factors, no general comparison of chlorine with chlorine dioxide can be made. In a real sense, it is like comparing apples and oranges.

D. Conclusions

Two important conclusions can be made from these field trials.

1. ClO_2 can effectively prevent/control/remove bacterial biofilm. This conclusion has significant implications for those cooling water systems for which cooling is critical [Pickrell et al. 1999].

2. ClO_2 can effectively disinfect cooling water systems. This conclusion has significant implications for those cooling systems where pathogenic bacteria such as Legionella, reside [Miller and Simpson 1999]. A number of organizations have realized the effectiveness of ClO_2 for control of Legionella, including the Joint Commission on Accreditation of Healthcare Organizations (JCAHO), the Association of Water Technologies (AWT), the Building Services Research and Information Association (BSRIA), the Health and Safety Commission (HSC), the Health and Safety Executive (HSE), and the American Society of Healthcare Engineering (ASHE).

The reaction chemistry of ClO_2 is not the reaction chemistry of Cl_2 or bleach. Therefore, old rules of thumb regarding measured residuals, points of application, dosages, and effectiveness do not apply. This has significant implications when it comes to selecting a vendor to apply ClO_2. The end user should select a company that has extensive experience with ClO_2. It is not good for end users to allow a water treatment company to use their cooling system as a training tool. Many large, well-established water treatment suppliers have argued against the use of ClO_2 in the past but have now been forced to accept it as a viable alternative. The increasing number of organizations that promote it and the increasing number of successful case histories [Robeck 1981; Simpson et al. 1993; Simpson 1994; Simpson et al. 1994] have overcome objections to its use. Care should be taken when selecting a ClO_2 supplier.

In summary, ClO_2 works when nothing else does. If one is going to hunt rabbits, then any gun will work. If one is going to hunt elephants, one needs a slightly larger caliber. ClO_2 is the larger caliber. If a system is relatively clean, then anything will work. If the system is highly contaminated, a number of chemicals <u>might</u> work, but ClO_2 <u>will</u> work, every time.

E. References

[ASHE] ASHE (American Society of Healthcare Engineering): In terms of Risk Assessment, guidelines include maintaining and auditing those systems designed to limit Legionella amplification in cooling systems and potable water treatment systems, e.g., copper/silver or chlorine dioxide.

[AWT] AWT (Association of Water Technologies): The AWT position paper on Legionella control, suggests that ClO_2 be considered as an alternative to chlorine or bromine. Specifically, AWT members are encouraged to consider "...chlorine dioxide, a proven biocide effective against Legionella without posing the corrosion problems associated with halogen biocides...."

[BSRIA] BSRIA (Building Services Research and Information Association): In the UK the BSRIA has found chlorine dioxide to be very effective against biofilm and Legionella under a variety of conditions.

[Chen and Stewart 1996] Chen, X. and Stewart, P., "Chlorine Penetration into Artificial Biofilm Is Limited by a Reaction-Diffusion Interaction," *Environmental Science and Technology*, 30(6), 2078(1996).

[Clark and Langley 1990] Clark, J. and Langley, D., "Biofilm Control," United States Patent 4,929,365, May 29, 1990.

[de Beer et al. 1994] de Beer, D., Srinivasan, R., and Stewart, P., "Direct Measurement of Chlorine Penetration into Biofilms During Disinfection," *Applied and Environmental Microbiology*, 60(12), 4339(1994).

[Dixie 2000] Dixie Products and Testing, Serial Dilution Bottles, Houston, TX, 2000.

[Gill et al. 1994] Gill, J., Yuke, M., Donlar, R., and Gibbon, D., "Fouling of Film Forming Cooling Tower Fills – A Mechanistic Approach," CTI Annual Meeting, Houston, TX, Feb 13-16, 1994.

[Hobson and Fitchett 1992] Hobson, E. and Fitchett, G., "The Effect of Cooling Water on Cooling Tower Plastic Pack Fouling," International Water Conference, 1992.

[HSC] HSC (The Health and Safety Commission): The HSC of the UK recommends halogen, including chlorine dioxide, for control of Legionella in cooling towers and hot water systems.

[HSE] HSE (Health and Safety Executive): A risk assessment document published by the UK HSE states that the major use of ClO_2 in the UK is "for the treatment and control of Legionella, as a sterilizing agent for hospital surgical equipment, for use in water treatment and as a biocide for controlling microorganisms in water during fruit and vegetable processing."

[JCAHO] Joint Commission on Accreditation of Healthcare Organizations (JCAHO) EC 1.7.

[Kozelski 1983] Kozelski, K., "Field Experience with a Simple Cooling Water Biofilm Monitoring Device," TP IWC-83-46, International Water Conference, Pittsburgh, PA, October 24-26, 1983.

[Laxton et al. 2001] Laxton, G., Sylvester, C., and Cornelius, S., "ClO_2 Can Provide Improved Performance and Decreased Cost in Ammonia Plant Cooling Water Systems," Gulf Coast Ammonia Producers Conference, Baton Rouge, LA, April 9-11, 2001.

[Masschelein 1991] Masschelein, W., "Chlorine Dioxide," Chemical Oxidation, Technology for the Nineties, First International Symposium, Vanderbilt University, Nashville, TN, February 20-22, 1991.

[Meek 1967] Meek, G., "Cellular Cooling Tower Fill," *Heating, Piping and Air Conditioning*, 84(December 1967).

[Miller and Simpson 1999] Miller, J., and Simpson, G., "Chemical Control of Legionella," Association of Water Technologies Annual Meeting, Palm Springs, CA, October 26-30, 1999.

[Monjoie et al. 1993] Monjoie, M., Noble, R., and Mirsky, G., "Research of Fouling Film Fill," *CTI*

Journal, 15(1), 20(1993).

[Nickels et al. 2000] Nickels, A., Harriman, C., Nickels, A., Freedman, A., and Harris, A., "A Novel Method of Applying Chlorine Dioxide in a Recirculating Cooling Water System," AWT Annual Convention and Exposition, Water Technologies 2000, Honolulu, HI, October 31-November 4, 2000.

[Pickrell et al. 1999] Pickrell, W., Germer, M., Miller, J., and Simpson, G., "Plants Increase Production Through Use of Chlorine Dioxide," Practical Industrial Water Treatment Technology for the New Millennium, Houston, TX, January 27-29, 1999.

[Robeck 1981] Robeck, G., "Chlorine, Is There a Better Alternative?" *Science of the Total Environment*, 18, 235(1981).

[Robinette and Tvedt 1999] Robinette, D. and Tvedt, T., "Cooling Tower Film Fill Fouling: Causes, Cleanup Techniques and Operational Guidelines," Practical Industrial Water Treatment Technologies for the New Millennium, Houston, TX, Jan 27, 1999.

[Simpson 1994] Simpson, G., "Chlorine Dioxide Offers a Cooling Tower Option," *Water Technology*, 17(4), 80(April 1994).

[Simpson 1995] Simpson, G., "Biofilm: Removal and Prevention with Chlorine Dioxide," The Third International Symposium on Chlorine Dioxide: Process Water, Drinking Water and Water Waste Issues, New Orleans, LA, September 14-15, 1995.

[Simpson 2001] Simpson, G., "Control of Biofilm in Cooling Towers with ClO_2," Fourth International Symposium on Chlorine Dioxide, Las Vegas, NV, February 15-16, 2001.

[Simpson and Miller 2000] Simpson, G. and Miller, J., "Control of Biofilm with Chlorine Dioxide," AWT Annual Convention and Exposition, Water Technologies 2000, Honolulu, HI, October 31-November 4, 2000.

[Simpson et al. 1993] Simpson, G., Laxton, G., Miller, R., and Clements, W., "A Focus on Chlorine Dioxide: Biocide of Choice for 'Stressed' Cooling Water Systems," WaterTech 93, Houston, TX, November 10-12, 1993.

[Simpson 2005] Simpson, G., <u>Practical Chlorine Dioxide: Volume I – Foundations</u>, Greg D. Simpson and Associates, publishers, 2005.

[Simpson 2006] Simpson, G., <u>Practical Chlorine Dioxide: Volume II – Applications</u>, Chapter 16 – More Frequently Asked Questions, Greg D. Simpson and Associates, publishers, 2006.

[Simpson et al. 1994] Simpson, G., Laxton, G., Ferriola, J., and Hogg, R., "A Focus on Chlorine Dioxide: Increasing Mold Life in Electric Arc Furnace Mini Steel Mills," WaterTech '94, Houston, TX, November 9-11, 1994.

[Trulear and Wiatr 1988] Trulear, M. and Wiatr, C., "Recent Advances in Halogen Based Biocontrol," Technical Paper No. 19, Corrosion 88, St. Louis, MO, March 21-25, 1988.

[White 1986] White, G. C., <u>Handbook of Chlorination</u>, Van Nostrand Reinhold Company, New York, 1986.

[Xu et al. 1996] Xu, X., Stewart, P., and Chen, X., "Transport Limitation of Chlorine Disinfection of *Pseudomonas aeruginosa* Entrapped in Alginate Beads," *Biotechnology and Bioengineering*, 49, 93(1996).

[Yu et al. 1998] Yu, F., Dallmier, A., and McCoy, W., "Fill Fouling Control in Cooling Towers," *CTI Journal,* 19(2), 34(1998).

Disinfection of Influent Water with ClO_2

A. Introduction

One of the earliest, if not the earliest, applications for chlorine dioxide was to control taste and odor in potable water [Synan et al. 1944; Synan et al. 1945; Mounsey and Hagar 1946; Vincent et al. 1946; Aston 1947; Faber 1947; Ingols and Ridenour 1948; Palin 1948; Coote 1950; Harlock and Dowlin 1953; Ringer and Campbell 1955]. Chlorine dioxide has proven to be so successful for this application that its use continues to the present day [Medsker et al. 1968; Barnhart 1980; Brett and Ridgeway 1981; Lalezary et al. 1986; Walker et al. 1986; Myers and Thompson 1986; Myers et al. 1986; Dietrich et al. 1995; Dickens and Graham 1995; Froese et al. 1999].

In addition to the numerous immediate benefits arising from the use of ClO_2 for influent treatment, there are a number of significant but lesser known benefits which occur downstream. Using ClO_2 has proven beneficial for cooling towers and for ion exchange systems which use the treated water as makeup and certainly other benefits may be observed, but few reports have been made of such treatments.

The use of ClO_2 is experiencing a resurgence in many industrial applications. It is being used increasingly in pulp bleaching of the pulp and paper industry [Correll 1992]. Replacement of chlorine by chlorine dioxide in pulp bleaching solves numerous environmental problems facing modern paper mills. Many of the toxic species that appear in the plant effluent disappear when chlorine dioxide is used [Axegard 1986]. It is also being used increasingly in industrial cooling towers [Ambrogi 1993; Geraci et al. 1993; Simpson 1994; Simpson et al. 1993a; Simpson et al 1993b]. And it is also being used to control trihalomethanes in potable water [Aieta 1986] and odor control [Pitochelli and Mason 1994]. A more detailed description of potential applications can be found elsewhere [Simpson 2005a]

The resurgence in use of ClO_2 is due in part to a reduction in use of gaseous chlorine as a disinfectant. Gaseous chlorine has been used historically because it is relatively inexpensive and very effective as a disinfectant. Many plants are considering alternatives to gaseous chlorine, including bleach, ozone, bromine, or chlorine dioxide [Simpson et al. 1993a].

Chlorine dioxide has unique performance characteristics that in many cases make it the ideal choice for disinfection or microbiological control. The following case history documents the immediate benefits of replacement of chlorine bleach with ClO_2, as well as several of the benefits to downstream users of the water produced.

B. Bleach for Disinfection of Influent Water

CWA, Coastal Water Authority, provides water to many of the industrial facilities along the Houston Ship Channel. One Gulf Coast industrial facility disinfected, lime softened, and clarified this water, and used it for internal plant water needs, including makeup to the demineralizers and cooling towers. This facility, shown schematically in Figure 1, also provided treated water to several neighboring plants.

Chlorine and ammonia were added by CWA to form chloramine to provide some disinfection to the water upstream and to help downstream drinking water facilities meet their Ct (concentration X time) requirements [USEPA 1991]. (Ct is a factor that is specific for a given disinfectant and a specific system. The stronger the disinfectant, the lower the concentration (C) and the shorter the contact time (t) which are required for adequate disinfection.) The disinfecting ability of chloramine is poor, which is demonstrated by EPA regulations requiring potable water facilities to achieve primary disinfection by use of an oxidant other than chloramine. Chloramine in such systems is used only to provide a measurable residual in the distribution system. Although no Ct credit for chloramine is generally given in a municipal plant, in this instance the high residence time from the point where chlorine and ammonia were added to the influent line allowed downstream users some contribution to their Ct requirements. In fact, it was sufficient, when added to the normal disinfection program of the plants using this water downstream to produce potable water, to meet minimum regulatory standards.

Figure 5-1. Plant Overview

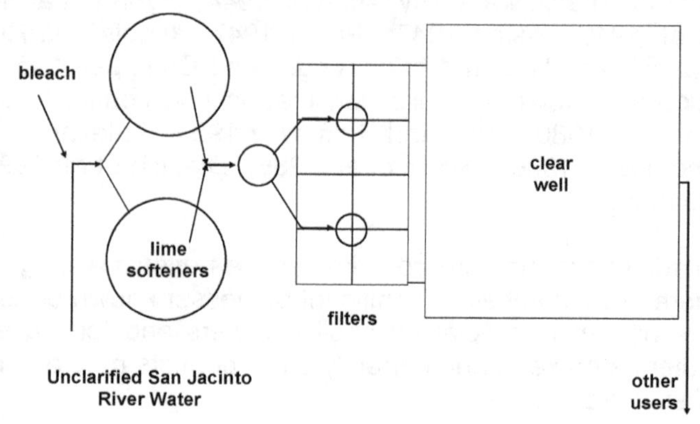

The presence of free ammonia or combined chlorine in the water provided a significant demand for bleach feed. Every mg/L of free ammonia requires 8—10 mg/L of chlorine to achieve breakpoint chlorination [White 1986a]. Thus, the bleach feed required to give a free chlorine residual in this industrial facility was significant. In addition, the degradation of the bleach required an even higher feedrate of bleach. The high pH, i.e., ~ 10, of the lime softened makeup also limited the disinfection ability of the hypochlorous acid, as at this pH the hypochlorous acid is essentially completely dissociated to hypochlorite ion, which is relatively ineffective as a disinfectant [White 1986b].

The excessive use of bleach at this facility led to a consideration of chlorine dioxide. Although chlorine dioxide costs more than bleach on a per-pound basis, differences in dosage requirements were thought to be significant enough to make chlorine dioxide economically attractive.

Even without the differences in dosages due to chloramines or ammonia, substantially less chlorine dioxide would normally be required. The primary reason is that chlorine dioxide, unlike chlorine or bleach, does not react with the vast majority of organics found in water. In addition, chlorine dioxide has a higher oxidation capacity in that it can accept up to 5 electrons in an oxidation-reduction reaction (although in most cases ClO_2 reacts to accept one electron to form chlorite ion), unlike chlorine which can accept only two. Thus chlorine dioxide is said to have roughly two and one half times the oxidation capacity of chlorine. The net result of these differences is that water generally shows a considerably higher demand for chlorine than the same water does for chlorine dioxide. For example, testing of San Jacinto River water in one plant along the Houston ship channel resulted in a demand of over 8 mg/L for chlorine and slightly less than 2 mg/L for chlorine dioxide.

C. The Impact of Bleach Replacement with ClO_2

The most direct effect of bleach replacement by chlorine dioxide was the improvement in the quality of the treated water leaving this facility. Microbiological counts while on bleach were < 10^1—10^2 cfu/ml, as measured with phenol red serial dilution bottles. However, filtering the treated water with a 0.45 μ Millipore filter resulted in the filter's having a slight greenish color while the facility was using bleach. The green color came from algae which were growing in the open million-gallon clear well due to the relative ineffectiveness of bleach on algae at this pH. In fact, one downstream customer that makes white polystyrene packing material had complained when algae growth was sufficient to turn their white polystyrene packing material green. Both the green color and the complaints disappeared when chlorine dioxide feed was implemented. Bacterial counts were further reduced to < 10^1 cfu/ml.

Another major effect of bleach replacement by chlorine dioxide was that the conductivity of the treated water decreased, due primarily to discontinuing the high bleach feedrate and removing the caustic soda and sodium chloride contained in the bleach. Bleach is essentially chlorine gas reacted with an excess of free caustic [Solvay 1960], as shown in Equation 5-1.

$$Cl_2 + NaOH\ (excess) \Rightarrow NaCl + HOCl\ (+\ excess\ NaOH) \qquad (5\text{-}1)$$

For every molecule of HOCl required to meet the demand of the water, one sodium ion and one chloride ion are also added, as shown by the chemical equation, Equation 5-1. This does not count the extra sodium ions coming from the excess caustic used for bleach stabilization. And, it does not count the increased concentration of sodium ions that would result from increased feed of bleach due to bleach degradation, which was significant in this case study.

Most of the water treated by this facility was utilized within its own plant as cooling tower makeup, process water, and feed to the demineralization system. As this plant also provided demineralized water to nearby plants, the demineralizer train used a substantial amount of the treated water. The decreased conductivity had a significant impact on demineralizer run lengths. Several days after the change from bleach to chlorine dioxide, demineralizer run lengths increased significantly, initially about 4%. Chlorine dioxide, produced by mixing an aqueous solution of sodium chlorite, industrial bleach, and hydrochloric acid, was fed at a substantially lower dosage than was bleach providing significant savings in terms of demineralization regeneration chemicals and helping minimize outages required for regeneration. Such results were important because the facility was near maximum capacity on the demineralizer trains.

The increase in run length led to a change from the three-chemical generation chemistry (HCl, bleach, and chlorite) to the two-chemical generation method (gaseous chlorine/chlorite). Thus, demineralizer run lengths were improved even further, shown in Figure 5-2, by eliminating sodium from the bleach used during the generation process.

Figure 5-2. Demineralizer Run Lengths

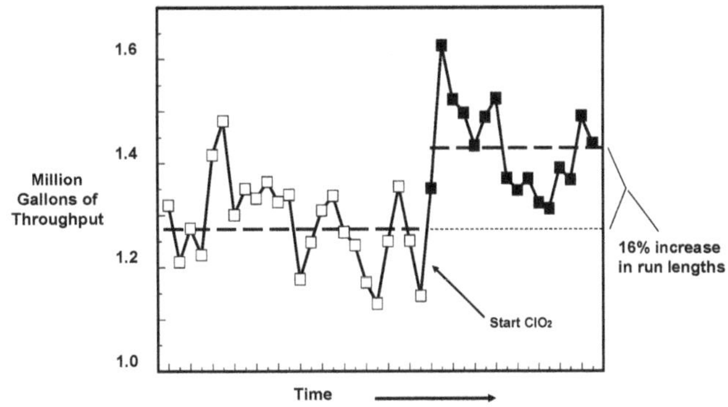

Another benefit the plant saw from replacement of bleach by chlorine dioxide was the reduced microbiological control chemicals required by the cooling towers. Cooling tower makeup at this facility typically contained about 0.2 mg/L chlorine dioxide, with about 0.8 mg/L chlorite ion. The presence of chlorine dioxide and chlorite had a substantial effect on algae, bacterial biofilm, and heterotropic bacteria present in the bulk water [Simpson 1995]. In one report, chlorite was shown to have a significant effect on bacterial regrowth in a potable water distribution system [Masschelein 1991]. While chlorite is not a very effective biocide, it can retard aftergrowth of bacteria in a system. In addition, the presence of chlorite also provides a significant amount of control over the formation of bacterial biofilm.

In addition to the increased demineralizer run lengths observed by this plant, other plants that purchased clarified water also noticed improvements. For example, algae on cooling tower support structures and decks disappeared or were greatly reduced. Increased demineralizer run lengths in other plants were noted. The alkalinity of at least one cooling system decreased due to the elimination of caustic soda that was present in the bleach, requiring a change in their chemical treatment program.

One nearby cooling tower which historically had problems with high efficiency PVC fill showed a substantial improvement in cooling efficiency [Thomas 1996].

D. The Economics of Replacing Bleach with ClO_2

Chlorine dioxide was a very cost-effective replacement for bleach. Bleach at feedrates of up to 1400 gallons per day was replaced by approximately 80—120 lbs. of chlorine dioxide per day with *improved* disinfection. Algae, which was present continually in the influent clarifiers/lime softeners with chlorine, were effectively controlled with ClO_2. In a straight dollar-to-dollar comparison, chlorine dioxide was slightly less expensive than bleach, shown in Figure 5-3.

Figure 5-3. Costs – Bleach vs. ClO_2

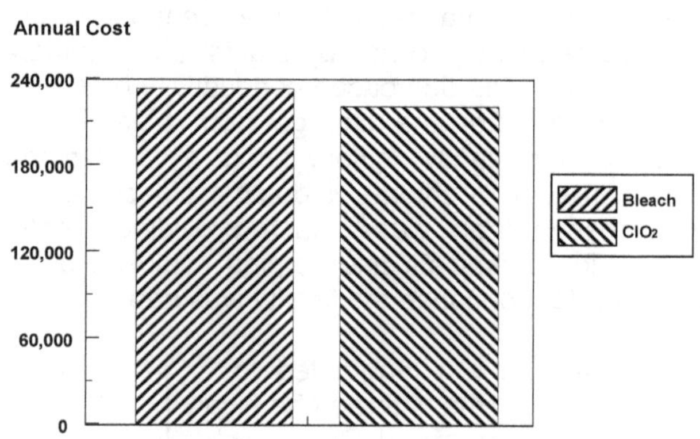

E. Soft-Dollar Economics

Cost comparisons among the various products used in plants are commonly made on a use-cost vs. use-cost basis. However, other factors should be considered when choosing between two products. A major factor is the consideration of "soft dollars." Soft dollars includes such things as extended run times between required equipment turn-arounds, improved heat transfer rates on heat exchangers, and reduced maintenance costs. Although soft dollars are hard to quantify, soft dollar savings are in many cases several times the increase in program costs that result from a change. In this case, actual use costs of chlorine dioxide were less than use costs of bleach. In addition, significantly less water was used in regeneration and rinsing of demineralizer resin. Further, significant savings resulted from the need for fewer demineralizer regenerant chemicals.

F. Conclusions

Several important points arise as a result of this case study.

- Chlorine dioxide, when used for influent clarification, can impact the entire plant. Such things as reduced microbiological control requirements in cooling systems, increased demineralizer run lengths and improvements in overall water chemistry may be observed.

- When a change is made in a treatment program, saved soft dollars may have a much greater impact than any increase in use-costs. In this case, the change from bleach to chlorine dioxide was economically attractive. Conditions of the influent water determined that use costs of chlorine dioxide were actually lower than that of bleach. As a result, soft-dollar economics provided even more of an incentive to consider a change of disinfectants.

- Plants that have problems with Asian clams, particularly in firewater systems, should consider replacing chlorine with chlorine dioxide for influent disinfection. Chlorine dioxide treatment of influent water would result in an overall improvement in microbiological control on the cooling systems. Its use would also eliminate difficult-to-handle and difficult-to-feed chemicals for treatment of Asian clams in firewater systems which use treated water for makeup.

G. References

[Aieta 1986] Aieta, E. and Berg, J. "A Review of Chlorine Dioxide in Drinking Water Treatment," *Journal of the American Water Works Association*, 78(6), 62(1986).

[Ambrogi 1993] Ambrogi, R., "Fouling Control Systems for Water Treatment," *Oebalia*, XIX, 355, 1993.

[Aston 1947] Aston, R., "Chlorine Dioxide Use in Plants on the Niagra," *Journal of the American Water Works Association*, 39(7), 687(1947).

[Axegard 1986] Axegard, P., "Substituting Chlorine Dioxide for Elemental Chlorine Makes the Bleach Plant Effluent Less Toxic," *TAPPI Journal*, 54(October 1986).

[Barnhart 1980] Barnhart, D., "Finding the Right RX for Tastes and Odors," *Pubic Works*, 95(June 1980).

[Brett and Ridgeway 1981] Brett, R. and Ridgeway, J., "Experiences with Chlorine Dioxide in Southern Water Authority and Water Research Centre," *Journal of the Institute of Water Engineers and Scientists*, 35(2), 135(1981).

[Coote 1950] Coote, R., "Chlorine Dioxide Treatment at Valparaiso, Indiana," *Water and Sewage Works*, 97(1), 13(1950).

[Correll 1992] Correll, A., "G-P Takes a Position on Chlorine Free Pulps," *Pulp and Paper*, 188(April 1992).

[Dickens and Graham 1995] Dickens, C. and Graham, P., "The Rupture of Algae During Abstraction from a Reservoir and the Effects on Water Quality," *Journal of Water Supply Research and Technology-Aqua*, 44(1), 29(1995).

[Dietrich et al. 1995] Dietrich, A., Hoehn, R., Dufresne, L., Buffin, L., Rashash, D., and Parker, B., "Oxidation of Odorous and Nonodorous Algal Metabolites by Permanganate, Chlorine and Chlorine Dioxide," *Water Science and Technology*, 31(11), 223(1995).

[Faber 1947] Faber, H., "A Theory of Chlorine Dioxide Taste and Odor Reduction," *Journal of the American Water Works Association*, 39, 691(1947).

[Froese et al. 1999] Froese, K., Wolanski, A., and Hrudey, S., "Factors Governing Odorous Aldehyde Formation as Disinfection By-Products in Drinking Water," *Water Research*, 33(6), 1364(1999).

[Geraci et al. 1993] Geraci, S., Ambrogi, R., Festa, V., and Piraino, S., "Field and Laboratory Efficacy of Chlorine Dioxide as Antifouling in Cooling Systems of Power Plants," *Oebalia*, XIX, 383(1993).

[Harlock and Dowlin 1953] Harlock, R. and Dowlin, R., "Chlorine and Chlorine Dioxide for Control of Algae Odors," *Water and Sewage Works*, 100, 74(February 1953).

[Hodgdon and Roach 1974] Hodgdon, R. and Roach, E., "Process for Defouling Anion Exchange Resins with Chlorine Dioxide," US Patent 3,787,339, January 1974.

[Ingols and Ridenour 1948] Ingols, R. and Ridenour, G. M., "The Elimination of Phenolic Tastes by Chloro-Oxidation," *Water and Sewage Works*, 95, 187(1948).

[Lalezary et al. 1986] Lalezary, S., Pirbazari, M., and McGuire, M., "Oxidation of Five Earthy-Musty Taste and Odor Compounds," *Journal of the American Water Works Association*, 62(March 1986).

[Masschelein 1991] Masschelein, W., "Chlorine Dioxide," Chemical Oxidation, Technology for the Nineties, First International Symposium, Vanderbilt University, Nashville, TN, February 1991.

[Medsker et al. 1968] Medsker, L., Jenkins, D., and Thomas, J., "An Earthy-Smelling Compound Associated with Blue-Green Algae and Actinomycetes," *Environmental Science and Technology*, 2(6), 461(1968).

[Mounsey and Hagar 1946] Mounsey, R. and Hagar, M., "Taste and Odor Control with Chlorine Dioxide," *Journal of the American Water Works Association*, 38, 1051(1946).

[Myers and Thompson 1986] Myers, G. and Thompson, A., "Controlling Taste and Odor with Chlorine Dioxide," *SW & Texas WWJ*, 4(March 1986).

[Myers et al. 1986] Myers, G., Thompson, A., Owen, D., and Baker, J., "Control of Trihalomethanes and Taste and Odor at Galveston County Water Authority," American Water Works Association Annual Conference, Denver, CO, June 22-26, 1986.

[Palin 1948] Palin, A., "Chlorine Dioxide in Water Treatment," *Journal of the Institute of Water Engineers*, 61(February 1948).

[Pitochelli and Mason 1994] Pitochelli, T. and Mason, J., "Large-Scale Field Applications of Chlorine Dioxide," Oxidative Treatment of Pollutants in Wastewater, Intertech Conferences, Philadelphia, PA, March 1994.

[Ringer and Campbell 1955] Ringer, W. and Campbell, S., "Use of Chlorine Dioxide for Algae Control at Philadelphia," *Journal of the American Water Works Association*, 47, 740(1955).

[Simpson 1994] Simpson, G., "Chlorine Dioxide Offers a Cooling Tower Option," *Water Technology*, 80(April 1994).

[Simpson 1995] Simpson, G., "Biofilm: Removal and Prevention with Chlorine Dioxide," The Third International Symposium on Chlorine Dioxide: Process Water, Drinking Water and Water Waste Issues, New Orleans, LA, September 1995.

[Simpson 2005a] Simpson, G., Practical Chlorine Dioxide: Volume I – Foundations, Chapter 1. History of Chlorine Dioxide, Greg D. Simpson & Associates, Publisher, 2005.

[Simpson 2005b] Simpson, G., Practical Chlorine Dioxide: Volume I – Foundations, Chapter 7. Frequently Asked Questions, Greg D. Simpson & Associates, Publisher, 2005.

[Simpson et al. 1993a] Simpson, G., Miller, R., Laxton, G., and Clements, W., "A Focus on Chlorine Dioxide: The 'Ideal' Biocide," Paper No. 472, Corrosion 93, New Orleans, LA, March 1993.

[Simpson et al. 1993b] Simpson, G., Miller, R., Laxton, G., and Clements, W., "A Focus on Chlorine Dioxide: Biocide of Choice for 'Stressed' Cooling Water Systems," WaterTech 93, Houston, TX, 1993.

[Solvay 1960] Solvay, "Chlorine Bleach Solutions," Solvay Division, Allied Chemical Company Bulletin 14, New York, Second Edition, 1960.

[Synan et al. 1944] Synan, J., MacMahon, J., and Vincent, G., "Chlorine Dioxide: A Development in the Treatment of Potable Water," *Water Works and Sewer*, 91, 423(1944).

[Synan et al. 1945] Synan, J., MacMahon, J., and Vincent, G., "A Variety of Water Problems Solved by Chlorine Dioxide Treatment," *Journal of the American Water Works Association*, 37, 869(1945).

[Thomas 1996] Thomas, K., private communication, April 1996.

[USEPA 1991] USEPA, "Guidance Manual for Compliance with the Filtration and Disinfection Requirements for Public Water Systems Using Surface Water Sources," 1991.

[Vincent et al. 1946] Vincent, G., MacMahon, J., and Synan, F., "Use of Chlorine Dioxide in Water Treatment," *American Journal of Public Health*, 1045(September 1946).

[Walker et al. 1986] Walker, G., Lee, F. and Aieta, E., "Chlorine Dioxide for Taste and Odor

Control," *Journal of the American Water Works Association,* 85(March 1986).

[White 1986a] White, G., The Handbook of Chlorination, Second Edition, Van Nostrand Reinhold Company, New York, 162(1986).

[White 1986b] White, G., The Handbook of Chlorination, Second Edition, Van Nostrand Reinhold Company, New York, 196(1986).

Disinfection of Influent Water with ClO_2

Control of Algae with ClO_2

A. What Are Algae and What Factors Influence their Growth?

Algae are comprised of several different groups of living organisms that capture light energy through a process called photosynthesis [Calvin 1956]. In photosynthesis, the algae use the energy from sunlight to convert inorganic substances into usable nutrients, producing oxygen in the process. Algae have been traditionally regarded as simple plants; they have no leaves, roots, flowers, or other organ structures that characterize higher plants.

Algae range from single-celled organisms to multi-cellular organisms, some with fairly complex differentiated forms. Most algae are photoautotrophic. That is, they have an absolute requirement for sunlight; some algae do not. Some unicellular species rely entirely on external energy sources.

Algae require, in general, sunlight, CO_2, and a few minerals. Phosphorous is one of the major nutrient elements required because phosphates are essential intermediates in photosynthesis [Lewin 1962]. This reliance on phosphorous is illustrated by one of the literature references to be described later in this chapter [Ringer and Campbell 1955].

Historically, the cyanobacteria have been classified as algae and have been referred to as the blue-green algae, although in some texts they are considered separately from algae. Cyanobacteria can be unicellular, colonial, or filamentous.

B. Algae-Related Potable Water Problems

In potable water applications, the most common problem is caused by free-floating algae, which can cause taste, odor, and appearance problems [Stigworth 1957; Lin 1976a; Lin 1976b; Hoehn 1988; Skulberg 1988; Juttner 1988]. A comprehensive study of tastes and odors was reported [Le Roux 1988]. Another report reviewed the source of aquatic odors [Persson 1988]. Treatment measures were described in other reports [Hrubec and De Kruijf 1983; Montiel 1983]. Although many chemicals can cause taste and odor (henceforth T & O) in potable water, the primary factors are biological [AWWARF 1976].

In addition to causing T & O problems, algae has been reported to also plug filters [Clare and Hopson 1975], and the metabolites of algae have been found to react with chlorine to form trihalomethanes (THMs) [Blanck et al. 1979].

The most common biological sources of those compounds that results in T & O problems are algae and Actinomycetes. Algae in surface water have been shown to be related to the numbers and types of bacteria in the same water [Overbeck and Babaenzien 1964; Olson et al. 1980]. One group demonstrated a positive correlation between coliform bacteria and Chlorella in a mountain stream [McFeters et al. 1995].

Two important T & O-causing chemicals produced by Actinomycetes and other organisms are geosmin and methyl isoborneol (MIB), tertiary alcohols whose structure makes them resistant to oxidation chemistries which are routinely applied for T & O improvement. These molecules have extremely low taste and odor threshold values, which makes them the foremost cause of off-taste and off-odor problems in potable water applications.

Open reservoirs are used in a number of cities to hold finished water prior to use. These open reservoirs are frequently contaminated by birds and airborne contamination, while sunlight promotes growth of algae. The problem of algae in open-finished water reservoirs was reviewed [Warf 1999].

Oxidation is the major chemical treatment process for taste and odor removal. Oxidizing agents utilized in the water industry include chlorine, chloramines, ozone, chlorine dioxide, and potassium permanganate, with chlorine gas (Cl_2) being the most popular oxidant/disinfectant used. It is cheap, easy to apply, provides a long-lasting residual concentration in the distribution system, and is efficient in destroying certain tastes and odors found in raw water. On the other hand it has several serious disadvantages: Firstly it produces chlorinated by-products with many trace organic compounds present in the water. These organohalogens may be odorous (for example, medicinal-smelling chlorophenols), or they may present a potential health risk (for example, trihalomethanes). Secondly, chlorine itself, if improperly dosed, gives an unpleasant chlorinous odor to the water. Thirdly, a number of odor-causing chemicals, particularly geosmin and MIB, are resistant to oxidation by chlorination.

C. Algae-Related Industrial Water Problems

In 1961, results of a two-year investigation of the relative amounts and types of algae most commonly found in cooling water were reported [Tehle 1961]. Particular note was made of the algae occurring in chromate-containing waters (Chromate was a common corrosion inhibitor used at that time.). Most of the algae were isolated in a pH range of 6.0—7.9. Samples were generally collected in towers where chlorine was not being used. Figure 6-1 shows the percentage of green algae found in the systems tested, while Figures 6-2 and 6-3 show the percentages of blue-green algae and diatoms present, respectively.

The most commonly found green algae were the unicellular forms Chlorella and Scenedesmus, and the filamentous Ulothrix.

Figure 6-1. Green Algae

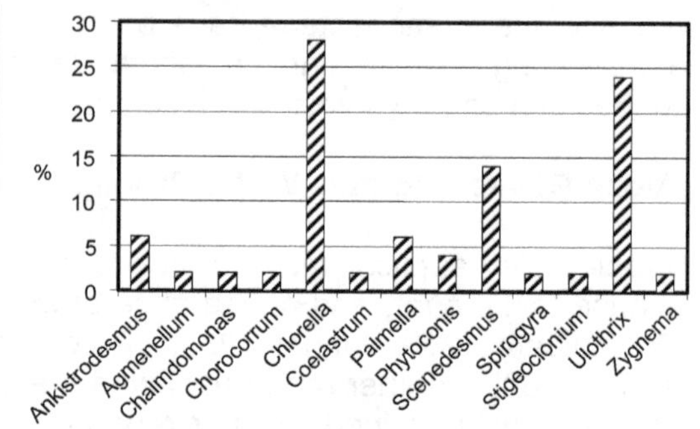

Figure 6-2. Blue-Green Algae

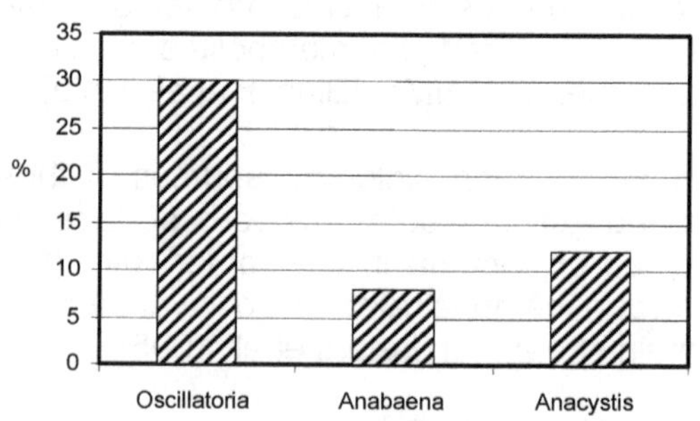

Figure 6-3. Diatoms

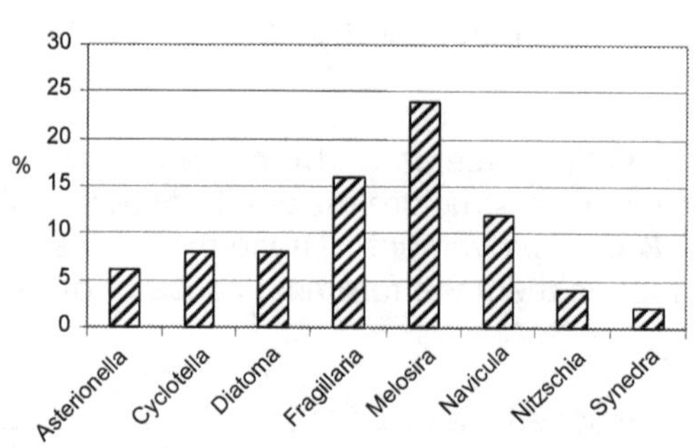

Only a few blue-green algae were found. Oscillatoria were the most prevalent of all the algae isolated. Oscillatoria have been found to cause corrosion in an open steel tank [Meyers 1947]. Algae have also been linked to a pitting type of corrosion [Ascolese 1977].

About 36% of the towers examined used chromate as a corrosion inhibitor. Several of the green algae and diatoms were found to be able to tolerate a wide range of chromate concentration, although chromate concentrations above about 60 mg/L were found to be inhibitory to many of the algae. As chromate is not being used currently, this data must be viewed with some caution.

Algae can form mats in internal sections of cooling towers that are exposed to water and sunlight. Algae grow on slats in conventional fill and prevent complete air/water contact required for cooling.

Algae, in general, are easier to control than bacteria in cooling systems [McCoy 1974]. Phosphate-based corrosion inhibitors can make algae control in such cooling systems more challenging than in those systems where another type of corrosion inhibitor is used.

Another common problem created by algae growth in cooling systems is their production of metabolic byproducts that provide a nutrient source to bacteria. For example, in 1987, 23.9% of the biofouling problems in power plant cooling systems were attributed to algae growth [Mussalli and Hillman 1987].

D. Common Methods for Algae Control

Physical Methods: Physical methods have been used in some cases to control algae.

- **Removal by Coagulation** Alum is often used as a coagulant and lime as a softener in the water treatment industry. Both of these chemicals remove phosphorus from water to some degree, thereby reducing the frequency and severity of algal blooms, since phytoplankton biomass is usually proportional to total phosphorus concentration [Schindler 1984].

- **Mechanical Removal** Mechanical removal or harvesting of aquatic organisms has been employed [Waldrop and Elfsiniotis 1998; Pinto et al. 2003].

- **Light Exclusion** The use of dye or covers to exclude light has been used with varying success to prevent light from reaching the water and allowing photosynthesis to occur. Covering open decks in recirculating cooling systems has been done with good success in a number of applications.

  Covering reservoirs in potable water applications may be impractical due to the size of the reservoir. In addition, sunlight inhibits the growth of nitrifying bacteria, which can create difficulties in systems treated with chloramines [Wolfe 1988].

Chemical Methods: Various chemicals have also been used in the control of algae.

- **Disruption of some Vital Cell Function** Certain chemicals have the ability to interrupt cell wall functions, disrupt enzyme functions, or inhibit photosynthesis. Such chemicals include copper sulfate, diquat, triazine compounds [Blackburn and Taylor 1976], alum [Kennedy and Cooke 1982] and lime [Prepas and Rigler 1982].

  In potable water, sudden algae kills may actually create other problems, most commonly the release of metabolites such as geosmin and methyl isoborneol (MIB). In any case, the metabolites released provide nutrient to other aquatic organisms.

- **Disinfection with Chlorine** Usually, water from reservoirs is treated with free chlorine to provide disinfection, sometimes several times before water reaches the customer. Disinfection is critical in conventional water purification schemes, and chlorine is the most commonly used disinfectant. Use of chlorine provides a residual concentration for the distribution system and destroys many of the harmful substances by oxidation. It is also relatively cheap and easy to apply. Despite these advantages, use of chlorine has some disadvantages, such as the creation of additional taste and odor problems and potential health problems resulting from disinfection by-products. This is true especially when pre-chlorination of raw water prior to the clarifier is done. Chlorination of algae results in cell lysis, releasing taste and odor producing substances and other organic compounds contained mainly within the cells [Wu and Juttner 1988]. The release into the water of these compounds may drastically increase the concentration of odorous compounds. The most common culprits, geosmin and methyl isoborneol (MIB), are very resistant to oxidation by chlorine, even at high dosages [Lalezary et al. 1986]. These compounds must be removed by other means.

  Chlorine disinfection of waters containing high algae levels coupled with increases in total organic carbon (TOC) concentrations attributable to algal growth, have been linked to elevated concentrations of trihalomethanes (THMs) [Pieterse 1989] and other disinfection by-products [Silverman 1983; Erb 1989; AWWA Committee Report 1983].

- **Use of Chloramines** In many water treatment plants chloramination has been seen as the ideal replacement for chlorination. However, a stronger disinfectant is typically used as the primary disinfectant, since chloramines do not provide sufficient Ct to act as the primary disinfectant, with chloramines being added primarily to provide a measurable residual in the distribution system. Additionally, there have been reports of nitrification problems in large, chloraminated covered reservoirs. One author suggested that the presence of sunlight inhibited growth of nitrifying bacteria, and its elimination by the covers may have aided the growth of nitrifying bacteria [Wolfe 1988]. Nitrification occurred in two covered finished-water reservoirs in Southern California following a change from free chlorine to chloramine disinfection.

E. Literature Review – Algae Control with ClO_2

The first use of ClO_2 was to control T & O in potable water in Niagra Falls, NY [Synan et al. 1944]. The approach was to provide disinfection with chlorine, followed by treatment with ClO_2 to oxidize organic components responsible for T & O problems. This approach was very effective.

After the first promising results with ClO_2, experienced by the municipalities of Niagra Falls and nearby cities, researchers at the city of Chicopee, Massachusetts, evaluated ClO_2 to control T & O problems that they had struggled with for some time [McCarthy 1945]. These workers noted the frequent high levels of hydrocarbon present in the raw water and the effectiveness of ClO_2 on slime on pipe. Apparently some of the slime was due to bacteria, emulsified hydrocarbon, or both. Whatever the source of slime, it was stripped by the ClO_2, resulting in some kerosene type T & O downstream. The workers found, as did Niagra Falls workers, that treatment with chlorine initially, followed by ClO_2 treatment, controlled T & O.

In 1945, a number of successful T & O applications, each with unique conditions, were reported [Synan et al. 1945]. The region and the type of problem is shown in Table 6-1.

Table 6-1. Location and Problem of Waters Successfully Treated with ClO_2	
Location	**Problem**
Niagra Falls	Phenols and algae
Eastern Pennsylvania	Organic with some phenol
Western Massachusetts	Decayed vegetation and algae, gasoline and kerosene
South Carolina	Algae
Maine	Highly colored with decayed vegetable matter, effluent from sulfite and kraft paper mills and town sewage

In the Niagra Falls application, use costs were actually reduced with the application of ClO_2.

Following the successful approach reported by Synan, Lawrence, Kansas treatment plant personnel evaluated ClO_2 and the results were reported in 1946 [Mounsey and Hagar 1946]. This potable water facility also suffered from periodic T & O problems.

A dam on the Kaw River created a somewhat stagnant condition in the river, and the municipal treatment plant processed a slip stream from this stagnant pool to make potable water.

While free chlorine, chloramine, alum, and activated carbon were sufficient to control T & O most of the time, occasionally conditions created a high level of microbiological growth, and control of T & O was lost. Use of ClO_2 alleviated this problem, and after optimizing, best results were achieved if the water had a high pH and if chlorine was used first, followed by application of ClO_2.

At about that same time, in another Niagra Falls municipality, workers reported that 0.5 mg/L ClO_2 was sufficient to destroy odor-causing compounds [Vincent et al. 1946].

In 1947, a summary report was made on the implementation of ClO_2 in several municipalities, including ones in Woodlawn, NY, Greenwood, SC, Tonawanda, NY, South Tonawanda, NY, Lockport, NY and Port Colborne, NY [Aston 1947]. T & O difficulties with phenolic tastes and odors, oil waste, and algae and vegetation decomposition problems were described. In addition, the removal of slime in the distribution pipes, along with removal of iron tubercules when hydrants were flushed, was described. (This phenomenon is also seen in industrial cooling systems.) The applications were deemed successful.

In Valparaiso, Indiana, two periods of odor episodes occur annually, one due to algae (June and July), and the other due to lake turnover [Coote 1950]. Algae taste and odors were attributed to Anabaena, Asterionella, Synura, Vorticella, Melosira, Ceratium, and Cyclops. During one severe algae odor episode, ClO_2 was evaluated. On June 16, 1948, the T & O of the processed water was very objectionable. The evaluation of ClO_2 began that afternoon. Two days later, June 18, 1948, the quality of the processed water was very much improved.

In a municipal treatment plant in McNeil Island, Washington, from mid September to mid-June, odors were not a problem but color was [Harlock and Dowlin 1953]. The rest of the year, algae, predominantly Anabaena, resulted in odors. These researchers found that use of chlorine alone would control algae odors if the reaction time was a minimum of 6 hours. However, use of ClO_2 in conjunction with chlorine significantly reduced this time.

In 1949, the city of Philadelphia used 1450 lbs of copper sulfate, fed in two separate doses, to control algae in their reservoir [Ringer and Campbell 1955]. Water entering the reservoir from another source was soft, had a slightly acidic pH, and thus was fairly corrosive, causing complaints from the area. To address this problem, a related facility began to use sodium hexametaphosphate for corrosion control in May of 1950. Some of this water entered the reservoir. The presence of the phosphate resulted in an explosive growth of algae, with 14—18 times the amount of copper sulfate being required to control the algae (28,000 lbs) than previously. This illustrates the direct dependence of algae on phosphate as a nutrient source. Even with this much copper sulfate, control was marginal. In 1951, 59,400 lbs of copper sulfate was used.

Over the next couple of years, workers evaluated ClO_2 in one basin or another or both, alone or coupled with copper sulfate. But use of copper sulfate was discontinued in May, 1954, and not started again for the remainder of the season. Results were exceptional. Their studies showed that ClO_2 was a more effective algaecide than copper sulfate, with the additional benefit of odor control.

Economically, the cost of ClO_2 was about 50% greater than the cost of copper sulfate. The authors noted that in 1951, a certain amount was spent for very poor control, while in 1954, about 50% more was spent for very good control. The extra cost was considered to be an acceptable tradeoff for the vast improvement in the quality of the finished waters.

On the other hand, in 1980, taste and odor complaints abounded as a result of the poor quality water from the Lee County, FL, municipal treatment facility [Wright 1980]. Two approaches were undertaken. The first was the installation of some granular activated carbon (GAC) filters and the second was to evaluate ClO_2. While both were effective, the plant found that they could use GAC filters alone and produce water of sufficient quality, and so ClO_2 was not employed.

ClO_2 is known to be inactivated by UV light. Results of one study conducted at The Los Angeles Aqueduct Filtration Plant [Zika et al. 1985] suggested that ClO_2 will decompose photolytically to chlorate. When stored in the dark over 6 hours, only 24% of the dosed chlorine dioxide in their study was converted to chlorite, while over 60% of the dosed chlorine dioxide remained. For the same period of time under fluorescent light, 40% of the dosed chlorine dioxide converted into chlorite. When the treated sample was stored in full daylight, the added chlorine dioxide decayed completely within 2 hours and the chlorite residual was less than 6% of the dosed chlorine dioxide after a six-hour exposure. The conclusion was that chlorine dioxide algal treatment should be conducted at night.

Many other reports concerning the effectiveness of ClO_2 towards algae have been reported and are too numerous to be described here [Clare and Hopson 1975; Hubbs et al. 1979; Steynberg et al. 1993; Malpas 1973; Sumner 1980; Walker et al. 1985; McKeon et al. 1986; Thompson and Ashe 1990; Sung 1990; Buffin et al. 1993; Dickens and Graham 1995; Dietrich et al. 1995; Steynberg et al. 1996a; Steynberg et al. 1996b; Stolarik and Liu 2000; Roach and Vickers 2001; Pan et al. 2002; Hunter 2003].

Many of the papers on use of ClO_2 to solve T & O problems are linked to specific strains of algae or Actinomycetes. In addition, a number of studies or review articles describe the use of chlorine dioxide in the successful control of musty odors, fishy tastes and odors from a variety of algae species [Medsker et al. 1968; Rav-Acha et al. 1995; Yongyi et al. 1995; Bryant et al. 1992; Warf 1999].

F. Summary

Chlorine dioxide provides a very cost effective disinfectant for the control of algae, in both potable water applications and industrial water applications.

G. References

[APHA 1930] APHA, "Aftergrowths in Water Distribution Systems," *American Journal of Public Health*, 43, 485(1930).

[Ascolese 1977] Ascolese, C. R., "Biofilm Formation: Attendant Problems, Monitoring, and Treatment in Industrial Cooling Systems," 27th Annual Liberty Bell Corrosion Course, Philadelphia, PA, September 14, 1977.

[Aston 1947] Aston, R., "Chlorine Dioxide Use in Plants on the Niagra," *Journal of the American Water Works Association*, 39(7), 687(1947).

[AWWA Committee Report 1953] American Water Works Association, Committee Report, "Potable-Water Storage Reservoirs," *Journal of the American Water Works Association*, 45(10), 1079(1953).

[AWWA Committee Report 1983] American Water Works Association, Committee Report, "Deterioration of Water Quality in Large Distribution Reservoirs (Open Reservoirs)," *Journal of the American Water Works Association*, 75(6), 313(1983).

[AWWARF 1976] AWWARF, Handbook of Taste and Odor Control Experiences in the U.S. and Canada, Denver, CO: American Water Works Association Research Foundation, 1976.

[Barnhart 1980] Barnhart, D., "Finding the Right RX for Tastes and Odors," *Pubic Works*, 95(June 1980).

[Blackburn and Taylor 1976] Blackburn, R. and Taylor, J., "Aquazine: A Promising Algaecide for Use in Southeastern Waters," *Proceedings of the Soil Weed Science Society*, 29, 365(1976).

[Blanck et al. 1979] Blanck, C., Arguello, M., Glaze, W., Rawley, R., Brett, R., and Calverley, R., "THM Levels at 100 US Plants; THM Levels at 14 US Utilities; THM Levels in East Texas; THM Levels at an English Utility; THM Research at Fairfax County, Virginia," American Water Works Association Research Foundation Water Quality Research News, 2, 6(November 1979).

[Bloesch 1997] Bloesh, D.F., "Harmful Algal Blooms in Coastal Waters Options for Prevention, Control, and Mitigation," NOAA Coastal Ocean Program Decision Analysis Series No. 10, Silver Spring, MD [Washington, D.C.]: National Oceanic and Atmospheric Administration Coastal Ocean Office; National Fish and Wildlife Foundation, XII, 49(1997).

[Brett and Ridgeway 1981] Brett, R. and Ridgeway, J., "Experiences with Chlorine Dioxide in Southern Water Authority and Water Research Centre," *Journal of the Institute of Water Engineers and Scientists*, 35(2), 135(1981).

[Bryant et al. 1992] Bryant, E., Fulton, G., Budd, G., Van Hoof, F., "Identifying and Characterizing Effects of Disinfection Byproducts," Chapter 3, Disinfection Alternatives for Safe Drinking Water," Van Nostrand Reinhold, 1992.

[Buffin et al. 1993] Buffin, L., Hoehn, R., Dietrich, A., Rashash, D., "Effectiveness of Chlorine, Chlorine Dioxide, and Permanganate for the Treatment of Cucumber, Grassy, and Fishy Odors in Water Supplies," Proceedings 1993 Water Quality Technology Conference; Part II, Miami, FL, 1859(November 7-13, 1993).

[Calvin 1956] Calvin, M., "The Photosynthetic Carbon Cycle," *Journal of the American Chemical Society*, 78, 1895(1956).

[Cherry 1962a] Cherry, A.K., "Rx for Tastes and Odors—$KMnO_4$ and Activated Carbon," *Water Works Engineering*, 3, 182(1962).

[Cherry 1962b] Cherry, A., "Use of Potassium Permanganate in Water Treatment," *Journal of the American Water Works Association*, 54, 417(1962).

[Clare and Hopson 1975] Clare, L. and Hopson, N., "Algae Problems in Eastern Lake Erie," *Journal of the American Water Works Association*, 67(3), 131(March 1975).

[Clement 1998] Clement, B., "Open, Finished-Water Reservoirs in the US and Puerto Rico," USEPA Region VIII, 1998.

[Coote 1950] Coote, R., "Chlorine Dioxide Treatment at Valparaiso, Indiana," *Water and Sewage Works*, 97(1), 13(1950).

[Dickens and Graham 1995] Dickens, C., and Graham, P., "The Rupture of Algae During Abstraction from a Reservoir and the Effects on Water Quality," *Journal of Water Supply Research and Technology-Aqua*, 44(1), 29(1995).

[Dietrich et al. 1995] Dietrich, A., Hoehn, R., Dufresne, L., Buffin, L., Rashash, D., and Parker, B., "Oxidation of Odorous and Nonodorous Algal Metabolites by Permanganate, Chlorine and Chlorine Dioxide," *Water Science and Technology*, 31(11), 223(1995).

[Erb 1989] Erb, T., "Implementation of Environmental Regulations for Improvements to Distribution Reservoirs in Los Angeles," American Water Works Association Annual Conference, Los Angeles, CA, 1989.

[Faber 1947] Faber, H. A., "A Theory of Chlorine Dioxide Taste and Odor Reduction," *Journal of the American Water Works Association*, 39, 691(1947).

[Ficek 1983] Ficek, K., "Raw Water Reservoir Treatment with Potassium Permanganate," 74th Annual Meeting Illinois Section American Water Works Association, Chicago, IL, 1983.

[Field and Giesbrecht 1985] Field, T. and Giesbrecht, G., "Taste And Odour Control Facilities For Regina/Moose Jaw Water Supply," 37th Annual Convention of the Western Canada Water and Sewage Conference, Winnipeg, Canada, September 18-20, 1985.

[Froese et al. 1999] Froese, K., Wolanski, A., and Hrudey, S., "Factors Governinig Odorous Aldehyde Formation as Disinfection By-Products in Drinking Water," *Water Research*, 33(6), 1364(1999).

[Harlock and Dowlin 1953] Harlock, R. and Dowlin, R., "Chlorine and Chlorine Dioxide for Control of Algae Odors," *Water and Sewage Works*, 100, 74(February 1953).

[Hoehn 1988] Hoehn, R., "Biological Causes of Taste and Odors in Drinking Water Supplies," *Water Quality Bulletin*, 13(2), 46(1988).

[Hrubec and De Kruijf 1983] Hrubec, J. and De Kruijf, H., "Treatment Methods for the Removal of Off-Flavours From Heavily Polluted River Water in the Netherlands--A Review," *Water Science and Technology*, 15(6/7), 301(1983).

[Hu et al. 2003] Hu, W., Liu, P., and Pei, H., "Characteristic and Mechanism of Inactivating Algae with O_3 and ClO_2," *Chinese Science Bulletin*, 48(9), 862(2003).

[Hubbs et al. 1979] Hubbs, S., Siria, J., Goers, M., "Use of Alternate Disinfectants at Louisville for Trihalomethane Control," Proceedings American Water Works Association, Controlling Organics in Drinking Water Seminar, San Francisco, CA, June 24, 1979.

[Hunter 2003] Hunter, K., "Control of Algae in Potable Water Supply—Bemm River Shadecloth Trial," Innovations in Water: Ozwater Convention & Exhibition Proceedings, Perth, Western Australia, Australia, April 6-10, 2003.

[Ingols and Ridenour 1948] Ingols, R. and Ridenour, G., "The Elimination of Phenolic Tastes by Chloro-Oxidation," *Water and Sewage Works*, 95, 187(1948).

[Juttner 1988] Juttner, F., "Biochemistry of Biogenic Off-Flavour Compounds in Surface Waters," *Water Science and Technology*, 20(8/9), 107(1988).

[Kennedy and Cooke 1982] Kennedy, R. and Cooke, G., "Control of Lake Phosphorus with Aluminum Sulfate: Dose Determination and Application Techniques," *Water Research*, 18, 389(1982).

[Krasner and Means 1986] Krasner, S. and E., "Means, Returning Recently Covered Reservoirs to Service: Health and Aesthetic Considerations," *Journal of the American Water Works Association*, 78(3), 94(1986).

[Kull et al. 2004] Kull, T., Backlund, P., Karlsson, K., and Meriluoto, J., "Oxidation of the Cyanobacterial Hepatotoxin Microcystin-LR by Chlorine Dioxide: Reaction Kinetics, Characterization, and Toxicity of Reaction Products," *Environmental Science and Technology*, 38(22), 6025(2004).

[Lalezary et al. 1986] Lalezary, S., Pirbazari, M., and McGuire, M., "Oxidation of Five Earthy-Musty Taste and Odor Compounds," *Journal of the American Water Works Association*, 78(3), 62(March 1986).

[Le Roux 1988] Le Roux, J., "Tastes and Odours in Drinking Water," *Municipal Engineering*, 19(7), 19(1988).

[LeChevallier et al. 1997] LeChevallier, M., Norton, W., and Atherholt, T., "Protozoa in Open Reservoirs," *Journal of the American Water Works Association*, 89(9), 84(1997).

[Lewin 1962] Lewin, R., ed., Physiology and Biochemistry of Algae, Academic Press, NY, 1962.

[Lin 1976a] Lin, S., "Sources of Tastes and Odors in Water, Part 1," *Water and Sewage Works*, 123(6), 101(1976).

[Lin 1976b] Lin, S., "Sources of Tastes and Odors in Water. Part 2," *Water and Sewage* Works, 123(7), 64(1976).

[Malpas 1965] Malpas, J., "Use of Chlorine Dioxide in Water Treatment," *Effluent and Water Treatment Journal*, 1965(July 1965).

[Malpas 1973] Malpas, J., "Disinfection of Water Using Chlorine Dioxide," *Water Treatment Examiners*, 22, 209(1973).

[McCarthy 1945] McCarthy, J., "Chlorine Dioxide for the Treatment of Water Supplies," *Journal of the New England Water Works Association*, 59, 252(1945).

[McCoy 1974] McCoy, J., The Chemical Treatment of Cooling Water, Chemical Publishing Company, NY, 1974.

[McFeters et al. 1995] McFeters, G., Stewart, P., Yu, F., De Beer, D., Srinivasan, R., Chen, X., Huang, C., and Wentland, E., "Spatial Patterns of Physiological Activities and Biocide Penetration Within Biofilms," Proceedings 1995 Water Quality Technology Conference; Part II, New Orleans, LA, November 12-16, 1995.

[McKeon et al. 1986] McKeon, W., Muldowney, J., Aptowicz, B., "The Evolution of a Modified Disinfection Strategy to Reduce Trihalomethane Formation," 1986 Annual Conference Proceedings, American Water Works Association; Water -- Key to Life, Denver, CO, June 22-26, 1986.

[Means and McGuire 1978] Means, E. and M. McGuire, "An Early Warning System for Taste and Odor Control," *Journal of the American Water Works Association*, 78(3), 77(1978).

[Medsker et al. 1968] Medsker, L., Jenkins, D., and Thomas, J., "An Earthy-Smelling Compound Associated with Blue-Green Algae and Actinomycetes," *Environmental Science and Technology*, 2(6), 461(1968).

[Montiel 1983] Montiel, A., "Municipal Drinking Water Procedures for Taste and Odour Abatement—A Review," *Water Science and Technology*, 15(6/7), 279(1983).

[Morra 1980] Morra, J., "A Review of Water Quality Problems Caused by Various Open Distribution Storage Reservoirs," *Journal of the New England Water Works Association*, 94(4), 316(1980).

[Mounsey and Hagar 1946] Mounsey, R. and Hagar, M., "Taste and Odor Control with Chlorine Dioxide," *Journal of the American Water Works Association*, 38, 1051(1946).

[Muchmore 1973] Muchmore, C., "Algae Control in Water Supply Reservoirs," Illinois Institute for Environmental Quality, Document No. 73-9. 1973.

[Mussali and Hillman 1987] Mussalli, Y. and Hillman, R., "Guidelines on Macrofouling Control Technologies," EPRI CS-5271, June 1987.

[Myers 1947] Myers, H., "The Role of Algae in Corrosion," *Journal of the American Water Works Association*, 39, 322(1947).

[Myers and Thompson 1986] Myers, G. and Thompson, A., "Controlling Taste and Odor with Chlorine Dioxide," *SW & Texas WWJ*, 4(March 1986).

[Myers et al. 1986] Myers, G., Thompson, A., Owen, D., and Baker, J., "Control of Trihalomethanes and Taste and Odor at Galveston County Water Authority," American Water Works Association, 1986 Annual Conference, Denver, CO, June 22-26, 1986.

[Nickels et al. 2000] Nickels, A., Harriman, C., Nickels, A., Freedman, A., and Harris, A., "A Novel Method of Applying Chlorine Dioxide," AWT Annual Conference, Nov 4, 2000.

[Olin 1972] Olin Corporation, "The Use of Sodium Chlorite for Control of Algae," unpublished report, 1972.

[Olson et al. 1980] Olson, B., Silverman, G., and Nagy, L., "Microbiological Survey of Two Uncovered Reservoirs in Southern California," Symposium on Surface Water Impoundments, Minneapolis, MN, 1980.

[Overbeck and Babaenzien 1964] Overbeck, J. and Babaenzien, H., "Bacterien und Phytoplankton eines Kleingewassers in Jahreszyklus," *Zeitschrift fur allgemeine Mikrobiologie*, 4, 59(1964).

[Palin 1948] Palin, A., "Chlorine Dioxide in Water Treatment," *Journal of the Institute of Water Engineers* (England), 61 (February 1948).

[Pan et al. 2002] Pan, S., Randtke, S., deNoyelles, F., Graham, D., "Occurrence, Biodegradation, and Control of Geosmin and MIB in Midwestern Water Supplies," 2002 Annual Conference Proceedings; American Water Works Association, New Orleans, LA, June 16-20, 2002

[Persson 1988] Persson, P., "Aquatic Off-Flavours—Past, Present, and Future," *Water Science and Technology*, 20(8/9), 283(1988).

[Pieterse 1989] Pieterse, A.., "Preliminary Observations on the Removal of Phytoplankton from Vaal River Water at the Balfonte Purification Plant," <u>Environmental Quality and Ecosystem Stability</u>, E. Spanier, Y. Steinberger, and M. Luria, Editors, 437(1989).

[Pinto et al. 2003] Pinto, L., Chandrasena, N., Dharmasena, C., "Managing Aquatic Weeds for Enhancing Nutrient Removal in the Constructed Wetland of Rouse Hill Recycled Water Plant," Innovations in Water: Ozwater Convention & Exhibition Proceedings, Perth, Western Australia, April 6-10, 2003.

[Pluntze 1974] Pluntze, J., "Health Aspects of Uncovered Reservoirs," *Journal of the American Water Works Association*, 67(8), 432(1974).

[Prepas and Rigler 1982] Prepas, E. and Rigler, F., "Improvements in Quantifying the Phosphorus Concentration in Lake Water," *Canadian Journal of Fisheries and Aquatic Science*, 39, 822(1982).

[Rav-Acha et al. 1995] Rav-Acha, Ch., Kummel, M., Salamon, I., and Adin, A., "The Effect of Chemical Oxidants on Effluent Constituents for Drip Irrigation," *Water Research*, 29(1), 119(1995).

[Ringer and Campbell 1955] Ringer, W. and Campbell, S., "Use of Chlorine Dioxide for Algae Control at Philadelphia," *Journal of the American Water Works Association*, 47(8), 740(1955).

[Roach and Vickers 2001] Roach, D. and Vickers, J., "Treatment of Nueces River Water Using Conventional Pretreatment and Microfiltration," 2001 Membrane Technology Conference Proceedings; American Water Works Association, San Antonio, TX, March 4-7, 2001

[Schindler 1984] Schindler, D., "Eutrophication and Recovery in Experimental Lakes: Implications for Lake Management," *Science*, 184, 897(1984).

[Silverman 1983] Silverman, G., Nagy, L., and Olson, B., "Variations in Particulate Matter, Algae, and Bacteria in an Uncovered Finished Drinking Water Reservoir," *Journal of the American Water Works Association*, 75(4), 191(1983).

[Simpson 1998] Simpson, G., "Irradiation of Aqueous ClO_2 and Speciation of Products," unpublished report, 1998.

[Skulberg 1988] Skulberg, O., "Chemical Ecology and Off-Flavour Substances," *Water Science and Technology*, 20(8/9), 167(1988).

[Steynberg et al. 1993] Steynberg, M., Geldenhuys, M., Guclielmi, M., Grobler, S., and Maree, B., "The Influence of Water Quality on the Efficiency of Chlorine Dioxide as Pre-Oxidant and Algicide in the Production of Potable Water," WRC Report No 281/1/94, December 1993.

[Steynberg et al. 1996] Steynberg, M., Guglielmi, M., Geldenhuys, J., Pieterse, A., "Chlorine and Chlorine Dioxide: Pre-Oxidants Used as Algocide in Potable Water Plants," *Journal of Water Supply Research and Technology-Aqua*, 45(4), 162(1996).

[Steynberg et al. 1996a] Steynberg, M., Pieterse, A., and Geldenhuys, J., "Improved Coagulation and Filtration of Algae as a Result of Morphological and Behavioural Changes Due to Pre-Oxidation," *Journal of Water Supply Research and Technology-Aqua*, 45(6), 292(December 1996).

[Stigworth 1957] Stigworth, E., "Control of Odor and Taste in Water Supplies," *Journal of the American Water Works Association*, 49(12), 1507(1957).

[Stolarik and Liu 2000] Stolarik, G. and Liu, P., "Chlorine Dioxide for Algae and DBP Control: The Los Angeles Experience," Proceedings Water Quality Technology Conference, Salt Lake City, UT, November 5-9, 2000.

[Sumner 1980] Sumner, A., "British Treatment of a Variable Surface Supply," American Water Works Association Research Foundation Water Quality Research News, 5, 12(February 1980).

[Sung 1990] Sung, W., "Taste and Odor Episodes at Wachusett Reservoir," *Journal of the New England Water Works Association*, 104(3), 165(September 1990).

[Suzuki and Gordon 1978] Suzuki, K. and Gordon, G., "Stoichiometry and Kinetics of the Reaction Between Chlorine Dioxide and Sulfur (IV) in Basic Solutions," *Inorganic Chemistry*, 17(11), 3115(1978).

[Synan et al. 1944] Synan, J., MacMahon, J., and Vincent, G., "Chlorine Dioxide, A Development in the Treatment of Potable Water," *Water Works and Sewage*, 91, 423(1944).

[Synan et al. 1945] Synan, J., MacMahon, J., and Vincent, G., "A Variety of Water Problems Solved by Chlorine Dioxide Treatment," *Journal of the American Water Works Association*, 37, 869(1945).

[Taylor 1933] Taylor, M. C., "Weed Eradication," US Patent 1,930,781, December 19, 1933.

[Tehle 1961] Tehle, E., "Here's What Grows in Cooling Tower Water," *The Oil and Gas Journal*, 84(Jan 16, 1961).

[Thompson and Ashe 1990] Thompson, J. and Ashe, C., "Removal of Total Oxidant Species of Chlorine Dioxide Disinfection by Granular Activated Carbon," 1990 Annual Conference Proceedings; American Water Works Association, Cincinnati, OH, June 17-21, 1990.

[USEPA 1998] USEPA, National Primary Drinking Water Regulations: Interim Enhanced Surface Water Treatment Rule; Final Rule. Federal Register, 63(241), 69478(1998).

[Vincent et al. 1946] Vincent, G., MacMahon, J., and Synan, F., "Use of Chlorine Dioxide in Water Treatment," *American Journal of Public Health*, 1045(September 1946).

[Vogel and Bowman 1998] Vogel, N. and Bowman, C., "Environmental Pollution Concerns Come to Forefront," *Sacramento Bee*. Sacramento, A1, 1998.

[Waldrop and Elfsiniotis 1998] Wardrop, D., Elefsiniotis, P., "An Evaluation of Weed and Algae Control Methods in Winnipeg's Stormwater Retention Basins," 1998 Western Canada Water & Wastewater Association 50th Annual Conference Proceedings, Calgary, Alberta, Canada, October 25-28, 1998.

[Walker et al. 1985] Walker, G., Lee, F., Aieta, E., "The Control of Taste and Odor Compounds in Drinking Water: The Edmonton Experience," Proceedings American Water Works Association Annual Conference; Managing a Priceless Resource, Washington, DC, June 23-27, 1985.

[Walker et al. 1986] Walker, G. S., Lee, F. P, and Aieta, E. M., "Chlorine Dioxide for Taste and Odor Control," *Journal of the American Water Works Association*, 85(March 1986).

[Warf 1999] Warf, C., "Chlorine Dioxide: An Attractive Alternative for Algae Control in Open, Finished-Water Reservoirs," 1999 Annual Conference Proceedings; American Water Works Association, Chicago, IL, June 20-24, 1999.

[Wolfe 1988] Wolfe, R., Means, E., Davis, M., and Barrett, S., "Biological Nitrification in Covered Reservoirs Containing Chloraminated Water," *Journal of the American Water Works Association*, 80(9), 109(1988).

[Wright 1980] Wright, R., "Eliminate Taste and Odor Problems Caused by Algae," *Water and Waste Engineering*, 67(May 1980).

[Wu and Juttner 1988] Wu, J. and Juttner, F., "Effect of Environmental Factors on Geosmin Production by Fischerella muscicola," *Water Science and Technology*, 20(8/9),143(1988).

[Yougyi 1995] Yongyi, W., Zhanpeng, J., and Guixiasheng, "Studies on Bacterial and Algaecidal Ability of Chlorine Dioxide," *Water Treatment*, 10, 347(1995).

[Zika et al. 1985] Zika, R. G., Moore, C. A., Gidel, L. T., Cooper, W. J., "Sunlight Induced Photodecomposition of Chlorine Dioxide," Chapter 82, Water Chlorination: Chemistry, Environmental Impact & Health Effects, Jolley, et al., Eds., Lewis Publishers, Inc., Chelsea, MI, 5, 1041(1985).

Enhancing Filter Performance with ClO_2

A. Introduction

A problem common to all water filters is microbiological fouling. Microbiological growth in filter media attaches to and binds the media together, making it difficult or impossible to remove the fouling material by backwashing alone. Such filters include those used in potable water and side stream filters found on cooling systems, both of which are dealt with in this chapter.

For cooling towers, side stream filtration offers numerous benefits, some of which were described in another report [Melancon 2004] including a reduction in turbidity which resulted in a substantial increase in cycles in one plant. This increase in cycles translated to significant reductions in cooling water chemical treatment costs, energy costs, and water costs.

While the benefits of filtration are well-documented, their limitations are also well known. The limitations are generally less related to the design of the filtration equipment or of the technology, than of the ability of conventional oxidants to control the fouling microbiological growth that limits the effectiveness of filtration equipment. With the increase in microbiological loading on such a filter and the documented inability of conventional oxidants such as chlorine [Chen and Stewart 1996; de Beer et al 1994; Marshall and Walker 1990; Xu et al. 1996] or bromine [White 1986; Franklin et al. 1991; Trulear and Wiatr 1988] to penetrate biofilm, for side stream filters the filter media becomes a "site of reinnoculation" where bacteria can be rapidly reintroduced into the cooling system. These limitations have resulted in many side stream filters being taken out of service.

For small package potable water plants, although it is the effluent turbidity which is regulated, the common practice is for operators to use differential pressure (DP) to initiate the backwashing event. Use of DP as trigger for the backwashing event provides some safeguard, as backwashing would be done well before the turbidity reaches its control limit. The fouling material backwashed from the filter media is generally composed of a high percentage of microbiological growth.

For package filters, backwashing is done until the filter is relatively clean. This process generally requires a great deal of water. It is not uncommon for approximately 30% of the potable water processed through the package plant to be required in filter backwashing event. Thus, backwashing of small package potable water plants typically results in a high cost to process the water, in terms of the rework that must be done and the chemicals used to treat the water.

In a recent field evaluation on a pilot unit in a potable water facility, an investigation was done to measure the impact that low ClO_2 concentrations fed immediately prior to the filters had on particle counts in the filter effluent [Hammond et al. 2004]. Although it was not a goal of this evaluation, differential pressure data was taken that showed some surprising results that impact the time between backwashing of filters, both small package potable water plants for small communities and side stream filters of cooling systems.

B. Case History

In this field evaluation on pilot filters in a potable water facility, an investigation was done to measure particle counts in the 2—5 µ, 5—10 µ, and 10—15 µ ranges in filtered water effluent [Hammond et al. 2004]. As filters in potable water facilities are generally controlled using turbidity (regulations require < 0.3 NTU), there has been some concern over the possibility of pathogens such as Giardia cysts (~ 10—12 µ) and Cryptosporidium oocysts (~ 4—6 µ) passing through the filter undetected, as the size of these pathogens are sufficiently small that they do not impact turbidity. Thus turbidity cannot be used as an indication of their removal. Because these organisms are particularly resistant to inactivation by chemical disinfectants, removal by filtration is the primary means of control.

Ozone has been found to destabilize particles and, when fed immediately before a filter, the particle counts of the filter effluent decrease. A number of reports of this phenomenon have been made. While chlorine dioxide has shown some of the same characteristics in terms of particle destabilization, only a couple of related investigations were found in the literature. The purpose of the field evaluation was to investigate the impact of ClO_2, when applied just prior to a GAC filter, on the size and numbers of particles exiting the filters. The results showed significant particle reduction, which averaged 23% at one potable water facility and about 13% at another facility.

Although the primary goal of the investigation was to determine the impact of chlorine dioxide on particle counts, filter differential pressure data was also taken. One surprising result was the impact small amounts of ClO_2 had on the rate of increase of differential pressure of the filter. And while the data was obtained in a potable water facility, the results are easily translatable to side stream filters in industrial cooling systems.

Experimental: Figure 7-1 shows a schematic of the pilot unit. In the pilot unit, four filters are in parallel. Each filter has 36" of bituminous-based GAC which lies atop 12" sand. The system is equipped so that each filter can be dosed separately. Raw water feed to the potable water facility is treated with chlorine to disinfect the water, but the chlorine feedrate is controlled so that no free chlorine reaches the GAC filters.

Figure 7-1. Pilot Unit Schematic

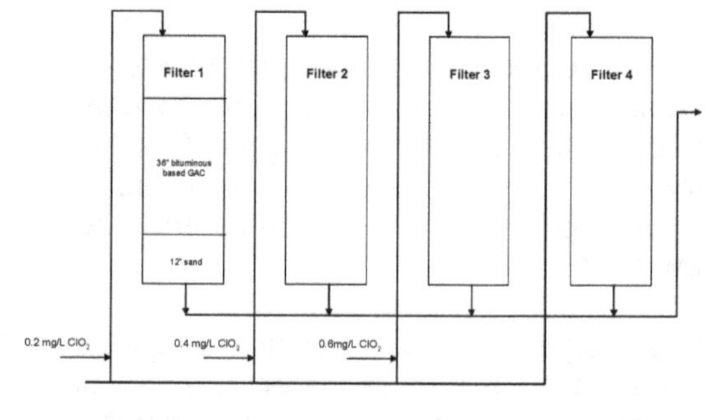

Chlorine dioxide, prepared by a method described in other work [Hammond et al. 2004], was dosed at about 0.2, 0.4, and 0.6 mg/L to Filters 1—3, respectively, with the number 4 Filter being held as the control, or untreated, filter. The ClO_2 dosages varied only slightly (± 0.1 mg/L) during the test. No other chemicals were added during the test.

Figure 7-2. DP - Filter 2 - 0.4 mg/L ClO_2

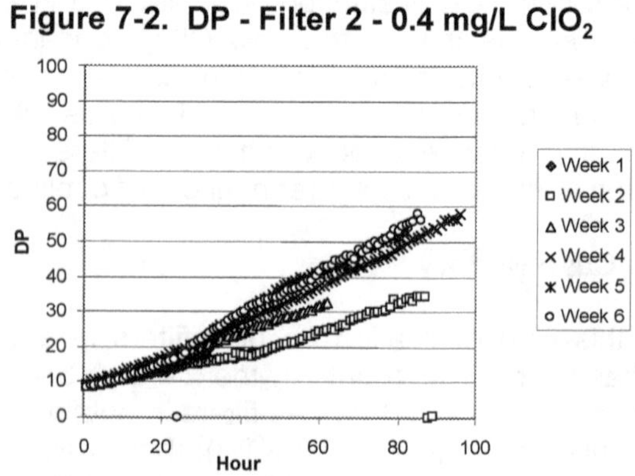

The general protocol followed was to treat the first 3 filters of the pilot unit over about a week's period of time. At the end of the week, feed of ClO_2 was stopped and the unit was shut down for the weekend. Backwashing each filter was done at the beginning of each week, at which time feed of ClO_2 was continued. The water going to the untreated filter had been well disinfected, with ClO_2 being added at the raw water line to meet the chlorine dioxide demand of the water. No ClO_2 was present in the water going to the untreated filter. Because the raw water had been well disinfected upstream, a minimal number of bacteria survived to see the untreated filter. The test lasted for about 6 weeks.

Results: The raw differential pressure data for Filter 2 (~ 0.4 mg/L ClO_2) and the untreated filter are shown in Figures 7-2 and 7-3, respectively. Clearly, the differential pressure increased over the duration of the trial for the control filter (Figure 7-3) at a more rapid rate than for Filter 2 (Figure 7-2). To illustrate this further, Figure 7-4 shows a comparison of the differential pressure data obtained during the sixth week of the test. This data shows that the differential pressure of the untreated filter increased at a higher rate than those treated with ClO_2. Surprisingly, there does not appear to be much difference between the filter treated with ~ 0.2 mg/L ClO_2 and the one treated with ~ 0.6 mg/L ClO_2. This raises the question as to how low ClO_2 can be dosed and still control the rate of increase in differential pressure.

Figure 7-3. DP - Untreated Filter

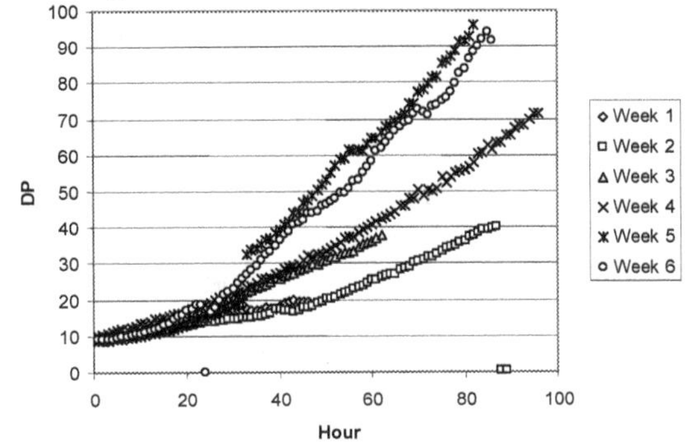

By plotting the difference between the DP of a treated filter and the untreated filter, the data show a cumulative effect. Figure 7-5 shows the plot of this ΔDP for Filter 1, which received the lowest applied [ClO_2].

The data show that during the first partial week of treatment, there was very little difference between the rate of change of the differential pressure of Filter 1 vs. that of the control filter.

Figure 7-4. DP Comparison - Week 6

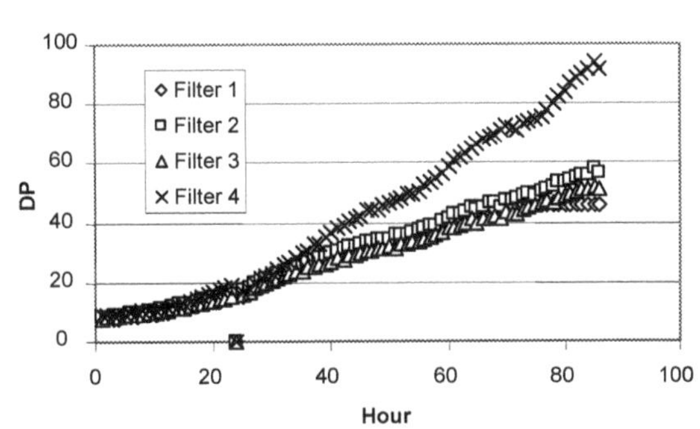

During week 2, except for what is considered some spurious data during the middle of the week, the data begins to tail downward toward the end of the week, indicating a slightly higher differential pressure for the control filter vs. that of Filter 1.

During the third week, initially there is little difference between the differential pressure of Filter 1 vs. that of the control filter, except that toward the end of the week, the data indicate a more rapid increase in differential pressure for the control filter than that of Filter 1.

During the fourth week, a noticeable change starts to occur during the middle of the week. From the middle of the week onward, there is an increasing difference between the differential pressure of the ClO_2-treated filter and the untreated filter.

During the fifth week, on about the end of the second day of treatment, there is a rapid increase in differential pressure between the ClO_2-treated filter and that of the control.

By week six, by the end of the first day or beginning of the second day, there is an increasing differential pressure change of the ClO_2-treated filter vs. that of the control filter.

These results were surprising, particularly the fact that the results appeared to be cumulative with time.

Figure 7-5. Delta DP (Filter 1 - Control)

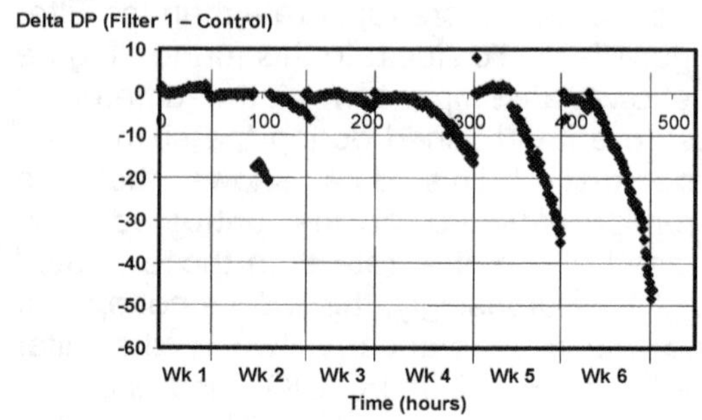

Anecdotal Information: In the Southwest, where the wind blows with some regularity, many side stream filters in many plants have been taken out of service, because the bacterial and algal biofilm holds the media together and prevents adequate backwashing. Thus, the side stream filters become laden with bacteria and become sites of reinnoculation during backwashing and routine operation.

In the early to mid-1990s, a number of these out-of-service side stream filters were cleaned and placed back into service by use of ClO_2. The process included initiating the backwash procedure and filling the side stream filter with ClO_2 of about 1000 mg/L and allowing it to sit for an hour or so before resuming the backwashing procedure. In this way the media was rapidly cleaned, allowing the filter to be placed back into service [Averett 2000].

In more recent work, the application of ClO_2 to a Parkson DynaSand® Filter in a desalination plant was evaluated. The DynaSand® filter is a continuous self-cleaning filter where air is bubbled through a small riser in the center of the filter. The air causes the sand to scrub against itself and abrade other sand particles and provide a scrubbing action to remove biological growth. Application of ClO_2 resulted in enhanced cleanup of the filter media. In addition, the differential pressure of the replaceable cartridge filter immediately downstream of the sand filter was maintained well below the manufacturer's upper limit where filter replacement was required. This filter helps to ensure protection of the downstream reverse osmosis unit. During the 2 week test on a pilot unit, replacement of the cartridge filter was not required, yet in the plant, replacement was being done every 3 days [Simpson et al. 2005].

C. Conclusions

Several conclusions can be made from this work.

1. The most obvious conclusion that can be reached from this data is that ClO_2 can provide significant reduction in the rate of microbial fouling of filters, when applied just prior to the filters.

2. A more significant, yet perhaps less obvious conclusion is that the effect of ClO_2 appears to be cumulative. ClO_2 can clean up aged biofilm, if sufficient turbulence (e.g., backwashing of filters or even cooling water exchangers [Simpson et al. 1993; Simpson 1995; Pickrell et al. 1999]) can be done. And although the data shown here is for continuous use of ClO_2, even intermittent use shows a similar trend, although the effect takes longer [Simpson et al. 1994].

3. A third conclusion is that the effect appears to be dosage independent over the range tested (0.2—0.6 mg/L). One question that arises is how low the dosage can be reduced and the effect still be observed. The answer is likely to be "quite low." The basis for this conclusion comes from work on cooling systems, where ClO_2 was applied on an intermittent basis at a rate of 0.01—0.05 mg/L with good control, even in stressed systems [Hobro 2000; Nickels et al. 2000; Simpson 2001].

4. For small package potable water treatment plants, backwashing of potable water filters is normally done automatically based on differential pressure. The use of ClO_2 immediately before such a filter, even at very low dosages, will greatly extend the run length between required backwashing of the filter. Since about 30% of the water produced through such a package plant is required to backwash the filter, the use of small amounts of ClO_2 before such a filter can greatly reduce the overall cost to process the water.

Chlorine dioxide is a niche oxidant, the use of which was well publicized during the anthrax-contaminated Hart Senate Office Building cleanup, which was successfully done using gaseous ClO_2. Its unusual properties, such as its ready penetration and inactivation of biofilm bacteria, make it an ideal candidate for applications such as enhancing the operation of both side stream cooling towers and potable water filters.

D. References

[Averett 2000] Averett, B., Biox, Inc., personal communication (2000).

[Chen and Stewart 1996] Chen, X. and Stewart, P., "Chlorine Penetration into Artificial Biofilm Is Limited by a Reaction-Diffusion Interaction," *Environmental Science & Technology*, 30(6), 2078(1996).

[de Beer et al 1994] de Beer, D., Srinivasan, R., and Stewart, P., "Direct Measurement of Chlorine Penetration into Biofilms During Disinfection," *Applied and Environmental Microbiology*, 60(12), 4339(1994).

[Franklin et al. 1991] Franklin, M., Nivens, D., Vass, A., Mittelman, M., Jack, R., Dowling, N., and White, D., "Effect of Chlorine and Chlorine/Bromine Biocide Treatments on the Number and Activity of Biofilm Bacteria and on Carbon Steel Corrosion," *Corrosion*, 47(2), 128(1991).

[Hammond et al. 2004] Hammond, D., Dean, A., and Simpson, G., "The Impact of Chlorine Dioxide on Particle Counts, Turbidities, and Differential Pressures of Potable Water GAC Filters," Southwest Regional Annual Meeting, American Water Works Association, Little Rock, AR, October 10-12 (2004).

[Hobro 2000] Hobro, D., Halox Corporation, personal communication (2000).

[Marshall and Walker 1990] Marshall, A. and Walker, G., "Increased Awareness of Biological Control and Its Implications," Paper No 354, Corrosion 90, Las Vegas NV, April 23-27, 1990.

[Melancon 2004] Melancon, C., "Cooling Tower Side Stream Filtration 101," *CTI Journal*, 25(2), 30(2004).

[Nickels et al. 2000] Nickels, A., Harriman, C., Nickels, A., Freedman, A., and Harris, A., "A Novel Method of Applying Chlorine Dioxide in a Recirculating Cooling Water System," AWT Annual Convention and Exposition, Water Technologies 2000, Honolulu, HI, October 31-November 4 (2000).

[Pickrell et al. 1999] Pickrell, W., Germer, M., Miller, J., and Simpson, G., "Plants Increase Production Through Use of Chlorine Dioxide," Practical Industrial Water Treatment Technology for the New Millennium, Houston, TX, January 27-29 (1999).

[Simpson 1995] Simpson, G., "Biofilm: Removal and Prevention with Chlorine Dioxide," The Third International Symposium on Chlorine Dioxide: Process Water, Drinking Water and Water Waste Issues, New Orleans, LA, September 14–15 (1995).

[Simpson 2001] Simpson, G., "Control of Biofilm in Cooling Towers with Chlorine Dioxide," Fourth International Symposium on Chlorine Dioxide, Las Vegas, NV, February 15-16 (2001).

[Simpson et al. 1993] Simpson, G., Laxton, G., Miller, R., and Clements, W., "A Focus on Chlorine Dioxide: Biocide of Choice for 'Stressed' Cooling Water Systems," WaterTech 93, Houston, TX, November 10-12 (1993).

[Simpson et al 1994] Simpson, G., Laxton, G., Ferriola, J., and Hogg, R., "A Focus on Chlorine Dioxide: Increasing Mold Life in Electric Arc Furnace Mini Steel Mills," WaterTech '94, Houston, Texas, November 9-11 (1994).

[Simpson et al. 2005] Simpson, G., Alarid, V., Matchim, D., and Hale, E., "From Seawater to Potable Water: Challenges and the Role of Chlorine Dioxide," unpublished results (2005).

[Trulear and Wiatr 1988] Trulear, M. and Wiatr, C., Recent Advances in Halogen Based Biocontrol, Technical Paper No. 19, Corrosion 88, March 21-25, St. Louis, MO (1988).

[White 1986] White, G. C., Handbook of Chlorination, Van Nostrand Reinhold Company, NY, (1986).

[Xu et al. 1996] Xu, X., Stewart, P., and Chen, X., "Transport Limitation of Chlorine Disinfection of Pseudomonas aeruginosa Entrapped in Alginate Beads," *Biotechnology and Bioengineering*, 49, 93(1996).

Control of *Legionella pneumophila* with ClO_2

A. Introduction

In 1999, Dr. Jennifer Miller and I presented a paper entitled "Chemical Control of Legionella" at the AWT Convention in Palm Springs [Miller and Simpson 1999]. This paper was a very comprehensive review article that discussed the known mechanisms of growth for Legionella and the relative effectiveness of alternative biocides, both oxidizing and non-oxidizing, on each of these mechanisms. Although the paper contained some very significant and highly relevant conclusions, the paper did not receive the attention it deserved. The reason for this was that the paper was much too comprehensive, with a great deal of supporting information, i.e., the conclusions were lost in the noise.

Therefore, this chapter revisits this topic in a much simpler format. The conclusions come first, with the supporting information following, should the reader wish to investigate in more detail the literature claims.

B. Chemical Control of Legionella – Conclusions

Regarding the mechanism of growth of Legionella, the key conclusions from this review are as follows:

1. **Legionella do not apparently multiply well in bulk water. They are not competitive bacteria, i.e., they have not been found to produce an antibiotic that will allow them to predominate in a diverse microbiological community. They do not grow well in the presence of other bacteria.**
2. **Legionella have been found to grow within protected microniches of bacterial biofilm.**
3. **Legionella have been found to grow within protected microniches of algal biofilm.**
4. **Legionella have been found to amplify within certain protozoa.**

Therefore, in addition to the normal bulk water disinfection, any chemical used to control Legionella must be able to do the following:

1. **The chemical must be effective on biofilm control and removal.**
2. **The chemical must be effective on algae.**
3. **The chemical must be able to control protozoa.**

The following sections review the relationship between Legionella and bulk water, biofilm, algae, and protozoa. In addition, the relative effectiveness of various chemicals on bulk water disinfection, control of biofilm, control of algae, and most importantly, control of protozoa, are discussed, from the perspective of the scientific literature.

In summary, for control of bacteria in the bulk water, all chemicals are relatively effective. When any of the other three items above are considered, the chemicals can be ranked according to their effectiveness. The order of ranking changes, generally, depending on whether biofilm, algae or protozoa is considered.

When all of these factors are considered, ClO_2 becomes the clear choice for control of Legionella. Supporting data are found in the following sections.

C. About Legionella

Discovery: In July 1976, the Pennsylvania Department of Health was inundated by reports of a strange illness. By August 2, department personnel concluded that all of those incidents involved persons who had stayed at the Bellevue-Stratford Hotel in Philadelphia July 21-24, 1976, or had been near the hotel during that time. Most had attended the 58th annual convention of the American Legion's Pennsylvania Chapter that was held at the hotel. Illness struck 221 persons. Ultimately there were 34 deaths [Fraser and McDade 1979]. Investigators traced the illness, which had been dubbed "Legionnaires' disease" by the press to a previously unknown bacterium, now known as *Legionella pneumophila*.

The discovery of Legionnaires' disease enabled researchers to reopen several cases in which the cause of illness could not previously be identified. Frozen specimens that had been retained were removed from long-term storage and re-tested. The results confirmed that Legionella was the causative agent. The earliest case identified occurred in 1947 and involved a soldier at Fort Bragg, North Carolina [McDade et al. 1979]. Another early manifestation where 3 or more persons contracted the disease occurred in 1957 in a meat packing plant [Osterholm et al. 1983]. The first known hospital outbreak occurred in 1965 when 81 patients at St. Elizabeth's Hospital in Washington, D.C. developed pneumonia with 14 fatalities [Thacker et al. 1978].

Outbreaks: Since those original incidents, a number of Legionella outbreaks have been identified [Addiss et al. 1989; Clepper 1995; Dondero et al. 1980; England et al. 1982; Garbe et al. 1985; Keller et al. 1996; Monohan 1978]. As a result, numerous investigations have been undertaken in an attempt to determine the mechanism of bacterial growth [Lee and West 1991], the mechanism of disease transmission [Muder et al. 1986], the number of Legionella required to promote development of the disease [O'Brien and Bhopal 1993], and the most effective control strategies.

Physiology: Legionella are rod-shaped, gram negative bacteria. Research has shown that they require very specific conditions, i.e., the presence of iron and specific amino acids, for growth [Tesh and Miller 1981; Wadowsky and Yee 1985; Lee and West 1991]. Therefore, it is thought that Legionella may survive, grow and multiply within well-defined specific microniches that contain the required nutrients, such as a bacterial biofilms [Lee and West 1991; Marro et al. 1993] or algal mats [Hume and Hann 1984; Tison et al. 1980; Pope et al. 1982]. These specific requirements for growth of Legionella explain the ubiquitous nature of the organism in the environment and the difficulty of culturing the bacteria in the laboratory [Fliermans et al. 1981a; Fliermans et al. 1981b].

Mechanism of Infection: Legionnaires' disease is thought to occur when a person with a compromised immune system inhales an aerosol contaminated with infective bacteria [Garbe et al. 1985; Lee and West 1991; Muder et al. 1986; Nagington and Smith 1980]. The infected persons white blood cells engulf the bacteria. White blood cells that engulf the bacteria provide the Legionella an environment conducive for growth. The Legionella proliferate and pneumonia develops. Those with healthy immune systems have no trouble staving off infection. Those with compromised immune systems, i.e., organ-transplant patients, people with AIDS, and cancer patients receiving chemotherapy, may develop Legionnaires' disease or Pontiac fever, a less virulent manifestation of the infection [Nagington and Smith 1980].

One remaining unanswered question is the number of bacteria required for development of an infection [O'Brien and Bhopal 1993; Berendt et al. 1980], as the exact composition of the infective aerosol is not yet known. What is known is that the infective aerosols may be contaminated by individual Legionella, clumps or packets of Legionella [Fliermans et al. 1981a; Fliermans et al. 1981b], Legionella-contaminated biofilm fragments [Potera 1999; Costerton 1998], Legionella-infected vacuoles or vesicles [Lee and West 1991; Rowbotham 1986], or Legionella-infected amoeba. Single Legionella are not thought to be the causative agent because white blood cells appear to be able to inactivate single bacteria. Amoeba are not likely to be the infective transport vehicles for Legionella because they are generally too large to get into the alveoli of the lung, the site of infection [Lee and West, 1991; Rowbotham 1986]. Of the possible transport mechanisms, it is the Legionella-packed foecal vesicle which appears to be able to carry the largest number of bacteria to the site of infection [Barker and Brown 1994; Holden et al. 1984]. An amoebal vesicle of 10 μm diameter, with 90% of the space occupied by bacteria measuring 0.32 X 0.6 μm could contain approximately 10^4 bacterial cells [Barker and Brown 1994].

Although there has been no general consensus regarding what constitutes an infectious dose of Legionella in water, a correlation between health risk and Legionella counts in cooling water has been made [Morris and Feeley 1990; Miller and Kenepp 1993].

Where Legionella Have Been Found: Legionella-contaminated aerosols have been found to originate from a number of sources, including cooling towers [Addiss et al. 1989; Bhopal et al. 1991; Dondero et al. 1980; England et al. 1982; Garbe et al. 1985; Keller et al. 1996; Monohan 1978; Lee and West 1991; Muder et al. 1986], evaporative condensers [Cordes et al. 1980], shower heads [Yee and Wadowsky 1982; Muder et al. 1986; Dennis et al. 1984], whirlpool spas [Henke and Seidel 1986; Witherell et al. 1984], hospital hot water systems [Fields et al. 1989; Tobin et al. 1980], medical respiration devices [Gorman et al. 1980], decorative fountains [Hady et al. 1993], and grocery store misting machines [Mahoney et al. 1992].

Some have suggested that Legionella are not free-living, aquatic bacteria [Rowbotham 1980], because they do not proliferate in bulk cooling water [Lee and West 1991; CTI 1997; Skaliy and McEachern 1979]. Others have suggested that conditions in cooling water may promote the growth of Legionella [Yamamoto et al. 1992; Yee and Wadowsky 1982].

D. Legionella, Bulk Water, and Control

Although Legionella have been found in bulk cooling water, research clearly indicates that controlling Legionella in the bulk water is not adequate to prevent a cooling tower from becoming a site of infection [Lee and West 1991]. In order to insure that a cooling system is free of Legionella, a number of things have to be done, including cleaning the tower (although in itself, this is insufficient to control Legionella) [Broadbent et al. 1994; Broadbent 1995].

Legionella do not appear to be able to produce antibiotics (these antibiotics would inhibit growth of other bacteria) that would allow them to predominate in a diverse biological community [Rowbotham 1980]. Thus, mechanisms of growth other than their proliferation in bulk water must account for their accumulating to numbers great enough to become a health threat.

Despite the diversity of opinions as to whether Legionella can grow in the bulk water or not, the fact remains that Legionella can be present in the bulk water. It is therefore important to be able to control Legionella that are present in the bulk water, whatever the source. The chemical control strategy chosen will depend on the origin, use, and fate of the water to be treated. This chapter focuses on chemical control strategies for cooling water systems.

Laboratory Studies vs. Field Studies - Biocidal Efficacy: A considerable number of studies have been completed to identify biocides that are effective against this organism in the bulk water. Much of this test work was conducted in the laboratory with Legionella cultured from broth or pure cultures, challenging Legionella with various biocides, both oxidizing [Wang et al. 1979; Hollis, and Smalley 1980; Skaliy et al. 1980; Dominque et al. 1988; McCoy and Wireman 1989] and non-oxidizing [Wang et al. 1979; Hollis and Smalley 1980; Skaliy et al. 1980; Grace et al. 1981; McCoy et al. 1986].

Researchers began to take note of the gap between laboratory results and field results of testing of biocidal efficacy against bacteria, including Legionella. Several researchers reported [Berg et al. 1981; Berg et al. 1982] that the methodology used to culture bacteria in a laboratory can result in bacteria that have significant differences in their resistance to disinfection. In particular, they found that laboratory-cultured bacteria are easier to kill than those found in nature. Their data suggested that bacteria cultured in chemostats can simulate those found in natural aquatic environments with regard to their resistance to disinfection.

In 1982 a report a compared the results of laboratory testing of biocidal efficacy against Legionella vs. actual field results from tests done on a number of cooling towers [England et al. 1982]. Their results showed that there did not appear to be any correlation between the presence of Legionella and use of biocides that have been shown in laboratory tests to be effective toward this organism.

In another report [Kurtz et al. 1982] fourteen cooling towers were surveyed to determine whether factors could be identified that might indicate the presence or absence of Legionella. No such correlation could be found. A tributyl tin oxide / quaternary ammonium compound, a chlorinated phenol, and a solid chlorine release compound were used as biocides in this investigation. The tin/quat was selected because previous work showed it to be effective towards Legionella.

Dosages of the biocides (tin/quat – 10 mg/L, chlorinated phenol – 100 mg/L, and the solid chlorine compound – maximum recorded residual – 1.2 mg/L) were found to be inadequate to control Legionella. No other chemicals were used during the trial.

Since tin/quat had been shown to be effective at controlling laboratory cultured Legionella and quat alone had been shown to be relatively ineffective, conclusions were that the tin compound, tributyl tin oxide (TBTO) was the effective agent [Soracco and Pope 1983].

The results suggest that cellular responses were related to the initial dosage of TBTO, and not the measured residual. At low or intermediate doses (10^4 to10^7 molecules per cell) the primary result of addition of TBTO was to slow down bacterial metabolism. At higher doses ($>10^7$ molecules per cell) the TBTO had a bactericidal effect. Their results suggest that when a tin/quat biocide is used, it appears that the TBTO plays the major role in the disinfection process.

In 1984, another group reported their results of a study wherein they tested the relative resistance of Legionella to inactivation by disinfectants. They confirmed that the dilute nutrient conditions used when culturing Legionella in continuous process promoted the growth of a Legionella phenotype that was generally much more difficult to inactivate than those cultured in laboratory environments [Berg et al. 1984a].

In their continuing work to investigate the effect growth conditions had on resistance to biocide by Legionella, one group demonstrated that ClO_2 was superior to chlorine for control of Legionella [Berg et al. 1984b].

Another group reported the results of a study involving two biocides, a chlorinated phenolic thioether and a polymeric quaternary ammonium chloride compound [Kurtz et al. 1984]. Each biocide was used in a number of cooling towers on a weekly basis. These towers were tested, just before dosing at monthly intervals. Every tower treated with the phenolic biocide showed positive for Legionella before the treatments began. None of the towers treated with the phenolic biocide (200 mg/L at 4 hours after dosing) showed positive for Legionella during treatment. Apparently, the phenolic biocide was very effective against Legionella. In contrast, Legionella were present in 24 of 35 systems treated with the quat (50 mg/L).

In their laboratory tests, 100 mg/L of the phenolic biocide reduced counts of Legionella by > 9 $\log_{10}$ within 3 hours, and the trophozoite form of *A. palestinesis* from 10^5/ml to below detectable levels in 1 hour. A similar dosage of the quat did not materially affect the counts or the protozoa. The authors suggest that it is important to use a biocide that is effective not only on Legionella, but also on protozoa.

In another study bromochlorodimethylhydantoin (BCDMH) was evaluated in a cooling tower [Fliermans and Harvey 1984]. In that field trial BCDMH was added to maintain a free halogen residual of 0.2—0.5 mg/L for 1 month. It was found to be ineffective for control of Legionella. Then, BCDMH was applied for 5 days at a dosage of 10 times the manufacturer recommendation, and it still did not reduce Legionella.

Another report described an investigation involving several biocides for their effectiveness at controlling various strains of Legionella, bacteria, and fungi in a working cooling tower [Elsmore 1986]. The results suggest that methylenebis thiocyanate applied once per week at 10 mg/L per treatment was ineffective at controlling Legionella. 2-Bromo-2-nitro-propane-1, 3-diol, applied twice weekly at 70 mg/L was effective at controlling Legionella. Although other biocides had been found to be effective in his laboratory experiments, they did not effectively control Legionella in this system.

One researcher noted that a number of biocides have been shown to be effective against Legionella in the laboratory, but that translation of this to the field has been difficult. This was attributed to a lack of a suitable standard method for evaluation of biocides against Legionella [Elsmore 1993].

In another report the results of several field trials for control of Legionella in cooling towers were described [Bentham and Broadbent 1995]. Sixteen cooling towers were treated with three commercially available biocides: bromonitropropanediol (BNPD), a chlorinated phenolic thioether (CPTE), and bromochlorodimethylhydantoin (BCDMH). The results of their use are shown in Table 8-1.

Table 8-1. Effect of Biocide Use on Legionella, Total Bacteria, and Protozoa

Biocide	**Effectiveness of Biocide[#]**		
	***Legionella* (# successful/# towers)**	**Total Bacteria**	**Protozoa**
Chlorinated phenolic thioether	8/9	NR[H]	NE[*]
Bromonitropropanediol	5/11	NR[H]	NE[*]
Bromochlorodimethylhydantoin	9/9	$\leq 10^3$	NE[*]

[#] Effectiveness was measured by reduction of Legionella below detectable levels.
[H]NR – Not Reduced
[*]NE – Not Effective

Recently a report was issued describing the results of work comparing various biocides, both oxidizing and non-oxidizing, in terms of their efficacy against Legionella in model cooling towers [Thomas et al. 1999]. The data suggested that oxidants work best, but only if a continuous residual is applied. Non-oxidizers were less effective and did not provide lasting control. Some non-oxidizing biocides were not effective.

Laboratory Work with Amoeba-Extracted Legionella: Various researchers have reported that Legionella that multiply within amoeba and are released after the amoeba lyses, are very motile, aggressive, and resistant to biocides [Barker et al. 1993; King et al. 1988; Lee and West 1991]. Thus researchers began to investigate the differences between Legionella extracted from amoeba and those grown from cultures.

One group reported the results of their work on the susceptibility of Legionella grown under several conditions to inactivation by several biocides [Barker et al. 1993]. The biocides tested were polyhexamethylenebiguanides (PHMB), 5-chloro-N-methylisothiazolone (CMIT; the active ingredient in Kathon), Fentichlor and tetradecyltrimethylammonium bromide (quat). Legionella grown inside amoeba were found to be significantly more resistant to biocide action than Legionella grown in broth, with the single exception of Fentichlor, after 6 hours of contact time.

Both PHMB and Fentichlor showed good reduction of Legionella grown from broth. Quat and CMIT were somewhat less effective. The results suggested that amplification of Legionella by protozoa might account for the differences between laboratory tests and results obtained from cooling systems.

E. Legionella, Biofilm, and Control

The Significance of Bacterial Biofilm: Biofilm have been shown to be a major source of Legionella in the natural environment [Bentham and Broadbent 1995; Marrao et al. 1993]. Biofilm has also been implicated as a major harbor for other pathogenic bacteria [Costerton 1995; Costerton and Lappin-Scott 1995; Costerton and Anwar 1994] and in particular is a problem in the health care industry [Potera 1999].

Some have suggested that biofilm in cooling towers may be a major source of Legionella [Bentham et al. 1993]. Bacteria associated with a biofilm have been found to be of considerably greater importance than those found in the bulk water, particularly as they impact corrosion and fouling [Kurtz et al. 1982; Videla et al. 1990]. Control of corrosion in biofouled cooling systems may require the removal of the biofilm to control the corrosion, and not just inactivation of bacteria within the biofilm [Roe et al. 1996].

Biofilm Control Strategies: In addition to these problems associated with a bacterial biofilm, the bacteria in microbial biofilm have been shown to have significantly greater resistance to bacterial disinfection than those found in the bulk water.

The fact that sessile bacteria exhibit a considerably greater resistance to non-oxidizing biocides than those found in the bulk water has been known for some time [Kajdasz et al. 1984].

In general, oxidizing biocides are more effective than non-oxidizing biocides in biofilm control and removal. However, chlorine has been found to relatively ineffective at controlling bacterial biofilm, even in a well-disinfected system such as a potable water treatment system [LeChevallier et al. 1990].

A general observation is that the weaker the oxidant, the more likely it is to be able to penetrate a biofilm and activate bacteria within it [Costerton et al. 1999]. Thus bromine, chlorine dioxide and chloramines would be likely to be better at controlling biofilm bacteria than chlorine.

Chlorine penetrates a bacterial biofilm very slowly. It reacts with biofilm fragments at the surface of the biofilm and is inactivated well before it can penetrate and disinfect the bacteria inside [Chen and Stewart 1996; de Beer et al. 1994; Xu et al. 1996]. In a cooling tower, the effectiveness of chlorine is a competition between the rate at which chlorine penetrates the biofilm and the production of biofilm by bacteria. This depends upon available nutrient. In cooling towers with high nutrient loading (i.e., systems where the rate of biofilm formation > the rate of biofilm removal), oxidants such as chlorine and bromine are ineffective [Trulear and Wiatr 1988].

Despite this, there is some controversy as to the effectiveness of bromine at biofilm control. Bromine has been shown to be significantly better than chlorine with regard to biofilm control by some researchers [Giusto 1990] while others have found there to be little difference between chlorine and bromine [Franklin et al. 1991].

There is also some controversy surrounding the effectiveness of ozone on biofilms. Some have reported that ozone can effectively control or remove biofilms [Kaur et al. 1988; Kaiga et al. 1987], while other reports indicate that ozone is relatively ineffective on biofilms [Garey 1976; Fitzsimmons et al. 1991; Mangum and McIlhenney 1975; Marine Research Inc. 1976]. Clearly the general observation cited previously would predict that ozone is so reactive it would be consumed readily at the surface of the biofilm, making it difficult to penetrate with any efficiency [Costerton et al. 1999].

Chloramines have been found to be able to penetrate a biofilm and inactivate bacteria better than chlorine or chlorine dioxide on biofilm grown on granular activated carbon [LeChevallier 1990]. However, chloramines are ineffective at removing the biofilm [Chen et al. 1993].

Of the oxidizing biocides, ClO_2 has been shown to be the most effective for control, prevention, and removal of established biofilms [Simpson 1995].

Walker and Morales, [Walker and Morales 1997] reported the results of their evaluation of one stabilized ClO_2 product (activated) for the control of biofilms. They found that 1 mg/L ClO_2 applied continuously for 18 hours was sufficient to reduce the planktonic population by 99.92%, while 1.5 mg/L ClO_2 was required to reduce the biofilm. The stabilized product alone (not activated) at 20 mg/L had no effect on bacterial populations.

Some have suggested that this resistance of sessile bacteria to inactivation by biocides may be a function of a slower growth rate of sessile bacteria when compared to the growth rate of planktonic bacteria [Brown and Gilbert 1993; Brown et al. 1988; Gilbert et al. 1990].

Inactivation of Biofilm-Associated Legionella: The efficacy of one biocide that had previously been identified as being effective towards Legionella was tested [Coughlin and Caplan 1987]. Its effectiveness towards both planktonic and sessile bacteria was measured via a Modified Robbins Device. They found bromonitropropanediol, (BNPD) to be very effective at inactivating biofilm Legionella, although the biocide did not remove the biofilm.

Two biocides, 2-Bromo-4-nitropropane-1,3-diol (Bronopol) and 5-chloro-2-methyl-4-isothiazolin-3-one and 2-methyl-4-isothiazolin-3-one (Kathon) were tested by another group against Legionella within a biofilm [Wright et al. 1991]. Their results demonstrated that in 9—12 hours, the biocides were effective against the planktonic population, but that > 48 hours were required to reduce the number of culturable bacteria to below detectable levels in the sessile population.

Green and Pirrie [Green and Pirrie 1993] described a laboratory apparatus that was used for the generation of Legionella biofilm and the testing of biocidal efficacy toward sessile populations of Legionella. They evaluated glutaraldehyde, a blend of glutaraldehyde and a quaternary amine, and bromonitropropanediol.

Their results showed that bromonitropropanediol was almost totally ineffective against both planktonic *Legionella bozemanii* and sessile Legionellas and other aerobic bacteria. Glutaraldehyde and the glutaraldehyde/quaternary ammonium compound blend performed well against sessile Legionellas, reducing numbers of both *Legionella bozemanii* and *Legionella pneumophila* to below detectable levels at > 250 mg/L with 3 hour contact time.

In 1994 one group reported their results of an investigation of bromochlorodimethylhydantoin (BCDMH), on a Legionella-containing biofilm [Walker et al. 1994]. At a biocide concentration of 1—2 mg/L, planktonic bacteria were dramatically reduced. Only a 1-log drop in viable bacteria was observed in the biofilm. At 1 mg/L BCDMH, however, no Legionella were recovered either from the bulk water or from the biofilm, indicating they are much more susceptible to inactivation by BCDMH than other biofilm bacteria.

BCDMH was ineffective at removing the biofilm. At 1 mg/L, it reduced planktonic populations of aerobic bacteria significantly, but only a 1-log drop in biofilm bacteria was observed.

Another group reported the results of their use of one stabilized ClO_2 product to treat a hospital water system that was found to contain Legionella [Walker et al. 1995]. Repeated hyperchlorination failed to eliminate Legionella and other organisms from the system. Continuous use of ClO_2 was implemented. No *Legionella pneumophila* were recovered from the water system after the ClO_2 trial. This product was quite effective at biofilm removal. However there were some small patches of biofilm remaining that showed the presence of viable bacteria, though none were *Legionella pneumophila*.

Recently the results of an investigation into a number of biocides, both oxidizing and non-oxidizing, for control of Legionella in model cooling tower systems were reported [Thomas et al. 1999]. Their system included Legionella, heterotropic bacteria and cyst-forming amoeba. They found that the faster acting biocides such as dibromonitropropionamide (DBNPA) (8mg/L active) and glutaraldehyde (27 mg/L active) initially reduced Legionella populations, but did not totally eliminate them or provide long lasting control.

The oxidizing biocides, chlorine, bromine and ozone, were found to be effective as long as a free residual of oxidant was present.

F. Legionella, Algae, and Control

Certain algae are known to provide suitable nutrients for proliferation of Legionella. In 1980 the results of a survey of several aquatic habitats near Aiken, S. C. was reported [Tison et al. 1980]. Legionella was found to be present in algal mats. The results suggested that *Legionella pneumophila* was apparently utilizing metabolites produced by the algae as a food source. The authors suggested that the relationship between *Legionella pneumophila* and certain blue-green algae could be responsible for the apparent widespread distribution of the bacterium in nature.

Another report showed that Legionella suspended in extracts of *Fischerella sp.* were significantly more stable than when suspended in tryptose-saline solution employed in previous experiments [Berendt 1981]. In addition, this stability disappeared after dialysis, which suggested that a relatively small molecule was involved. The data suggested that algal extracts might play a role in the transmission of the bacteria in aerosols.

In 1982, research showed that cultures of several species of alga added to *Legionella pneumophila* grown in dilute nutrient solution resulted in a burst of new growth of that organism. This demonstrated that strains of algae other than *Fischerella* support growth of *Legionella pneumophila* [Pope et al. 1982].

Another group isolated three extracts from Fischerella. Two of the extracts stimulated respiration of Legionella, but one did not [Bohach and Snyder 1983]. This confirmed earlier work that algae could stimulate growth of Legionella.

In 1984 research was reported that demonstrated that Legionella grew better in the presence of green algae than in the absence of it [Hume and Hann 1984].

Chemical Control of Algae and Associated Legionella: In 1983 results of testing of nine strains of algae and nine strains of Legionella for their susceptibility to inhibition by a variety of commercially available microbiocides were reported [Soracco et al. 1983]. The responses ranged from ineffective to effective at 1/100 the manufacturers' recommended slug doses. Those biocides which were found to be the most effective for control of both algae and Legionella at dosages of 1/10 that recommended by the manufacturer were didecyl dimethyl ammonium chloride, TBTO/chlorophenate, TBTO/n-alkyl-1, 3-propylenediamine mono-propionate, N-n-alkyl-1, 3-propylenediamine monoacetate/n-alkyl dimethyl benzyl ammonium chloride, TBTO/n-alkyl dimethyl benzyl ammonium chloride.

Of the oxidizers, it is ClO_2 that has been used specifically for control of algae [Mounsey and Hagar 1946; Vincent et al. 1946; Harlock and Dowlin 1953]. The first recorded use of ClO_2 in potable water treatment was to eliminate taste and odor problems that are caused by algae [Aston 1947]. Use cost of ClO_2 for control of algae has been reported to be less than that of copper sulfate [Ringer and Campbell 1955].

G. Legionella, Protozoa, and Control

The interesting relationship between Legionella and certain protozoa has been one of the more thoroughly researched areas. Research suggests that survival, growth and amplification of Legionella in nature may require the presence of certain protozoa [Anad et al. 1983; Barker and Brown 1994; Holden et al. 1984; Kilvington and Price 1990; King et al. 1988; Fields et al. 1989].

Protozoa can be thought of as microscopic cows that graze on the biofilm, engulfing bacteria much like a snake engulfs a rodent. This process of ingesting a bacterium is known as phagocytosis. Unlike most bacteria, which are utilized as food source by certain protozoa, the Legionella can under specific conditions, after phagocytosis, multiply inside the protozoa [Barker and Brown 1994; Fields et al. 1984; Rowbotham 1980, Skinner et al. 1983, Holden et al. 1984].

A motile, feeding and replicating trophozoite form characterizes all free-living amoebae (FLA) with many species also able to produce a cyst stage. The work that has been done shows that encysted protozoa are considerably more difficult to destroy than the trophozoite form.

For example, if an Acanthamoeba becomes infected with Legionellae, the Acanthamoeba will try to encyst. If the internal multiplication of Legionella is at a faster rate than the Acanthamoeba can encyst, the amoeba can lyse, releasing many Legionella into the water [Rowbotham 1986; Oldham and Rodgers 1985; Rowbotham 1984]. Released amoeba-grown Legionella are motile [Barker and Brown 1994] and stay that way for about 24 hours, when they become non-motile. Thus infected cysts appear to be a product of a race between the amoeba to encyst and the Legionella to kill it. If the trophozoite successfully encysts prior to lysis due to Legionellae, the cyst can be packed with non-motile bacteria and stay that way for months [Rowbotham 1986].

If the amoeba encysts, the bacteria are protected further from harsh conditions and may survive for extended periods of time [Lee and West, 1991; Rowbotham 1986; Anad et al. 1983; King et al. 1988; Skinner et al. 1983; Barker et al. 1993]. For example, Legionella within protozoa are much more resistant to biocides, surviving chlorine levels as high as 50 mg/L [Kilvington and Price 1990; King et al. 1988].

Since amoeba in cooling towers can be a source of virulent Legionella, it is instructive to look at work done on the effect of biocides on various protozoa.

Non-Oxidizing Biocides for Inactivation of Protozoa: Only a few studies have investigated the inactivation of various protozoa with cooling water biocides.

In one report the amoebicidal properties of chlorine, chlorine dioxide, ozone, and deciquam 222 (didecyldimethylammonium bromide) were reported [Cursons et al. 1980]. They found that Naegleria were significantly more sensitive to biocides than Acanthamoeba. Based on their results, they concluded that:

didecyldimethylammonium bromide (Deciquam 222), > ClO_2 > ozone > Cl_2

In other work polyhexamethylene biguanide at 10 mg/L was lethal to Naegleria and Acanthamoeba in trophozoite form [Dawson et al. 1983].

Kilvington tested a variety of biocides and contact lens disinfectants against pathogenic free-living amoeba [Kilvington 1990]. He determined the minimum amoebicidal or cysticidal concentration (MAC or MCC), that is, the lowest concentration of test solution that resulted in complete lysis or degeneration of the trophozoites or cysts, respectively. Both Acanthamoeba and Naegleria were tested. The results are shown for 2-hour exposure time at pH 8.0 in Table 8-2.

Table 8-2. Minimum Amoebicidal or Cysticidal Concentration of Biocides

Biocide	MAC and MCC mg/L			
	Acanthamoeba		Naegleria	
	Trophozoite	Cyst	Trophozoite	Cyst
Peracetic acid	15	150	8	8
Chlorinated phenolic thioether	10	80	2	20
Isothiazolin	2	150	<1	2
Polyhexamethylene biguanide	10	5	20	2000
Bromonitropropanediol	200	>10,000	50	25
Methylenebis thiocyanate	3	>1,000	5	<1
Chlorine	2	>50	2	4

These results show two things: that some protozoa, particularly in their encysted form, are very resistant to some biocides, and that different protozoa exhibit markedly different resistance to different biocides.

Another group reported that Legionella taken from inside amoeba were more difficult to kill than ones grown from culture [Barker et al. 1992]. One possible explanation is that an amoeba-imposed iron restriction might alter the phenotypic response of the Legionella released to the environment.

A report in 1993 described the results of testing of two species of amoebae, Acanthamoeba and Cochliopodiuim, which were isolated from the cooling tower of an air conditioning system [Srikanth and Berk 1993]. These were examined for effects of low-level exposure to 4 cooling tower biocides: a thiocarbamate, tin-quat, quat alone, and isothiazolin. Two amoeba obtained from a commercial supply, were also tested.

The two strains of amoeba isolated from cooling towers were more resistant to the biocides than those obtained commercially. In addition, the results suggested that cooling tower amoebae might adapt to biocides. One alarming observation the authors made was that not only were the amoeba not inhibited by the biocides at the concentrations tested, but they also reproduced faster in the presence of low concentrations of the biocide.

A 1996 report described the survival of several protozoa in cooling tower biocides [Sutherland and Berk 1996]. The results, shown in Table 8-3, demonstrate the extreme resistance certain

Table 8-3. Minimum Lethal Concentration of Biocides for Protozoa

	Biocide concentration (mg/L) against:						
Organism	**Tetrahymena**	**Colpoda**		**Vannella**	**Acanthamoeba**		**MRMD**
Biocide		**feeding stage**	**cyst stage**		**feeding stage**	**cyst stage**	
Thiocarbamate	244	977	31,250	3,906	3,906	125,000	10—30
Isothiazolin	31	31	7,813	122	244	31,250	35—219
QAC	122	61	488	61	61	62,500	8—80
TBT/QAC	15	15	31	61	31	122	156

protozoa have for certain non-oxidizing biocides, and that protozoa may survive the manufacturers recommended maintenance dosages (MRMD) of certain biocides.

Oxidizing Biocides for Inactivation of Protozoa: A significantly greater number of studies were done to investigate the effectiveness of oxidants on protozoa [Chang 1978; Cursons et al. 1980; de Jonckheere and van de Voorde 1976; Kilvington and Price 1990; Chen et al. 1985; Ferguson and Andrews 1992; Hoehn 1976; Hoff 1986; Korich et al. 1990; Lykins et al. 1990; Lykins et al. 1989; Leahy et al. 1987; Peeters et al. 1989; Sproul et al. 1983; Wallis et al. 1990]. Most of this work has been in the drinking water area and thus the oxidants are generally chlorine, chloramine, ozone and chlorine dioxide. In addition, much of this work was done at neutral pHs where chlorine works well, but that most comfort cooling systems operate well into the alkaline region, where chlorine is less effective.

A 1972 report indicated that 10 mg/L chlorine was not sufficient to destroy Naegleria in a contaminated swimming pool [Anderson and Jamieson 1972].

Another report showed that HOCl was able to eradicate both Naegleria and Acanthamoeba at 0.5 mg/L [Derreumaux et al. 1974].

Other workers reported that Naegleria were destroyed in 30 minutes with 2 mg/L chlorine [de Jonckheere et al. 1976]. They also reported that at 40 mg/L chlorine, some Acanthamoeba survived after 3 hours, showing that Acanthamoeba were significantly more difficult to destroy than Naegleria.

They also demonstrated that *Acanthamoeba culbertsoni* cysts were not destroyed by up to 1.0 mg/L bromine, up to 5.0 mg/L iodine, or up to 5.0 mg/L iodophore after 24 hours exposure. In addition, up to 1.0 mg/L bromine had no effect after 24 hours on Naegleria, although both iodine and iodophore were effective under the conditions studied.

The relative difficulty of disinfecting organisms was investigated [Hoehn 1976]. In decreasing order of resistance to disinfection the organisms are as follows:

bacterial spores > protozoan cysts > viruses > vegetative bacteria

He noted that in general protozoan cysts are second only to bacterial spores with regard to their resistance to disinfectants.

Another worker reported that Naegleria were destroyed by both chlorine and iodine [Chang 1978]. The effectiveness of each oxidant varied with concentration. That is, above about 3.4 mg/L, iodine was found to be more effective, but below about 3.1, chlorine was found to be more effective.

A number of researchers [Chen et al. 1985; Hoff, 1986; Leahy et al. 1987] reported the results of their investigation regarding the effectiveness of various oxidants at inactivating amoebic cysts. They found generally that $O_3 > ClO_2 > Cl_2$.

The effectiveness of ozone or ClO_2 for inactivation of *Crytosporidium parvum* oocysts was reported [Peeters et al. 1989]. Both ozone and ClO_2 were effective at inactivating these oocysts.

A 1990 report showed that cysts of *Acanthamoeba polyphaga* produced from infected trophozoites were found to protect Legionella from at least 50 mg/L of free chlorine [Kilvington and Price 1990].

An indication of the relative difficulty of inactivating certain amoeba is shown in the Table 8-4. This report gives estimates of Ct values for common potable water disinfectants [Korich et al. 1990].

Table 8-4. Ct Estimates for *Cryptosporidium parvum* Oocyst

Ct value	Disinfectant			
	Ozone	ClO_2	Cl_2	Monochloramine
	5 - 10	78	7200	7200

It is clear from this that certain amoebic cysts are very resistant to both chlorine and monochloramine.

Other workers have confirmed the variable resistance of different cysts [Lykins 1990].

In a plant test, ozone and ClO_2 were tested for inactivation of Giardia cysts. Results showed that ClO_2 worked as well as ozone, and that even the 1 mg/L limit for ClO_2 is sufficient over a contact time of a few minutes to effectively control this pathogen [Wallis et al. 1990].

In summary, it is clear from these data that certain protozoa commonly found in cooling systems, notably Acanthamoeba, are extremely resistant to disinfection by chlorine particularly in the encysted form. The Legionella that proliferate within such amoeba are also protected from very high levels of disinfectants.

While chlorine is clearly inadequate for control of these cysts, use of ozone or ClO_2 may offer a better option, in that reports from the literature indicate that both oxidants are significantly more effective than chlorine for their control. In addition, some reports have suggested that non-oxidizing biocides such as didecyldimethylammonium bromide may be effective.

H. General Conclusions

The studies conducted to date on efficacy of certain biocides against Legionella show a considerable "gap" between laboratory results and field test results. Even in laboratory testing, some studies indicate a particular biocide to be very effective against Legionella while other studies indicate it is less effective. The source of this controversy may be the result of how the Legionella were cultured for the study. Bacteria in general, and Legionella specifically, have been shown to respond significantly differently to biocides, depending upon how they are grown. Those organisms grown in dilute nutrient solutions, similar to those found in nature, are much more resistant to biocides than those grown in nutrient-rich environments. In addition the response of Legionella to biocides depends on whether the bacteria are planktonic or located within a microniche of a bacterial biofilm.

Although there is evidence that certain algal extracts support growth of Legionella, it is less clear whether algae protect Legionella in some way from the effects of biocides. It is likely that such is the case.

Finally, one of the major explanations for the gap between laboratory testing of biocidal efficacy and field results is the increased biocidal resistance of protozoa-grown Legionella. Research has shown that Legionella that are released to the environment from the lysing of a host protozoa have a high degree of motility and increased biocidal resistance. These Legionella lose their motility within about 24 hours after being released to the environment. Whether they lose their resistance to biocides at this same time is unclear.

All of the studies conducted to date provide valuable data that are pieces of the Legionella puzzle. When viewed together, these data provide a clearer picture of the necessary steps required to adequately control Legionella. As the relationships between Legionella and bacteria, bacterial biofilm, algae, and protozoa are explored, it becomes clear that investigating any one of these areas, without considering the others, will not provide an accurate picture of a real system. To insure that biocidal efficacy studies translate to the field, the systems used to test biocidal efficacy must incorporate all of these factors: bulk water bacteria, bacterial biofilm, algae, and protozoa.

I. References

[Addiss et al. 1989] Addiss, D., Davis, J., LaVenture, M., Wand, P., Hutchinson, M., and McKinney, R., "Community-Acquired Legionnaires' Disease Associated with a Cooling Tower: Evidence for Longer-Distance Transport of *Legionella pneumophila*," *American Journal of Epidemiology*, 130(3), 557(1989).

[Anad et al. 1983] Anad, C., Skinner, A., Malic, A., and Kurtz, J., "Interaction of L. pneumophila and a Free Living Amoebae (*Acanthamoeba palestinensis*)," *Journal of Hygiene (Cambridge)*, 91, 167(1983).

[Anderson and Jamieson 1972] Anderson, K., and Jamieson, A., "Primary Amebic Meningo-Encephalitis," *Lancet* I: 902(1972).

[Aston 1947] Aston, R., "Chlorine Dioxide Use in Plants on the Niagara," *Journal of the Water Works Association*, 39(7), 687(1947).

[Barker and Brown 1994] Barker, J., and Brown, M., "Trojan Horses of the Microbial World: Protozoa and the Survival of Bacterial Pathogens in the Environment," *Microbiology*, 140, 1253(1994).

[Barker et al. 1992] Barker, J., Brown, M., Collier, P., Farrell, I., and Gilbert, P., "Relationship between *Legionella pneumophila* and *Acanthamoeba polyphaga*: Physiological Status and Susceptibility to Chemical Activation," *Applied and Environmental Microbiology*, 58(8), 2420(1992).

[Barker et al. 1993] Barker, J., Brown, M., Gilbert, P., Collier, P., and Farrell, I., "The Physiological Status of *Legionella pneumophila* and Its Susceptibility to Chemical Inactivation," Legionella: Current Status and Emerging Perspectives, Barbaree, Breiman and Dufour, eds., American Society for Microbiology (1993).

[Bentham and Broadbent 1995] Bentham, R., and Broadbent, C., "Field Trial of Biocides for Control of Legionella in Cooling Towers," *Current Microbiology*, 30, 167(1995).

[Bentham et al. 1993] Bentham, R., Broadbent, C., and Marwood, L., "The Influence of the Sessile Population in the Legionella Colonization of Cooling Towers," Legionella: Current Status and Emerging Perspectives, Barbaree, Breiman and Dufour, eds., American Society for Microbiology (1993).

[Berendt 1981] Berendt, R., "Influence of Blue-Green Algae (Cyanobacteria) on Survival of *Legionella pneumophila* in Aerosols," *Infection and Immunity*, 32(2), 690(1981).

[Berendt et al. 1980] Berendt, R., Young, H., Allen, R., and Knutsen, G., "Dose-Response of Guinea Pigs Experimentally Infected with Aerosols of *Legionella pneumophila*," *The Journal of Infectious Diseases*, 141(2), 186(1980).

[Berg et al. 1981] Berg, J., Matin, A., and Roberts, P., "Growth of Disinfection Resistant Bacteria and Simulation of Natural Aquatic Environments in the Chemostat," Water Chlorination, Environmental Impact and Health Effects, Jolley, et. al., eds., 4(81), 1137(1981).

[Berg et al. 1982] Berg, J., Matin, A., and Roberts, P., "Effect of Antecedent Growth Conditions on Sensitivity of Escherichia coli to Chlorine Dioxide," *Applied and Environmental Microbiology*, 44(4), 814(1982).

[Berg et al. 1984a] Berg, J., Hoff, J., Roberts, P., and Matin, A., "Disinfection Resistance of *Legionella pneumophila* and *Escherichia coli* Grown in Continuous and Batch Culture," Water Chlorination: Chemistry, Environmental Impact and Health Effects, Jolley, et. al., eds., 5(47), 603(1984).

[Berg et al. 1984b] Berg, J., Hoff, J., Roberts, P., and Matin A., "Growth of *Legionella pneumophila* in Continuous Culture and its Sensitivity to Inactivation by Chlorine Dioxide," Legionella, C. Thornsberry et al., editors, American Society for Microbiology, Washington, DC, (1984).

[Bhopal et al. 1991] Bhopal, R., Fallon, R., Buist, E., Black, R., and Urquhart, J., "Proximity of the Home to a Cooling Tower and Risk of Non-Outbreak Legionnaires' Disease," *British Medical Journal*, 302, 378(1991).

[Bohach and Snyder 1983] Bohach, G. and Snyder, I., "Cyanobacterial Stimulation of Oxygen Uptake by *Legionella pneumophila*," *Applied and Environmental Microbiology*, 46, 528(1983).

[Bova et al. 2004] Bova, G., Sharpe, P., and Keane, T., "Evaluation of Chlorine Dioxide in Potable Water Systems for Legionella Control in a Acute Care Hospital Environment," International Water Conference, IWC-04-11, October 2004.

[Broadbent 1995] Broadbent, C., "Cooling Towers, the Environment, and Public Health..." *I.P.E.A. National Journal,* 10(2), 17(1995).

[Broadbent et al. 1994] Broadbent, C., Bentham, R., and Marwood, L., "A Model to Explain Autumn Outbreaks of Legionnaires Disease Associated with Small Cooling Towers, *Australian Institute of Refrigeration Air Conditioning and Heating Journal*, 19(1994).

[Brown and Gilbert 1993] Brown, M., and Gilbert, P., "The Sensitivity of Biofilms to Antimicrobial Agents," *Journal Applied Bacteriology Symposium Supplement*, 74, 87S(1993).

[Brown et al. 1988] Brown, M., Allison, D., and Gilbert, P., "Resistance of Bacterial Biofilms to Antibiotics: A Growth Rate Related Effect?" *Journal of Antimicrobial Chemotherapy*, 22, 777(1988).

[Chang 1978] Chang, S. L., "Resistance of Pathogenic Naegleria to Some Common Physical and Chemical Agents," *Applied and Environmental Microbiology*, 35(2), 368(1978).

[Chen and Stewart 1996] Chen, X. and Stewart, P., "Chlorine Penetration into Artificial Biofilm is Limited by a Reaction-Diffusion Interaction," *Environmental Science & Technology*, 30(6), 2078(1996).

[Chen et al. 1985] Chen, Y., Sproul, O., and Rubin, A., "Inactivation of *Naegleria gruberi* Cysts by Chlorine Dioxide," *Water Research*, 19, 783(1985).

[Chen et al. 1993] Chen, C., Griebe, T., and Characklis, W., "Biocide Action of Monochloramine on Biofilm Systems of *Pseudomonas aeruginosa*," *Biofouling*, 7, 1(1993).

[Clepper 1995] Clepper, I., "Legionnaires' Disease Strikes in Minnesota; Second Recent Outbreak," *Air Conditioning, Heating and Refrigeration News*, 1(August 14, 1995).

[Cordes et al. 1980] Cordes, L., Fraser, D., Skaliy, P., Perlino, C., Elsea, W., and Mallison, G., "Legionnaires' Disease Outbreak at an Atlanta, Georgia, Country Club: Evidence for Spread from an Evaporative Condenser," *American Journal of Epidemiology*, 111, 425(1980).

[Costerton 1995] Costerton, J., "Overview of Microbial Biofilms," *Journal of Industrial Microbiology*, 15, 137(1995).

[Costerton 1998] Costerton, W., Center for Biofilm Engineering, Montana State University, private communication (1998).

[Costerton and Anwar 1994] Costerton, J. and Anwar, H., "*Pseudomonas aeruginosa*: The Microbe and Pathogen," Pseudomonas aeruginosa Infections and Treatment, A. Baltch and R. Smith (eds), Marcel Dekker, Inc., Chapter 1, 1(1994).

[Costerton and Lappin-Scott 1995] Costerton, J. and Lappin-Scott, H., "Introduction to Microbial Biofilms," Microbial Biofilms, H. Lappin-Scott and J. Costerton, eds., 1(1995).

[Costerton et al. 1999] Costerton, J., Camper, A., Stewart, P., Zelver, N., and Dirckx, M., "The Problem: Not Just Bacteria – Bacterial Biofilms," *The Analyst*, Summer, 18(1999).

[Coughlin and Caplan 1987] Coughlin, M. and Caplan, G., "Microbicidal Efficacy of BNPD against *Legionella pneumophila*," TP87-18, Cooling Tower Institute Annual Meeting, New Orleans, LA, February 25-27 (1987).

[CTI 1997] Cooling Tower Institute "1997 Educational Seminar on Legionella Control," *CTI Journal,* 18(2), 70(1997).

[Cursons et al. 1980] Cursons, R., Brown, T., and Keys, E., "Effect of Disinfectants on Pathogenic Free-living Amoebae: In Axenic Conditions," *Applied and Environmental Microbiology*, 40, 62(1980).

[Dawson et al. 1983] Dawson, M., Brown, T., and Till, D., "The Effect of Baquacil on Pathogenic Free-Living Amoeba (PFLA) 1. In Axenic Conditions," *New Zealand Journal of Marine and Freshwater Research*, 17, 305(1983).

[de Beer et al. 1994] de Beer, D., Srinivasan, R., and Stewart, P., "Direct Measurement of Chlorine Penetration into Biofilms during Disinfection," *Applied and Environmental Microbiology*, 60(12), 4339(1994).

[de Jonckheere and van de Voorde 1976] de Jonckheere, J. and van de Voorde, H., "Differences in Destruction of Cysts of Pathogenic and Nonpathogenic Naegleria and Acanthamoeba by Chlorine," *Applied and Environmental Microbiology*, 31(7), 294(1976).

[Dennis et al. 1984] Dennis, P., Wright, A., Rutter, D., Death, J., and Jones, B., "*Legionella pneumophila* in Aerosols from Shower Baths," *Journal of Hygiene (Cambridge)*, 93, 349(1984).

[Derreumaux et al. 1974] Derreumaux, A., Jadin, J., Willaert, E., and Moret, R., "Action du chlore sur les amibes de l'eau," *Ann Soc. Belbe Med. Trop.*, 54, 294(1974).

[Dominique et al. 1988] Dominque, E., Tyndall, R., Mayberry, W., and Pancorbo, O., "Effects of Three Oxidising Biocides on *Legionella pneumophila* Serogroup 1," *Applied and Environmental Microbiology*, 54, 741(1988).

[Dondero et al. 1980] Dondero, T., Rendtorff, R., Mallison, G., Weeks, R., Levy, J., Wong, E., and Schaffner, W., "An Outbreak of Legionnaires' Disease Associated with a Contaminated Air Conditioning Cooling Tower," *New England Journal of Medicine*, 302, 365(1980).

[Elsmore 1986] Elsmore, R., "Biocidal Control of Legionellae," *Israel Journal of Medical Science,* 22, 647(1986).

[Elsmore 1993] Elsmore, R., "Microbiocides and the Control of Legionella," Legionella: Current Status and Emerging Perspectives, Barbaree, Breiman and Dufour, eds., American Society for Microbiology, (1993).

[England et al. 1982] England, A., Fraser, D., Mallison, G., Mackel, D., Skaliy, P., and Gorman, G., "Failure of *Legionella pneumophila* Sensitivities to Predict Culture Results from Disinfectant-Treated Air-Conditioning Cooling Towers," *Applied and Environmental Microbiology*, 43, 240(1982).

[Ferguson and Andrews 1992] Ferguson, M. and Andrews, R., "Evaluation Methodology for Selection of Alternative Disinfectants using the CT Concept," Proceedings of the Annual Convention Wht Can. 44th, Water Wastewater Assoc., (1992).

[Fields et al. 1984] Fields, B., Shotts, E., Feeley, J., Gorman, G., Martin, W., "Proliferation of *Legionella pneumophila* as an Intracellular Parasite of the Ciliated Protozoan *Tetrahymena pyriformis*," *Applied and Environmental Microbiology*, 47(3), 467(1984).

[Fields et al. 1989] Fields, B., Sanden, G., Barbaree, J., Morrill, W., Wadowsky, R., White, E., and Feeley, J., "Intracellular Multiplication of *Legionella pneumophila* in Amoebae Isolated from Hospital Hot Water Tanks," *Current Microbiology* 18, 137(1989).

[Fitzsimmons et al. 1991] Fitzsimmons, T., Lindteigen, W., and Roeder, J., "Ozone Water Treatment Pilot Testing in a Comfort Cooling Tower," Paper No. 208, Corrosion 91, March 11-15, Cincinnati, OH, (1991).

[Fliermans and Harvey 1984] Fliermans, C. and Harvey, R., "Effectiveness of 1-Bromo-3-Chloro-5,5-Dimethylhydantoin Against *Legionella pneumophila* in a Cooling Tower," *Applied and Environmental Microbiology*, 47(6), 1307(June 1984).

[Fliermans et al. 1981a] Fliermans, C., Cherry, W., Orrison, L., Smith, S., Tison, D., and Pope, D., "Ecological Distribution of *Legionella pneumophila*," *Applied and Environmental Microbiology*, 41, 9(1981).

[Fliermans et al. 1981b] Fliermans, C., Cherry, W., Orrison, L., Thacker, L., "Isolation of *Legionella pneumophila* from Nonepidemic-related Aquatic Habitats," *Applied and Environmental Microbiology*, 37, 1239(1981).

[Franklin et al. 1991] Franklin, M., Nivens, D., Vass, A., Mittelman, M., Jack, R., Dowling, N., and White, D., "Effect of Chlorine and Chlorine/Bromine Biocide Treatments on the Number and Activity of Biofilm Bacteria and on Carbon Steel Corrosion," *Corrosion*, 47(2), 128(1991).

[Fraser and McDade 1979] Fraser, D. and McDade, J., "Legionellosis," *Scientific American* 241(4), 82(1979).

[Garbe et al. 1985] Garbe, P., Davis, B., Weisfeld, J., Markowitz, L., Miner, P., Garrity, F., Barbaree, J., Reingold, A., "Nosocomial Legionnaires' disease, Epidemiological

Demonstration of Cooling Towers as a Source," *Journal of the American Medical Association*, 254(4), 521(1985).

[Garey 1976] Garey, J., "A Comparison of the Effectiveness of Ozone and Chlorine in Reducing Slime Growth Within Condensers of an Electric Generating Station Using Seawater as a Coolant," Proceedings of the International. Ozone Institute, June 2-4, Chicago, (1976).

[Gilbert et al. 1990] Gilbert, P., Collier, P., and Brown, M., "Influence of Growth Rate on Susceptibility to Antimicrobial Agents: Biofilms, Cell Cycle, Dormancy, and Stringent Response," *Antimicrobial Agents and Chemotherapy*, 34(10), 1865(1990).

[Giusto 1990] Giusto, M., "An Alternative Oxidizer, Bromine Offers a Northeast Chemical Plant Improved Corrosion and Fouling Control," Technical Paper No. TP90-08, Cooling Tower Institute, 1990 Annual Meeting, Houston, TX, (1990).

[Gorman et al. 1980] Gorman, G., Yu, V., Brown, A., Hall, J., Corcoran, L., Martin, W., Bibb, W., Morris, G., Magnussen, M., and Fraser, D., "Isolation of Pittsburgh Pneumonia Agent from Nebulizers used in Respiratory Therapy," *Annals of Internal Medicine,* 93(4), 572(1980).

[Grace et al. 1981] Grace, R., Dewar, N., Barnes, W., and Hodges, G., "Susceptibility of *Legionella pneumophila* against Three Cooling Tower Microbicides," *Applied and Environmental Microbiology*, 41(1), 233(1981).

[Green and Pirrie 1993] Green, P. and Pirrie, R., "A Laboratory Apparatus for the Generation and Biocide Efficacy Testing of Legionella Biofilms," *Journal of Applied Bacteriology*, 74(4), 388(1993).

[Hady et al. 1993] Hady, W., Mullen, R., Mintz, C., Shelton, B., Hopkins, R., and Daikos, G., "Outbreaks of Legionnaires' Disease Linked to a Decorative Fountain," *American Journal of Epidemiology*, 138, 555(1993).

[Harlock and Dowlin 1953] Harlock, R. and Dowlin, R., "Chlorine and Chlorine Dioxide for Control of Algae Odors," *Water and Sewage Works*, 100, 74(1953).

[Henke and Seidel 1986] Henke, M., and Seidel, K., "Association between *Legionella pneumophila* and Amoebae in Water," *Israel Journal of Medical Science*, 22(9), 690(1986).

[Hoehn 1976] Hoehn, Robert C., "Comparative Disinfection Methods," *Journal of the American Water Association*, Water Technology/Quality, 302(1976).

[Hoff 1986] Hoff, J., "Inactivation of Microbial Agents by Chemical Disinfectants," EPA 600/S2-86/067, (1986).

[Holden et al. 1984] Holden E., Winkler, H., Wood, D., and Leinbach, E., "Intracellular Growth of *Legionella pneumophila* Within *Acanthamoebae castellanii Neff*," *Infection and Immunity*, 45(1), 18(1984).

[Hollis and Smalley 1980] Hollis, C. and Smalley, D., "Resistance of *Legionella pneumophila* to Microbiocides," *Developments in Industrial Microbiology*, 21, 265(1980).

[Hume and Hann 1984] Hume, R. and Hann, W., "Growth Relationships of *Legionella pneumophila* with Green Algae (Chlorophyta)," Legionella, Thornsberry, et. al., eds, Proceedings of the 2nd International Symposium," American Society of Microbiology, Washington, DC, 323(1984).

[Kaiga et al. 1987] Kaiga, N., Seki, T., Iyasu, K, "Ozone Treatment in Cooling Water Systems," Proceedings 8th Ozone World Congress, Vol. 2 (Zurich, Switzerland: Intl. Ozone Assoc.), G1(1987).

[Kajdasz et al. 1984] Kajdasz, R., Young-Bandala, L., and Einstman, R., "Biocidal Efficacy with Respect to Sessile and Planktonic Organisms," Paper No. IWC-84-60, International Water Conference, Pittsburgh, PA, October 22-24 (1984).

[Kaur et al. 1988] Kaur, K, Bott, T., Leadbeater, B., "Ozone as a Biocide in Industrial Cooling Water Systems," in Ozone in Water Quality Management (Burgess Hill, Sussex, England: International Ozone Association, British Secretariat, 147(1988).

[Keller et al. 1996] Keller, D., Hajjeh, R., Demaria, A., Fields, B., Pruckler, J., Benson, R., Kludt, P., Lett, S., Mermel, L., Giorgio, C., and Breiman, R., "Community Outbreak of Legionnaires'

Disease: An Investigation Confirming the Potential For Cooling Towers To Transmit Legionella Species," *Clinical Infectious Diseases*, 22, 257(1996).

[Kilvington 1990] Kilvington, S., "Activity of Water Biocide Chemicals and Contact Lens Disinfectants on Pathogenic Free-Living Amoebae," *International Biodeterioration,* 26, 127(1990).

[Kilvington and Price 1990] Kilvington, S. and Price, J., "Survival of *Legionella pneumophila* within Cysts of *Acanthamoeba polyphaga* following Chlorine Exposure," *Journal of Applied Bacteriology*, 68, 519(1990).

[King et al. 1988] King, C., Shotts, E. Jr., Wooley, R., and Porter, K., "Survival of Coliforms and Bacterial Pathogens within Protozoa during Chlorination," *Applied and Environmental Microbiology*, 54, 3023(1988).

[Korich et al. 1990] Korich, D., Mead, J., Madore, M., Sinclair, N., and Sterling, C., "Effects of Ozone, Chlorine Dioxide, Chlorine, and Monochloramine on *Cryptosporidium parvum* Oocyst Viability," *Applied and Environmental Microbiology*, 56, 1423(1990).

[Kurtz and Davis 1988] Kurtz, J. and Davis, V., "Efficacy of Bromine Biocide in Cooling-Water Systems," *The Lancet*, 304(1988).

[Kurtz et al. 1982] Kurtz, J., Bartlett, C., Newton, U., White, R., and Jones, N., "*Legionella pneumophila* in Cooling Water Systems: Report of a Survey of Cooling Towers in London and a Pilot Trial of Selected Biocides," *Journal of Hygiene (Cambridge)*, 88, 369(1982).

[Kurtz et al. 1984] Kurtz, J., Bartlett, C., and Newton, U., "Field Trial of Biocides in Control of Legionella in Cooling Water Systems," Legionella, Proceedings of the 2nd International Symposium, Thornsberry, et. al., eds, American Society of Microbiology, Washington, DC 340(1984).

[Langmark et al. 2005] Langmark, J., Storey, M., Ashbolt, N., and Stenstrom, T., "Accumulation and Fate of Microorganisms and Microspheres in Biofilms Formed in a Pilot-Scale Water Distribution System," *Applied and Environmental Microbiology*, 71(2), 706(2005).

[Leahy et al. 1987] Leahy, J., Rubin, A., and Sproul, O., "Inactivation of *Giardia muris* Cysts by Free Chlorine," *Applied and Environmental Microbiology*, 53(7), 1448(1987).

[LeChevallier 1990] LeChevallier, M., "Disinfection of Bacterial Biofilms," *Water Chlorination*, 6, 905(1990).

[LeChevallier et al. 1990] LeChevallier, M., Lowry, C., and Lee, R., "Disinfecting Biofilms in a Model Distribution System," *Journal of the American Water Works Association,* 87(1990).

[Lee and West 1991] Lee, J. and West, A., "Survival and Growth of Legionella Species in the Environment," *Journal of Applied Bacteriology Symposium Supplement,* 70, 121S(1991).

[Lykins et al. 1989] Lykins, B., Goodrich, J., Hoff, J., and Kothari, N., "Chlorine Dioxide for Drinking Water Disinfection," 1989 Montana Section of AWWA/ Montana WPC Association Joint Conference, Great Falls, MT, March 29-31 (1989).

[Lykins et al. 1990] Lykins, B., Goodrich, J., and Hoff, J., "Concerns with Using Chlorine Dioxide Disinfection in the USA," *Journal of Water Supply Research and Technology-Aqua* (London), 39(6), 376(1990).

[Mahoney et al. 1992] Mahoney, F., Hoge, C., Farley, T., Barbaree, J., Breiman, R., Benson, R., and McFarland, L., "Communitywide Outbreak of Legionnaires' Disease Associated with a Grocery Store Misting Machine," *Journal of Infectious Diseases*, 165, 736(1992).

[Mangum and McIlhenny 1975] Mangum, D. C., and McIlhenny, W. F., "Control of Marine Fouling in Intake Systems - A Comparison of Ozone and Chlorine, in Aquatic Applications of Ozone," Blogoslawski, W. and Rice, R., eds., International Ozone Institute, 138(1975).

[Marine Research Inc 1976] Marine Research Inc., "A Report on Possible Alternatives to Chlorination for Controlling Fouling in Power Station Cooling Water Systems," Falmouth, MA (1976).

[Marrao et al. 1993] Marrao, G., Verissimo, A., Bowker, R., and da Costa, M., "Biofilms as Major Sources of *Legionella sp.* in Hydrothermal Areas and their Dispersion into Stream Water," *FEMS Microbiology Ecology,* 12, 25(1993).

[McCoy and Wireman 1989] McCoy, W., and Wireman, J., "Efficacy of Bromochlorodimethylhydantoin against *Legionella pneumophila* in Industrial Cooling Water," *Journal of Industrial Microbiology*, 4(6), 403(1989).

[McCoy et al. 1986] McCoy, W., Wireman, J., and Lashen, E., "Efficacy of Methylchloroisothiazolone biocide against *Legionella pneumophila* in Cooling Tower Water," *Journal of Industrial Microbiology*, 1(1), 49(1986).

[McDade et al. 1979] McDade, J., Brenner, D., Bozeman, F., "Legionnaires' Disease Bacterium Isolated in 1947," *Annals of Internal Medicine*, 90, 659(1979).

[Miller and Kenepp 1993] Miller, R., and Kenepp, K., "Risk Assessment for Legionnaires' Disease Based on Routine Surveillance of Cooling Towers for Legionella," in Legionella: Current Status and Emerging Perspectives, Barbaree, Breiman and Dufour, eds., American Society for Microbiology (1993).

[Miller and Simpson 1999] Miller, J. and Simpson, G., "Chemical Control of Legionella," Association of Water Technologies Annual Meeting, Palm Springs, CA, October 26-30, 1999.

[Monohan 1978] Monohan, B., "A/C Cooling Tower May have Hidden Legionnaires' Disease," *Air Conditioning Heating & Refrigeration News*, (August 21), 1(1978).

[Morris and Feeley 1990] Morris, G., and Feeley, J., Abstr. American Society of Heating, Refrigeration, and Air Conditioning Engineers Annual Meeting, 76(1990).

[Mounsey and Hagar 1946] Mounsey, R. and Hagar, M., "Taste and Odor Control with Chlorine Dioxide," *Journal of the American Water Works Association*, 38, 1051(1946).

[Muder et al. 1986] Muder, R., Yu, V., and Woo, A., "Mode of Transmission of *Legionella pneumophila*: A Critical Review," *Archives of Internal Medicine*, 146, 1607(1986).

[Nagington and Smith 1980] Nagington, J. and Smith, D., "Pontiac Fever and Amoebae," *Lancet* ii, 1241(1980).

[O'Brien and Bhopal 1993] O'Brien, S. and Bhopal, R., "Legionnaires' Disease: the Infective Dose Paradox," *The Lancet*, 342, 5(1993).

[Oldham and Rodgers 1985] Oldham, L. and Rodgers, F., "Adhesion, Penetration and Intracellular Replication of *Legionella pneumophila*: An In vitro Model of Pathogenesis," *Journal of General Microbiology*, 131, 697(1985).

[Osterholm et al. 1983] Osterholm, M., Chin, T., Osborne, D., Dull, H., Dean, A., Fraser, D., Hayes, P., and Hall, W., "A 1957 Outbreak of Legionnaires' Disease Associated with a Meat Packing Plant," *American Journal of Epidemiology*, 117, 63(1983).

[Panikov et al. 1993] Panikov, N., Merkurov, A., and Tartakovskii, I., "Kinetic Studies of Legionella Interactions with Protozoa," Legionella, Current Status and Emerging Perspectives, Barbaree, Breiman and Dufour, eds., American Society for Microbiology 153(1993).

[Peeters et al. 1989] Peeters, J., Mazas, E., Masschelein, W., Maturana, I., and Debacker, E., "Effect of Disinfection of Drinking Water with Ozone or Chlorine Dioxide on Survival of *Cryptosporitium parvum* Oocysts," *Applied and Environmental Microbiology*, 55(6), 1519(1989).

[Pope et al. 1982] Pope, D., Soracco, R., Gill, H., and Fliermans, C., "Growth of *Legionella pneumophila* in Two-Membered Cultures with Green Algae and Cyanobacteria," *Current Microbiology*, 7, 319(1982).

[Potera 1999] Potera, C., "Forging a Link Between Biofilms and Disease," *Science*, 283, 183(1999).

[Ringer and Campbell 1955] Ringer, W. and Campbell, S., "Use of Chlorine Dioxide for Algae Control at Philadelphia," *Journal of the American Water Works Association*, 47, 740(1955).

[Roe et al. 1996] Roe, F., Lewandowski, Z., and Runk, T., "Simulating Microbiologically Influenced Corrosion by Depositing Extracellular Biopolymers on Mild Steel Surfaces," *Corrosion*, 52(10), 744(October 1996).

[Rowbotham 1980] Rowbotham, L., "Preliminary Report on the Pathogenicity of *Legionella pneumophila* for Freshwater and Soil Amoeba," *Journal of Clinical Pathology*, 33, 1179(1980).

[Rowbotham 1984] Rowbotham, T., "Legionellae and Amoebae," Legionella, Thornsberry, C., et. al., eds.,, Proceedings of the 2nd International Symposium, American Society of Microbiology, Washington, DC, 325(1984).

[Rowbotham 1986] Rowbotham, T., "Current Views on the Relationships between Amoebae, Legionella and Man," *Israel Journal of Medical Sciences,* 22, 678(1986).

[Simpson 1995] Simpson, G., "Biofilm: Removal and Prevention with Chlorine Dioxide," The Third International Symposium on Chlorine Dioxide: Process Water, Drinking Water and Water Waste Issues, New Orleans, LA, September 14–15, 1995.

[Skaliy and McEachern 1979] Skaliy, P., and McEachern, H., "Survival of the Legionnaires' Disease Bacterium in Water," *Annals of Internal Medicine*, 90, 662(1979).

[Skaliy et al. 1980] Skaliy, P., Thompson, T., Gorman, G., Morris, G., McEachern, H., and Mackel, D., "Laboratory Studies of Disinfectants against *Legionella pneumophila*," *Applied and Environmental Microbiology*," 40(4), 697(October 1980).

[Skinner et al. 1983] Skinner, A., Anand, C., Malic, A., and Kurtz, J., "Acanthamoebae and Environmental Spread of *Legionella pneumophila*," *The Lancet,* 289(1983).

[Soracco and Pope 1983] Soracco, R. and Pope, D., "Bacteriostatic and Bactericidal Modes of Action of Bis(tributyltin)oxide on *Legionella pneumophila*," *Applied and Environmental Microbiology*, 45(1), 48(January 1983).

[Soracco et al. 1983] Soracco, R., Gill, H., Fliermans, C., and Pope, D., "Susceptibilities of Algae and *Legionella pneumophila* to Cooling Tower Biocides," *Applied and Environmental Microbiology*, 45(4), 1254(1983).

[Sproul et al. 1983] Sproul, O., Chen, Y., Engel, J., and Rubin, A., "Comparison of Chlorine and Chlorine Dioxide for the Inactivation of an Amoebic Cyst," *Environmental Technology Letters*, 4, 335(1983).

[Srikanth and Berk 1993] Srikanth, S. and Berk, S., "Stimulatory Effect of Cooling Tower Biocides on Amoebae," *Applied and Environmental Microbiology*, 59(10), 3245(1993).

[Sutherland and Berk 1996] Sutherland, E. and Berk, S., "Survival of Protozoa in Cooling Tower Biocides," *Journal of Industrial Microbiology*, 16, 73(1996).

[Tesh and Miller 1981] Tesh, M. and Miller, R., "Amino Acid Requirements for *Legionella pneumophila* Growth," *Journal of Clinical Microbiology*, 13, 865(1981).

[Thacker et al. 1978] Thacker, S., Bennett, J., Tsai, T., Fraser, D., McDade, J., Shepard, C., Williams, K., Stuard, W., Dull, H., Eickhoff, T., "An Outbreak in 1965 of Severe Respiratory Illness Caused by Legionnaires' Disease Bacterium," *Journal of Infectious Diseases*, 138(4), 512(1978).

[Thomas et al. 1999] Thomas, W., Eccles, J., and Fricker, C., "Laboratory Observations of Biocide Efficiency against Legionella in Model Cooling Tower Systems," *American Society of Heating Refrigeration Air Conditioning and Engineering Transactions*, Technical Paper SE-99-3-4(1999).

[Tison et al. 1980] Tison, D., Pope, D., Cherry, W., and Fliermans, C., "Growth of *Legionella pneumophila* in Association with Blue-Green Algae (Cyanobacteria)," *Applied and Environmental Microbiology*, 39(2), 456(1980).

[Tobin et al. 1980] Tobin, J., Beare, J., Dunnill, M., Fisher-Hoch, S., French, M., Mitchell, R., Morris, P. and Muers, M., "Legionnaires' Disease in a Transplant Unit: Isolation of the Causative Agent from Shower Baths," *The Lancet* ii, 118(1980).

[Trulear and Wiatr 1988] Trulear, M. and Wiatr, C., "Recent Advances in Halogen Based Biocontrol," Technical Paper No. 19, Corrosion 88, March 21-25, St. Louis, MO (1988).

[UK Department of Health 1988] Department of Health (UK), "Report of the Expert Advisory Committee on Biocides," London, Her Majesty's Stationary Office, Wright, A., Chairman (1988).

[Videla et al. 1990] Videla, H., Mele, M., Silva, R., Bianchi, F., and Canales, C., "A Practical Approach to the Study of the Interaction Between Biofouling and Passive Layers on Mild Steel and Stainless Steel in Cooling Water," Paper No. 124, Corrosion 90, Las Vegas, NV, April 23-27, 1990.

[Vincent et al. 1946] Vincent, G., MacMahon, J., and Synan, F., "Use of Chlorine Dioxide in Water Treatment," *American Journal of Public Health*, 1045(1946).

[Wadowsky and Yee 1985] Wadowsky, R. and Yee, R., "Effect of Non-Legionellae Bacteria on the Multiplication of *Legionella pneumophila* in Potable Water," *Applied and Environmental Microbiology*, 49(5), 1206(1985).

[Walker and Morales 1997] Walker, J. and Morales, M., "Evaluation of Chlorine Dioxide (ClO_2) for the Control of Biofilms," *Water Science and Technology*, 35(11-12), 319(1997).

[Walker et al. 1994] Walker, J., Rogers, J., and Keevil, C., "An Investigation of the Efficacy of a Bromine Containing Biocide on an Aquatic Consortium of Planktonic and Biofilm Micro-Organisms Including *Legionella pneumophila*," *Biofouling*, 8, 47(1994).

[Walker et al. 1995] Walker, J., Mackerness, C., Mallon, D., Makin, T., Williets, T., and Keevil, C., "Control of *Legionella pneumophila* in a Hospital Water System by Chlorine Dioxide," *Journal of Industrial Microbiology*, 15, 384(1995).

[Wallis et al. 1990] Wallis, P., Roodselaar, A., Neuwirth, M., Roach, P., Buchanan-Mappin, J., and Mack, H., "Inactivation of Giardia Cysts in a Pilot Plant using Chlorine Dioxide and Ozone," Proceedings of the Water Quality Technology Conference, 17, 695(1990).

[Wang et al. 1979] Wang, W., Blaser, M, Cravens, J., and Johnson, M., "Growth, Survival and Resistance of Legionnaires' Bacterium," *Annals of Internal Medicine*, 90, 614(1979).

[White 1986] White, G. C., Handbook of Chlorination, Van Nostrand Reinhold Company, NY, (1986).

[Witherell et al. 1984) Witherell, L., Orciari, L., Spitalny, K., Pelletier, R., Stone, K., and Voght, R., "Disinfection of *Legionella pneumophila* Contaminated Whirlpool Spas," Legionella, Thornsberry, C., et. al., eds., Proceedings of the 2nd International Symposium, American Society of Microbiology, Washington, DC, 325 (1984).

[Wright et al. 1991] Wright, J., Russeska, I., and Costerton, J., "Decreased Biocide Susceptibility of Adherent *Legionella pneumophila*," *Journal of Applied Bacteriology*, 71, 531(1991).

[Xy et al. 1996] Xu, X., Stewart, P., and Chen, X., "Transport Limitation of Chlorine Disinfection of *Pseudomonas aeruginosa* Entrapped in Alginate Beads," *Biotechnology and Bioengineering*, 49, 93(1996).

[Yamamoto et al. 1992] Yamamoto, H., Sugiura, M., Kusunoki, S., Ezaki, T., Idedo, M., and Yabuuchi, E., "Factors Stimulating Propagation of Legionellae in Cooling Tower Water," *Applied and Environmental Microbiology*, 58(4), 1394(1992).

[Yee and Wadowsky 1982] Yee, R. and Wadowsky, R., "Multiplication of *Legionella pneumophila* in Unsterilized Tap Water," *Applied and Environmental Microbiology*, 43(6), 1330(1982).

Control of *Corbicula fluminea* (the Asian Clam) with ClO_2

A. An Overview of Asian Clams and the Problems They Cause

Morphology, Dispersal and Reproduction [Post et al. 2000]: The Asian clam (*Corbicula fluminea*) is a bivalve filter feeder, which feed by siphoning and exhausting water through a pair of small siphons, straining plankton and absorbing oxygen as the water passes over their gills.

A species not native to the US, the Asian clam is thought to have been introduced to the Pacific Northwest as a food source by immigrant Asian workers in the 1930s. It was first discovered in the US on the north shore of the Columbia River near Knappton, Washington in 1938 [Burch 1944; Lee et al. 1981]. It has since spread from coast to coast [Counts 1986] and is now found in 40 of the 48 contiguous United States [Post et al. 2000].

The Asian claim can reach its maximum size of almost two inches after 4—5 years. It can survive very low oxygen conditions (1—2 mg/L) for extended periods. It has a broad temperature tolerance (32°F—95°F), few natural predators, and a life cycle of a few years.

The Asian clam is hermaphroditic (capable of self fertilization) and can produce from 10,000—40,000 veligers (larval stage of a mollusk) per year per spawning adult. It can reach sexual maturity in 6 months, making it capable of releasing a second generation within a year. There are two spawning periods: a major one in the spring and a smaller one in the fall, resulting in a fairly long spawning season. This organism has a lengthy planktonic stage, which facilitates dispersal, as the size of veligers varies from 100—200 microns.

Asian clams are not like zebra mussels, in that zebra mussels live in the water attached to surfaces. Asian clams are benthic organisms, that is, they live in the muck [Smith et al. 1977; Cameron et al. 1987]. They are like mussels in that they filter fairly large volumes of water for food. It is during this filtration that they can be killed with chemicals.

Problems with the Asian Clam: The Asian clam forms dense populations up to 10,000 organisms per square foot. They are able to attach if the water velocities are as high as about 1.5 ft/sec, and are generally found in areas where silt or sand is present. Due to their size, their remarkable fecundity, and their rapid growth rate, like the zebra mussel, they can clog intake pipes of industrial plants [Smith et al. 1979] along with the pipes inside industrial facilities [Goss and Cain 1977; Harvey 1981; Ingram 1959; Leathrum 1983; McMahon 1982; West et al. 1983]. For example, adults and relic shells have been found in piles up to 10—14 feet in low flow areas in the corners of intake structures. Fouling by Asian clams has caused shutdowns in both nuclear [Harvey 1971] and conventional [Goss et al. 1977] power plants.

Impact of Asian Clams: Losses in electric generating unit availability due to condenser biofouling average about 3% according to industry surveys. Of this 3%, about 40% have been attributed to macrobiological fouling.

B. Chlorine for Control of the Asian Clam

Some work has been done to investigate the impact of various chemicals and temperature on the Asian clam.

One of the earliest reports describing work to control the Asian clam was done by the Tennessee Valley Authority [Goss et al. 1977]. The Asian clam was identified at TVA Shawnee Steam Plant in 1957 and since has spread to all of the TVA power plants.

Control of *Corbicula fluminea* (the Asian Clam) with ClO_2

In 1974, they began to suspect biofouling at Browns Ferry Nuclear Plant near Decatur Alabama, which began commercial operation in August 1974. In October of that year, significant numbers of clam shells were found. However, it was only after an Amertap system was installed for condenser tube cleaning, and the majority of the approximately 4500 Amertap balls remained in the condenser because tubes were blocked by clam shells, that the extent of the fouling became apparent. The fouling was so extensive that the condenser had to be dismantled and cleaned with brushes.

It was suspected that the majority of clams fouling condenser tubes at this facility, which had 3/8" vertical traveling screens, had grown inside the intake tunnel, as the intake system had been flooded for more than 2 years before the unit was started up.

These researchers studied the use of chlorine, ClO_2, and BrCl as possible chemical treatment strategies for control of the Asian clam and found that chlorine and BrCl could likely be used as an effective control chemical, but that ClO_2 would not be an effective mollusk control chemical. No discussion was provided as to how ClO_2 was prepared or measured, which subsequently has been found to be important [Simpson 2005a].

In later work, data suggested that the mortality of Asian clams was related to water temperature but not to chlorine exposure [Mattice et al. 1982].

In 28—32 day laboratory studies, data showed a marked temperature dependence upon the mortality of the Asian clam exposed to halogen [Doherty et al. 1986]. For example, continuous total residual chlorine (TRC) at low levels (0.2—1.0 mg/L) resulted in < 53% mortality of Asian clams when the temperature was < 16 °C. Under the same conditions, if the temperature was > 18 °C, the mortality was significantly greater. Their work suggested no essential difference between use of bromine and chlorine.

Results of their field trials, however, showed that in the spring, while ambient temperatures rose from 20 to 25 °C, mortality levels exceeded 90% after 28 days of exposure to 0.25 mg/L TRC. During the fall/winter when temperatures dropped from 20 to 12 °C, mortality levels did not exceed 23% after 28 days of exposure to elevated TRC levels (< 0.5 mg/L).

A summary of some of the results with chlorine is shown in Table 9-1.

Table 9-1. TRC, Temperature, Duration and % Mortality				
TRC (mg/L)	T (°C)	Duration (Days)	Mortality (%)	Reference
10—40	5	2	35—90	[Tilly 1976]
1.0	18	12	100	[Bernhard 1986]
1.0	23	8	100	[Bernhard 1986]
1.0	29	6	100	[Bernhard 1986]
1.0—1.5	24—26	19	98	[Pearson 1962]
0.5—0.75	26—28	12	100	[Pearson 1962]
0.2—0.5	25—27	21	100	[Pearson 1962]
0.5	9—39	28	100	[Cherry et al. 1986]
0.2	25	13	100	[Ramsay et al. 1988]
0.1	25	26	100	[Ramsay et al. 1988]
0.05	25	36	100	[Ramsay et al. 1988]
0.5	-	2.2	100	[Sickel 1976]
0.5	25—28	4—4.5	100	[Goss et al. 1977]

C. ClO_2 for Control of the Asian Clam

In a later work, researchers compared six disinfectants, including free chlorine, monochloramine, chlorine dioxide, potassium permanganate, cupric sulfate, and bromamine, the goal being to identify a chemical that would efficiently inactivate juvenile clams within 24 hours, without producing THMs [Cameron et al. 1987; Cameron et al. 1989a].

Permanganate and copper were found to be too weak to be effective under the arbitrary constraints. Monochloramine and ClO_2 were found to be effective, although monochloramine was chosen for economic reasons. However, their method of generation of ClO_2, the acid-chlorite method, is known to be the least efficient (most expensive) of the commonly used generation methods [Simpson 2005b].

In later work, this group investigated the effect of temperature and pH on the toxicity of monochloramine to the Asian clam [Cameron et al. 1989b]. As would be predicted, they found a fairly strong dependence of mortality on temperature. Temperature effects can be compensated for somewhat by adjustment of the concentration of monochloramine.

In more recent work, researchers suggested that ClO_2 requires lower dosages and shorter exposure times than chlorine. A ClO_2 concentration of 0.2 mg/L can achieve 100%mortality in 8 days, compared to > 3 weeks with chlorine [Post et al. 2000].

D. Summary

Asian clams are benthic organisms, that is, they live in the muck and sludge. As a result, they can be quite difficult to control in the sediment, but are more easily controlled in their free-floating veliger state [Smith et al. 1979]. They can be controlled effectively with chlorine, monochloramine or ClO_2.

In addition, as for the disinfection of potable water, there appears to be a concentration X time phenomenon for control of the Asian clam. That is, maintaining a lower concentration of a given chemical for an extended period of time seems to be able to control the organism effectively. Maintaining a higher concentration of a given chemical for a shorter time can also be effective. As several workers have determined, this depends to a great deal on temperature, as the Asian clam, once exposed to a toxic chemical, can close up and wait for the toxic environment to change. The ability of the Asian clam to remain closed is a definite function of temperature.

E. References

[Bernhard 1986] Bernhard, H., "Asiatic Clams Tested at Edison Stations," *Edison Elec. Biocurr.* 2, 12(1986).

[Burch 1944] Burch, J., "Check List of West North America Marine Mollusca from San Diego, Calif., to Alaska," Minutes Concho. Club South. Calif., 38, 11(1944).

[Cameron et al. 1987] Cameron, G., Symons, J., Spencer, S., and Ma, J., Comparison of Various Disinfectants for Control of the Asiatic Clam (*Corbicula fluminea*), Proceedings, American Water Works Association Annual Conference, 1987.

[Cameron et al. 1989a] Cameron, G., Symons, J., Spencer, S., Ma, J., "Minimizing THM Formation during Control of the Asiatic Clam: A Comparison of Biocides," *Journal of the American Water Works Association*, 81(10), 53(1989).

[Cameron et al. 1989b] Cameron, G., Symons, J., Bushek, D., and Kilkarni, R., "Effect of Temperature and pH on the Toxicity of Monochloramine to the Asiatic Clam," *Journal of the American Water Works Association*, 63, October 1989).

[Cherry et al. 1986] Cherry, D., Roy, R., Lechleitner, R., Dunhardt, P., Peters, G., and Cairns, J., "Corbicula Fouling and Control Measures at the Celco Plant, Virginia," Proceedings of the Second International Corbicula Symposium, 1986.

[Counts 1986] Counts, C., III., "The Zoogeography and History of the Invasion of the United States by *Corbicula fluminea* (Bivalva: Corbiculidae)," *American Malac. Bull. Special Ed.*, 2, 2(1986).

[Doherty et al. 1986] Doherty, F., Farris, J., Cherry, D., and Cairns, J., "Control of the Freshwater Fouling Bivalve *Corbicula fluminea* by Halogenation," *Archives of Environmental Contamination and Toxicology*, 15, 535(1986).

[Goss and Cain 1977] Goss, L. and Cain C., "Power Plant Condensor and Service Water System Fouling by Corbicula, the Asiatic Clam," in *Biofouling Control Procedures: Technology and Ecological Effects*, Jensen, L., ed, Marcel Dekker, New York, 1977.

[Goss et al. 1977] Goss, L., Jackson, J., Flora, H., Isom, B., Gooch, C., Murray, S., Burton, C., and Bain, W., "Control Studies on Corbicula for Steam-Electric Generating Plants," Proceedings, First International Corbicula Symposium, Fort Worth, TX, Oct 13-15, 1977.

[Harvey 1981] Harvey, R., "Recolonization of Reactor Cooling Water Systems by the Asiatic clam, *Corbicula fluminea*," *Nautilus*, 95(3), 131(1981).

[Ingram 1959] Ingram, W., "Asiatic Clams as Potential Pests in California Water Supplies," *Journal of the American Water Works Association*, 363(March 1959).

[Leathrum 1983] Leathrum, R., "Control of Corbicula Fouling in Industrial Cooling Water Systems," Second International Corbicula Symposium, Little Rock, AR, June 21-24, 1982.

[Lee eet al. 1981] Lee, N., Haag, W., and Jolley, R., "Cooling Water Pollutants: Bioaccumulation by Corbicula," Chapter 59, Water Chlorination, Environmental Impact and Health Effects, 4(2), 1981.

[Mattice et al. 1982] Mattice, J., McLean, R., and Burch, M., "Evaluation of Short-Term Exposure to Heated Water and Chlorine for Control of the Asiatic Clam (*Corbicula fluminea*)," Oak Ridge National Laboratory, Publication No 1748, January 1982.

[McMahon 1977] McMahon, R., "Shell Size-Frequency Distributions of *Corbicula manilensis* from a Clam-Fouled Steam Condensor," *Nautilus*, 91(2), 54(1977).

[Pearson 1962] Pearson, H., "Field Investigation on Control of Asiatic Clams," Metropolitan Water

District of Southern California Internal Report, 1962.

[Post et al. 2000] Post, R., Petrille, J., Lyons, L., "A Review of Freshwater Macrobiological Control Methods for the Power Industry," 20th Annual Electric Utility Chemistry Workshop, University of Illinois, Champaign, Illinois, May 9 2000.

[Ramsay et al. 1988] Ramsay, G., Tackett, J., and Moris, D., "Effect of Low-Level Continuous Chlorination on *Corbicula fluminea*," *Environmental Toxicology and Chemistry*, 7, 855(1988).

[Sickel 1976] Sickel, J., "An Ecological Study of the Asiatic Clam, *Corbicula manilensis* (Philippi, 1841) in the Altamaha Rivere, Georgia, with emphasis on Population Dynamics, Productivity and Control Methods," Ph.D. Thesis, Emory University, 1976.

[Simpson 2005a] Simpson, G., "The Chemistry of Chlorine(III)," Chapter 4, Practical Chlorine Dioxide: Volume I – Foundations, Greg D. Simpson Publishers, 2005.

[Simpson 2005b] Simpson, G., "Practical Generation Chemistry and Equipment," Chapter 6, Practical Chlorine Dioxide: Volume I – Foundations, Greg D. Simpson Publishers, 2005.

[Smith et al. 1979] Smith, A., Muia, R., Farkas, J., and Bassett, D., "Clams – A Growing Threat to Inplant Water Systems," *Plant Engineering*, 33, 165(1979).

[Tilly 1976] Tilly, L., "Clam Survival in Chlorinated Water," Report No. DP-1398, E. I. DuPont de Nemours & Company, Savannah River Laboratory, Aiken SC, 1976.

[West et al. 1983] West, R., Calloway, J., and Adams, C., "Corbicula Fouling in Cooling Water Systems at Arkansas Nuclear One, Units 1 & 2," Second International Corbicula Symposium, Little Rock, AR, June 21-24, 1983.

Sulfide/H_2S Oxidation with ClO_2

A. What Is Hydrogen Sulfide (H_2S)?

Hydrogen sulfide, H_2S, (also called hydrosulfuric acid, sulfuretted hydrogen, hepatic gas and sewer gas) is a colorless gas which, at concentrations less than about 150 ppm, smells like rotten eggs. The properties of this chemical are shown in Table 10-1. Hydrogen sulfide is produced as a metabolic byproduct by certain types of bacteria, most notably *Desulfovibrio desulfuricans*. These bacteria, commonly called sulfate-reducing bacteria (SRB), are strict anaerobes, that is, they do not thrive in an oxygenated environment, but in environments where oxygen is absent. They obtain their energy from the sulfate ion, SO_4^{-2}, hence their name (SRB).

Table 10-1. H_2S – Physical Properties	
Formula wt	34.08 g/mole
Melting point	-85.5 °C
Boiling point	-60.7 °C
Olfactory detection limit	0.002 mg/L
Specific gravity	1.185
pH of saturated solution	4.5
Explosion limits	4.3 to 46%
H_2S solubility in H_2O @10 °C	1 g/187 ml H_2O (5347 mg/L)
@20 °C	1 g/242 ml H_2O (4132 mg/L)
@30 °C	1 g/314 ml H_2O (3185 mg/L)
H_2S solubility @ 20 °C in	Mole fraction (mole H_2S/mole solvent)
Hexane	0.0341
Octane	0.0440
Dodecane	0.0513
Cetane	0.0578
Cyclohexane	0.0338
TLV	20 ppm
LC_{50}	713 ppm/Liter inhalation

H_2S is often smelled near open sewer gratings, because bacteria in municipal sewer pipes produce the gas. H_2S is also found in petroleum and natural gas, in a variety of industries, and in ground water in some parts of the country.

H_2S dissociates in aqueous solution as a function of pH, as shown in Equation 10-1.

$$H_2S \overset{k_1}{\Leftrightarrow} H^+ + HS^- \overset{k_2}{\Leftrightarrow} 2H^+ + S^{-2} \qquad (10\text{-}1)$$

Figure 10-1 shows a representation of the dissociation curve of H_2S with pH. The values for the dissociation constants of H_2S are approximate, as there is some variation in the values reported in the literature.

Figure 10-1. pH Dissociation Curve for H_2S

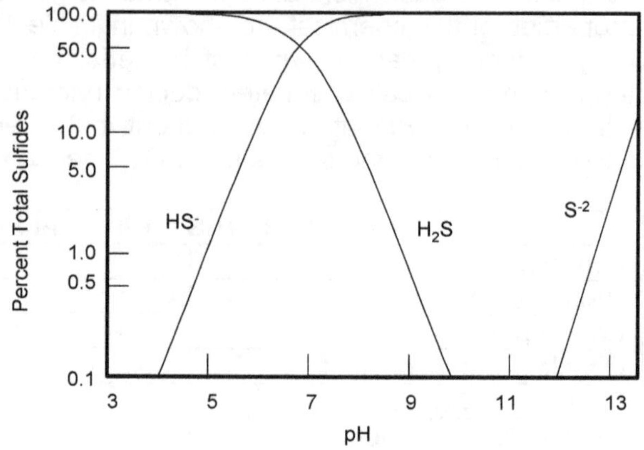

Reported values of the first ionization constant, pK_1, of H_2S have varied from 6.72 to 7.04 at 25.0 °C [Sillen and Martell 1964; Latimer 1952; Himmelblau and Loy 1961]. A pK_1 of 7.0 is commonly used.

The second ionization constant, pK_2, is more uncertain, with reported values that vary between 12.2 and 17.1 at 25 °C [Dohnalek and Fitzpatrick 1983].

In general, as the pH decreases below 7, the H_2S becomes the increasingly predominant form of sulfide present (Figure 10-1). At a pH of about 7, the reduced sulfur species are roughly divided between H_2S and HS^-. From a pH > 7 to ~ 11, HS^- comprises over 90% of the reduced sulfur species present. The bisulfide ion (HS^-) is believed to be the most reactive species [Bowers et al. 1966].

B. Problems Associated with H_2S

Human Exposure: H_2S odors can be detected by the human nose at levels as low as 0.0001 ppm [Pomeroy et al. 1969]. The exposure limits vary somewhat by industry, as shown in Table 10-2.

Table 10-2. Human Exposure Limits for H_2S

Agency	Industry	Exposure	Citation
OSHA	General Industry	20 ppm ceiling 50 ppm peak	29 CFR 1910.1000Z-2 Table
OSHA	Construction	10 ppm 15 mg/m^3 TWA	29 CFW 1926.55 Appendix A
OSHA	Shipyards	10 ppm 15 mg/m^3 TWA	29 CFW 1915.1000 Table Z-Shipyards
ACGIH*		10 ppm 15 mg/m^3 TWA 15 ppm, 21 mg/m^3 STEL	
NIOSH**		10 ppm, 15 mg/m^3 Ceiling IDLH – 100 ppm	

* ACGIH – American Council of Government Industrial Hygienists
**NIOSH – National Insitute for Occupational Safety and Health

In the gas phase, H_2S has a permissible exposure limit (PEL) of 20 ppm maximum concentration and a peak exposure limit of 50 ppm for no more than 10 minutes (Short Term Exposure Limit – STEL) for general industry.

H_2S is highly poisonous. Exposure to H_2S can cause eye irritation, dizziness, coughing, and headache. The physiological effects of H_2S exposure is shown in Table 10-3.

Table 10-3. Physiological Effect of H_2S Exposure	
Concentration (ppm)	Effect
0.0047	Odor threshold for 50% of humans. (Some can detect levels as low as 0.0001 ppm, or. 0.1 ppb)
10	Beginning of eye irritation.
50—100	Leads to eye damage.
150—250	Olfactory nerve is paralyzed after a few inhalations; the sense of smell disappears.
320—530	Lead to pulmonary edema with the possibility of death.
530—1000	Strong stimulation of the central nervous system and rapid breathing, leading to loss of breathing.
800	Lethal concentration for 50% of humans for 5 minute exposure.
>1000	Immediate collapse with loss of breathing, after a single breath.

A number of deaths resulting from H_2S exposure have occurred when workers enter poorly ventilated spaces such as deep wells, underground tanks or sewer systems. H_2S gas is heavier than air; thus it accumulates in low-lying areas.

Oil Production: H_2S is troublesome in oil production. The presence of sulfide creates conditions that support the proliferation of highly corrosive SRB. The corrosion process releases iron, and the iron reacts with sulfide to form iron sulfide. In many cases, a biofilm/FeS matrix results which fouls or plugs piping and thus limits production and causes corrosion. It is desirable to treat the sulfide downhole, so that when the oil reaches the surface the H_2S has been oxidized or complexed and the potential for creating a poisonous atmosphere is minimized. The desirability of using ClO_2 to treat the sulfide results in part from the fact that, under the right conditions, ClO_2 oxidizes sulfide to sulfate, with minimal production of elemental sulfur.

Petroleum Refining, Chemical Processing, and Natural Gas Production: H_2S is removed from petroleum during the refining process, chemical processing, or the production of natural gas. H_2S poses the same hazards in these facilities as it does in oil production. As limits tighten on the amount of sulfur which is allowed in refined petroleum products, there is a corresponding need to reduce, remove, or oxidize sulfur further and further upstream. Restrictions on the amount of sulfide allowed in pipelines require producers to treat the gas or crude to reduce sulfide. Sulfide scavenger chemistry has abounded to help meet these restrictions. Unfortunately, most sulfide scavengers do not destroy sulfide. They form fairly stable complexes with the sulfide, providing a short term fix to the problem. The complexed sulfide can later be released by heat, pressure, or other processing. So these scavengers do not solve the problem, they push the problem further downfield. Someone else has to deal with the problem later. In many cases the producer, the pipeline, and the refiner are separate business units of the same company, each trying to increase profitability of their business unit. In such a case as this, the best decision for the specific business unit is not the best decision for the parent company. That is, it would be best to oxidize the sulfide downhole, making it unnecessary for the pipeliner or refiner to have to deal with it later.

C. Literature Review – ClO_2 and Sulfide/H_2S

In a study of the kinetics and stoichiometry of the reaction between gaseous ClO_2 bubbled through an aqueous solution of Na_2S, the following reaction was proposed, [Ishi 1962]:

$$2Na_2S + 2ClO_2 \Rightarrow 2NaCl + Na_2SO_4 + S^\circ \qquad (10\text{-}2)$$

One group reported that chlorine dioxide make use of all of its oxidation capacity in this reaction, oxidizing sulfide to sulfate stoichiometrically [Synan and Malley 1975]. This chemistry was reported to be as follows:

$$8ClO_2 + 5H_2S + 4H_2O \Rightarrow 5H_2SO_4 + 8HCl \qquad (10\text{-}3)$$

Note that in this case, strong mineral acids are formed from this reaction. This has led to corrosion concerns when ClO_2 is used in some applications, including oil production.

In 1976, a patent was issued for the "Treatment of Sulfide-Bearing Waters with Chlorine Dioxide" [Smeck 1976]. This patent was specifically for the use of ClO_2 to remove sulfide in water without producing colloidal sulfur. This patent work demonstrated the complexity of the reaction of ClO_2 with various sulfur species, in that small changes in conditions resulted in substantial differences in the ClO_2 required for complete oxidation. He showed that the amount of chlorine dioxide required varied from 2.7—4 mg/L ClO_2/S^{-2}.

In another report, ClO_2 was shown to oxidize sulfide in an 8:5 molar ratio (3.6:1 on wt/wt basis) [Dussert 1990]. This work was in general agreement with that of Smeck. However, Dussert found that in acidic solution, SO_4^{-2} was produced, while in neutral or slightly basic solutions, the product distribution included both SO_4^{-2} and S°.

The reaction of ClO_2 and H_2S is not straightforward, but is a sum of many reactions, the reaction or reactions that predominate being dictated by conditions. The complexity of the reaction is illustrated by the possible valence states of oxychlorine compounds shown in Table 10-4 and the sulfur species shown in Table 10-5.

Table 10-4. Cl Species, Formula and Valence

Species	Formula	Valence
Chloride	Cl^-	-1
Hypochlorite	OCl^-	1
Chlorine Monoxide	Cl_2O	1
Chlorite	ClO_2^-	3
Chlorine Dioxide	ClO_2	4
Chlorate	ClO_3^-	5
Perchlorate	ClO_4^-	7

Table 10-5. S Species, Formula and Valence

Species	Formula	Valence
Dithionate	$S_2O_6^{-2}$	5
Bisulfites	$S_2O_5^{-2}$	4
Sulfite	SO_3^{-2}	4
Sulfur Dioxide	SO_2	4
Dithionite or Hydrosulfite	$S_2O_4^{-2}$	3
Polysulfides	S_x^{-2}, x = 4,5	-
Thiosulfate	$S_2O_3^{-2}$	2
Elemental Sulfur	S°	0
Hydrogen Sulfide	H_2S	-2

It is not possible for the reaction between chlorine dioxide and sulfide to be expressed as a single equation for all conditions, considering the multitude of possible reaction pathways. About the best than can be said is that it will take somewhere between 2 and 4 mg/L chlorine dioxide per mg/L sulfide, assuming nothing else is present in the water to consume chlorine dioxide, which is unlikely. Demand testing is highly recommended.

D. Experiment 1 – Reaction of ClO_2 with Sulfide

A petroleum refinery processed high sulfur crude, with a caustic solution being used to scrub the high levels of H_2S from the process. Once the caustic solution had been saturated with sulfide, the caustic was bled into the wastewater treatment system for further treatment (Many dispose of spent caustic in deep injection wells.). The refinery treated this combined stream with high levels of hydrogen peroxide.

The evaluation of a new crude with twice the sulfide level resulted in gaseous evolution of H_2S in the wastewater treatment plant. The very high sulfide levels resulted in the plant having to operate the peroxide pumps at their maximum pumping rate. Even with this, there was insufficient peroxide to completely oxidize all of the sulfide. Thus some sulfide remained which off-gassed when the pH of the stream was reduced during further processing of the wastewater. In addition, the extremely high federate of hydrogen peroxide resulted in significant corrosion downstream. Since the newer crude produced a very good end product, there was a desire to continue to use this crude and so alternate means of handling the sulfide had to be found.

A treatability study was done with ClO_2. As the safe concentration of ClO_2 is limited to about 3500 mg/L, the high levels of sulfide present required a dilution of the sample to allow treatment with ClO_2 of this concentration. Therefore, the wastewater stream was diluted 10 fold. Then, varying aliquots of 3500 mg/L ClO_2 solution were added and the solution measured for sulfide by the methylene blue method [APHA 1999]. The addition of ClO_2 did not result in the production of any precipitate. Although one can make a case for possible production of elemental sulfur on paper by balanced chemical equations, in real-world applications, production of elemental sulfur is not generally observed, primarily because at the depressed pH which results, a complete oxidation of sulfide to SO_4^{-2} occurs.

In addition to the concentration of ClO_2 dosed and the concentration of sulfide remaining after treatment, the pH was measured after addition of ClO_2, because ClO_2 is known to react with sulfide to form strong mineral acids, as shown below.

$$8ClO_2 + 5H_2S + 4H_2O \Rightarrow 5H_2SO_4 + 8HCl \qquad 3.2/1\ ClO_2/H_2S$$

The data obtained from this work is shown in the Table 10-6. The data have been adjusted to reflect the diluted concentrations used.

Table 10-6. Chlorine Dioxide for Sulfide Reduction

Sample	ClO_2 mg/L	pH final	Sulfide mg/L
1	0	10.3	~2700
2	1154	7.0	1625
3	1698	6.65	-
4	2727	3.1	625
5	3684	2.25	350

A plot of this data is shown in Figure 10-2. This plot shows a direct reduction in sulfide concentration with ClO_2. In addition, the ClO_2 depresses the pH significantly, and this pH depression, coupled with the high levels of ClO_2, can impact corrosion significantly.

The use of ClO_2 shows an almost linear reduction in sulfide, initially. In general, the data show that the addition of 2000 mg/L reduced sulfide from 2700 to about 800 mg/L. Thus the initial ratio of ClO_2 to sulfide was approximately 1:1 on a weight basis. To take the reaction to completion requires an estimated ratio of 3:1 on a weight basis.

E. Experiment 2 – Corrosive Effects of ClO_2 with Sulfide

The purpose of this experiment was to investigate the relationship that the concentration of ClO_2 and pH have on the corrosivity of an aqueous solution containing 25 mg/L sulfide.

A Central Composite design of experiments (DOE) approach was taken. The concentration of ClO_2 and pH were varied to explore the response surface of the interaction between these parameters as they relate to corrosion rate.

Figure 10-2. Sulfide Reduction with ClO_2

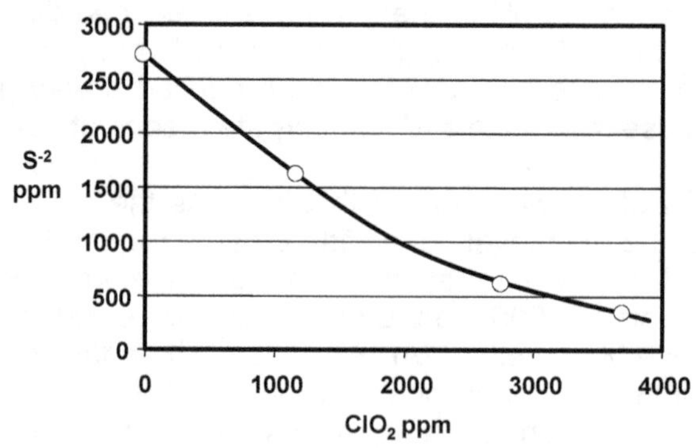

A 2-factor, 3-level design was chosen, so that curvature in the response surface could be explored. Prior work showed that at a pH range of 4 to 7, corrosion would likely reach a minimum in this region. Factors chosen are shown in Table 10-7.

Table 10-7. Central Composite Design - Factors

	Low	Medium	High
Variable	-1	0	+1
[ClO_2]	0	50	100
pH	4	5.5	7

The DOE resulted in the test matrix shown in Table 10-8.

Table 10-8. Design Matrix		
Run	$[ClO_2^-]$	pH
1	0	4
2	0	7
3	100	4
4	100	7
5	50	5.5
6	50	5.5
7	0	5.5
8	100	5.5
9	50	4
10	50	7

The tests were run in duplicate. To provide a reasonable brine solution, 1000 mLs of 3.5% brine was added to a 1000 mL vacuum Erlenmeyer flask. The sulfide solution was added, followed by pH adjustment and finally, the ClO_2 was added. Given the much greater oxidation potential of ClO_2 than molecular oxygen and the much greater concentration of ClO_2 in the solution than oxygen, the effects of oxygen were considered negligible and were ignored.

A Rohrback Cosasco Model 9030 Plus was used to record corrosion rates of duplicate samples, prepared independently. The corrosion rate was monitored over time. For a single sample, the maximum corrosion rate observed was recorded as the worst-case corrosion rate on new corrater carbon steel probes. The measured corrosion reached a maximum initially, while the surface was being conditioned to the materials with which it was in contact. Then, over time, the measured corrosion rate declined. The maximum corrosion rate was recorded. Although this corrosion rate data did not reflect actual corrosion rates which would be measured in a real system, the use of the maximum corrosion rate allowed a comparison with other corrosion rates obtained in a similar way under different experimental conditions. This is important because the corrosion process changes the surface of corrosion probes significantly, and thus use of fresh probes eliminates this source of variation. In this way, a very worse case scenario can be developed. Actual corrosion rates in real-world systems will likely be much less.

Experimental results are shown in Table 10-9.

Table 10-9. Corrosion Results (mpy)						
Run	Test 1	Test 2		Run	Test 1	Test 2
1	240	270		6	86	81
2	280	400		7	200	220
3	1000	1000		8	200	220
4	200	220		9	140	160
5	120	92		10	80	90

Table 10-10. Regression Analysis Table					
Y-hat Model		**Corrosion Rate**			
Factor	**Name**	**Coeff**	**P(2 Tail)**	**Tol**	**Active**
Const		25.714	0.5879		
A	ClO_2	99.167	0.0074	1	X
B	pH	-125.00	0.0015	1	X
AB		-223.75	0.0000	1	X
AA		246.07	0.0003	0.9722	X
BB		153.57	0.0091	0.9722	X
	R^2	0.8741			
	Adj R^2	0.8291			
	Std Error	109.7274			
	F	19.4406			
	Sig F	0.0000			
	F_{LOF}	522.3403			
	Sig F_{LOF}	0.0000			
	Source	**SS**	**df**	**MS**	
	Regression	1170333.7	5	234066.7	
	Error	168561.3	14	12040.1	
	$Error_{Pure}$	1175.0	11	106.8	
	$Error_{LOF}$	167386.3	3	55795.4	
	Total	1338895.0	19		

The data was recorded into the Design of Experiments software, DOE PRO XL, Air Academy Associates, Colorado Springs, CO. The data was analyzed and the Regression Analysis Table is shown (Table 10-10).

The table suggested that both the concentration of ClO_2 and pH were important parameters. The fairly high R^2 value suggested that the data were reasonably explained by the model.

A response surface plot of this data is shown below:

Figure 10-3. Y-hat Surface Plot of Corrosion Rate vs. pH and [ClO2]

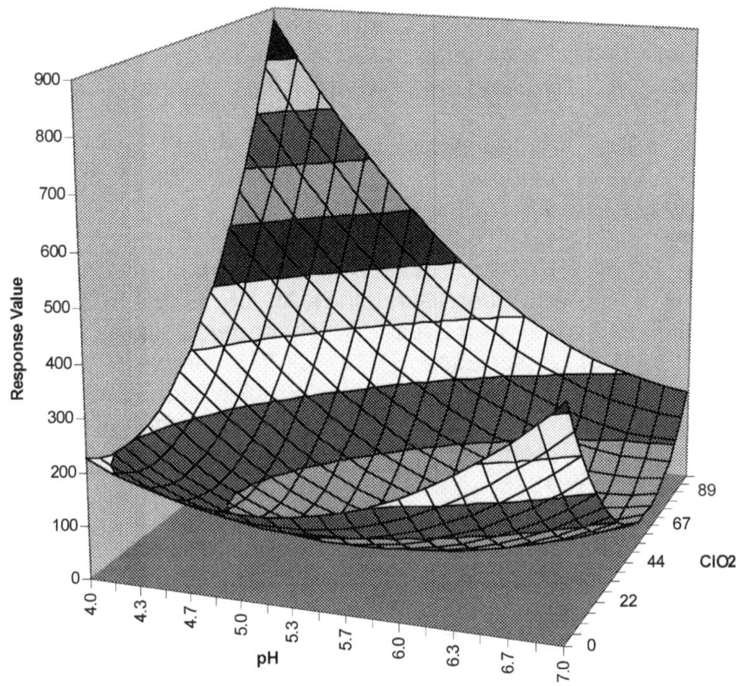

F. Discussion

These results showed that as the pH approaches the 5.7—6.3 range, the corrosion rate reaches a minimum at a ClO_2 to sulfide ratio of about 40:25 or 1.6:1 on a weight: weight basis. In addition, since the rate of change near the minimum corrosion rate is slow, the ClO_2 to sulfide ratio can vary considerably, e.g., from about 1:1 to 3:1. This is well within the range of Experiment 1.

This experiment suggests that both sulfide and ClO_2 are corrosive to carbon steel. However, the addition of ClO_2 to a sample containing sulfide actually results in a reduction in corrosion rate. A minimum is reached, and then, once sufficient ClO_2 has been added to oxidize the sulfide, further addition adds ClO_2 that is not reacted with sulfide, and the free ClO_2 results in increased corrosion.

This data shows that when an aqueous ClO_2 solution is added carefully to a solution of sulfide of known concentration, the corrosion rate can be controlled to a great extent. The above experiments show that in the absence of any corrosion inhibitors, there is substantial variability in the measured corrosion rate. Unfortunately, in most downhole applications, ClO_2 is shoved down hole, with little regard to concentration of the ClO_2. In such an application, the excessive ClO_2 can be quite corrosive, particularly at the high temperatures and pressures which exist downhole.

G. References:

[Bowers et al. 1966] Bowers, J., Fuller, M., and Packer, J., "Auto Oxidation of Aqueous Sulfide Solutions," *Chemistry and Industry*, No 2, 65(1966).

[Dohnalek, D., and FitzPatrick, J., "The Chemistry of Reduced Sulfur Species and Their Removal from Groundwater Supplies," *Journal of the American Water Works Association*, 298(1983).

[Dussert 1990] Dussert, B., "Destruction of Sulfide by Chlorine Dioxide," unpublished report, 1990.

[Himmelblau and Loy 1961] Himmelblau, D. and Loy, H., "The First Ionization Constant of Hydrogen Sulfide in Water," *Journal of Physical Chemistry*, 65, 264(1961).

[Ishi 1962] Ishi, G., *Kogyo Kagaku Zasshi*, 65, 1013(1962).

[Latimer 1952] Latimer, W., The Oxidation States of the Elements and Their Potentials in Aqueous Solution, Prentiss Hall, Inc., Englewood Cliffs, N.J., Second Edition, 1952.

[Patton 1986] Patton, C., Applied Water Technology, NACE, Houston, Texas. 1986.

[Pomeroy et al. 1969] Pomeroy, R. and Cruse, H. "Hydrogen Sulfide Odor Threshold," *Journal of the American Water Works Association*, 61(12), 677(December 1969).

[Sillen and Martell 1964] Sillen, L. and Martell, A., "Stability Constants of Metal Ion Complexes, Special Publication 17, *Journal of the Chemical Society*, London, England, 1964.

[Smeck, 1976] Smeck, C., "Treatment of Sulfide-Bearing Waters with Chlorine Dioxide," US Patent 4,077,879, March 7, 1978.

[APHA 1999] APHA, "Method 4500-S^{2-} D. Methylene Blue Method," *Standard Methods for the Examination of Water and Wastewater*, 20th Edition, American Public Health Association, American Water Works Association and Water Environment Federation, 1999.

[Synan and Malley 1975] Synan, J. F., and Malley, H. A., "Chlorine Dioxide: An Alternative to Chlorine for Disinfection in Water Systems," paper presented to the Engineering Panel of the Campden Food Preservation Research Association, Chipping Campden, Glos., England, Oct 21, 1975.

[Taradoire 1941] Taradoire, F., *Bulletin de la Societe Chimique de France*, 8(5), 860(1941).

[Woodward 1953] Woodward, E., "Chlorine Dioxide for Odor Control," *TAPPI Journal*, 36(5), 216(May 1953).

Particle Count Reduction with ClO_2

A. Introduction

Particle Counts vs. Turbidity as a Trigger for Filter Backwash: The inactivation, destruction, or removal of waterborne pathogens is a major goal of potable water treatment. Two of the most difficult organisms to inactivate are *Giardia lamblia* and *Cryptosporidium parvum*, due to their resistance to inactivation by oxidants. For example, Table 11-1 shows the Ct factor (concentration X contact time) for inactivation of Cryptosporidium for several common potable water disinfectants. The data show that 80 mg/L of chlorine or chloramine would have to be applied at 90 minutes just to achieve a 2-log reduction in Cryptosporidium.

Table 11-1. Concentration and Contact Time for 99% Inactivation of Cryptosporidium [Korich et al. 1990]

Disinfectant	Concentration (mg/L)	Contact time (min) for 99% inactivation
Ozone	1	5
ClO_2	1.3	60
Cl_2	80	90
Chloramine	80	90

As a result of the questionable efficacy of disinfection as a method of inactivating these organisms [Korich et al. 1990], the focus has been on their physical removal by filtration [Huck et al. 2001].

Turbidity of filtered water effluent is used to gauge the effectiveness of the filtration process. Surface water treatment plants are required by the EPA to reduce the filtered effluent turbidity to 0.3 NTU. Recent studies have questioned the use of turbidity as an indicator of removal of Giardia and Cryptosporidium [Huck et al. 2001] because the size of these organisms (Giardia ~ 10—12 µ and Cryptosporidium ~ 4—6 µ) is below that which would impact turbidity. Therefore, low turbidity water does not insure the absence of Giardia or Cryptosporidium [Hatukai et al. 1997].

It has been suggested that particles may act as surrogates for these pathogens, and that survival of these organisms can be linked to particle removal for a given plant [McTigue et al. 1998; Huck et al. 2001], despite the variability inherent in different particle counting technologies and the associated errors that arise from plant to plant [Lewis and Manz 1991; Chowdhury et al. 2000].

Since particle size measuring systems have a detection limit in the range of 2 µ, particle counts of the filter effluent has been suggested as a better indicator of removal of these pathogens [Hall and Croll 1997].

A number of studies to identify the factors that impact the stability of a particle in solution have been reported. Several important factors have been identified, including the type and concentration of oxidant, the concentration of divalent cations, and the nature and amount of natural organic matter present in the raw water [Grasso and Weber 1988; Dowbriggin and Singer 1989; Singer and Chang 1989; Edwards and Benjamin 1991; Chang and Singer 1991; Tobiason et al. 1992; Reckhow et al. 1993; Hatukai et al. 1997; Chandrakanth and Amy 1998; Swietlik et al. 2002].

Ozone for Particle Destabilization and Particle Count Reduction: Ozone, which has been found to have a significant impact on particle stability, has been the focus of several studies [Felix-Filho 1985; Reckhow et al. 1986; Grasso 1987].

A number of researchers have identified that, in the absence of divalent cations, ozone has no effect on particle stability [Felix-Filho 1985; Dowbriggin and Singer 1989]. In the presence of calcium, ozone was found to promote coagulation [Dowbriggin and Singer 1989], although some results show that high doses of ozone have a negative impact on flocculation in the absence of particles, calcium and other cations [Van Breeman et al. 1979; Jekel 1983; Johnson and Randtke 1983; Reckhow and Singer 1984].

While some have identified an optimum dosage range for ozone performance [Schalekamp 1979; Gurol and Pidatella 1983; Saunier et al. 1983], others have not [Scrivner et al 1980].

In one study, some of the effects normally attributed to ozone were shown to result from pH effects and precipitation of calcium carbonate [Edwards and Benjamin 1991].

Many researchers have reported that absorbed natural organic matter (NOM) increases particle stabilization [Jekel 1983; Gibbs 1983; Amy et al. 1987; Felix-Filho et al 1987; Jekel 1986]. Some have suggested that flocculation benefits from ozonation are dependent on the specific NOM being studied [Chandrakanth and Amy 1998].

Several researchers have found that the THM formation potential can sometimes be lowered with ozonation if the pH of the water is < 7.0 [Trussell and Umphres 1978; Rice 1980; Scrivner et al 1980; Reckhow and Singer 1984; Reckhow et al. 1986; Reckhow et al. 1993], while others have found that at pH > 7.0, the THM formation potential increases much of the time [Riley et al. 1978; Dore et al. 1978; Scrivner et al 1980; Reckhow and Singer 1984; Reckhow et al. 1993].

Within the past few years, workers have shown that ozone addition immediately prior to a filter has resulted in a significant reduction in particle counts out of the filters [Becker et al. 2003].

ClO_2 for Particle Destabilization and Particle Count Reduction: Although ClO_2 has shown an effect similar to that of ozone on microflocculation, only a few detailed studies have been reported which describe the impact of ClO_2 on particle reduction or organics bound to filter media. In two reports, use of ClO_2 was found to produce lower particle counts, a shorter ripening period, and a lower flocculant dosage when compared to chlorine [Sukenik et al. 1987; Hatukai et al. 1997].

In another report, the size of the organic molecules present in the water, and the impact ClO_2 had on the molecular weight distribution was investigated. Lower molecular weight molecules are less retained than the higher molecular weight components of NOM on granular activated carbon (GAC). In addition, the lower molecular weight NOM causes less ClO_2 demand [Swietlik et al. 2002].

To expand the knowledge of how ClO_2 impacts the size and number of particles passing through a filter and the filter effluent turbidities, this study was undertaken at the Mohawk and AB Jewell water treatment plants, for the City of Tulsa, Oklahoma.

B. Experimental

Mohawk Plant: The Mohawk plant, put into service in 1929 and renovated in May of 1998, processes water from Yahola Terminal Reservoir (source Spavinaw Lake) to provide ~ 47 MGD potable water to the northern/western part of Tulsa (population ~ 500,000). Treatment at Mohawk included use of alum (15—20 mg/L as dry), ~ 1.25 mg/L DADMAC polymer, and chlorine added just after the raw water pumps.

To simulate as nearly as possible conditions in the plant, water at the end of the sedimentation basis and just before filtration was pumped through each pilot plant. In this way, the only chemical added at the pilot plant was a small amount of aqueous chlorine dioxide. The filters had 60" of lignite-based GAC which lies atop 12" of sand.

Chlorine dioxide was made by a method of Pitochelli [Pitochelli 1999], whereby a small amount of dry chlorite, a solid chlorine donor, and a solid proton donor were thoroughly mixed in 1 liter of water to produce ~ 3000 mg/L ClO_2 with only a very small concentration of free chlorine (e.g., < 10 mg/L free chlorine). One of these batches was prepared periodically during the test, and diluted prior to use. The concentration of ClO_2 was measured directly using a Hach DR 4000 Spectrophotometer. The concentration was measured immediately after dilution at the beginning of a run, and at the end of a run, just prior to preparation of a new batch. The concentration of ClO2 decreased by about 15% over the life of the batch. Degradation was less at the Mohawk than at AB Jewell, as the solution was cooler and the working area was darker.

Filter effluent turbidity was measured continuously by the Hach 1720 Turbidimeter. Particles in the 2—5 µ, 5—10 µ, and total particle count ranges were counted on effluent water from the three pilot filters at the Mohawk plant using a Hach 2200 PC instrument, which was calibrated prior to use.

AB Jewell: The AB Jewell plant, put into service in 1971, drew water from Lynn Lane Reservoir (source Ologah Lake) to produce ~ 50 MGD potable water to southern/eastern Tulsa. Treatment at AB Jewell included use of alum (35—40 mg/L as dry) and ~ 2.0—2.25 mg/L DADMAC polymer, with chlorine dioxide (1.0—1.25 mg/L) being added to the raw water (gravity flow).

In a manner similar to that at Mohawk, water at the end of the sedimentation basis and just before filtration was pumped through each pilot plant. In this way, the only chemical added at the pilot plant was a small amount of aqueous chlorine dioxide.

At AB Jewell, the pilot filters were somewhat shallower than that at Mohawk, with 36" of bituminous-based GAC on top of 12" of sand. Chlorine dioxide was made by the same method used at Mohawk.

Turbidity was measured continuously by the Hach 1720 Turbidimeter. Particles in the 2—5 µ, 5—10 µ, and 10—15 µ ranges were counted on effluent water from the four pilot filters at the AB Jewell plant using a Hach 2200 PC instrument, which was calibrated prior to use.

C. Results

Mohawk: Figure 11-1 shows a plot of 2—5 µ particle counts over the 5-week trial period. Initially, there was a fairly wide split between the three filters, with Filter 1 (ClO_2 dosage ~ 0.3 mg/L) showing the lowest number of particles in this range. Filter 2 (ClO_2 dosage ~ 0.7 mg/L) showed the second lowest number and the control filter, Filter 3 (no ClO_2) showed the highest number of particles in the 2—5 µ range. This split was unexpected. Therefore, to see if the effect was related to [ClO_2] the ClO_2

Figure 11-1. 2.0—5.0 u Particles

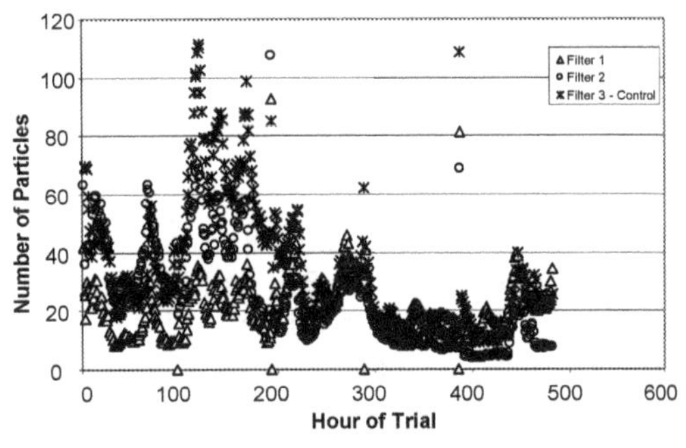

Figure 11-2. 2—5u (Filter 1 – Control)

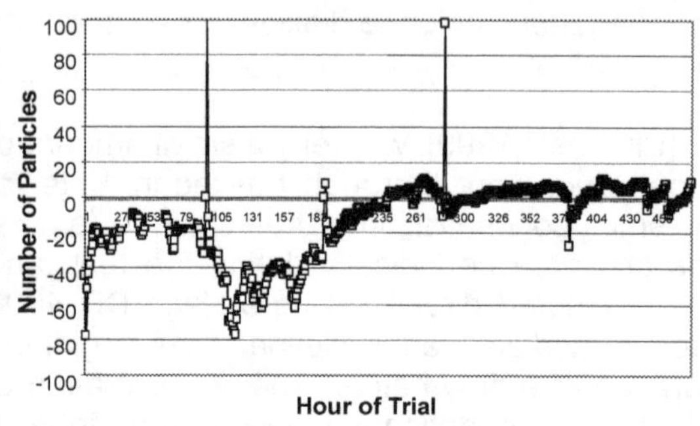

Figure 11-3. 2—5u (Filter 2 – Control)

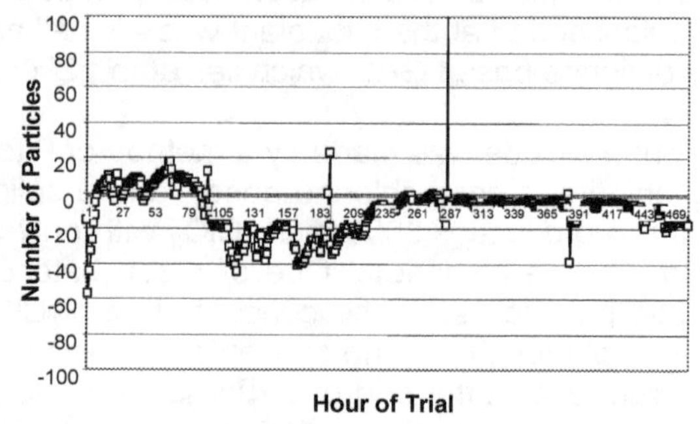

solution pumps for Filters 1 and 2 were reversed. Since this change in pumps and consequently dosage had no observable effect, the pumps were switched back after about a week and left alone for the rest of the test. This unusual trend led to speculations as to whether Filter 1 was in some way fundamentally different (e.g., different media) than the other two filters.

The plots all follow the same general trend. Since raw water changes are the primary cause of the variation, the data was normalized, i.e., each data point for the control filter was subtracted from the corresponding data point of each ClO_2-treated filter. This removes the overall variation due to changes in the raw water and allows observation of general trends over a large number of data points. The data for 2—5 μ particle deltas (Δ) for Filter 1 and 2 are shown in Figures 11-2 and 11-3, respectively. Any point falling below the 'zero' line indicates that there is improvement in particle reduction (i.e., the un-treated filter allows passage of a higher number of particles than the particular ClO_2-treated filter). While the use of this technique is questionable for a given data point, observing a large number of data points allowed general conclusions to be made.

Figure 11-4. 5—10u (Filter 1 – Control)

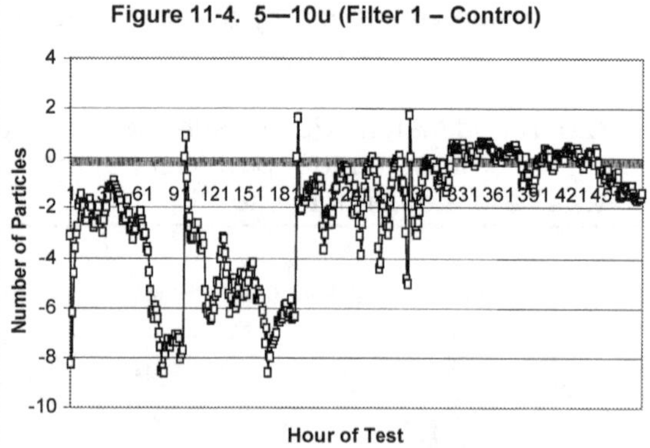

Figure 11-5. 5—10u (Filter 2 – Control)

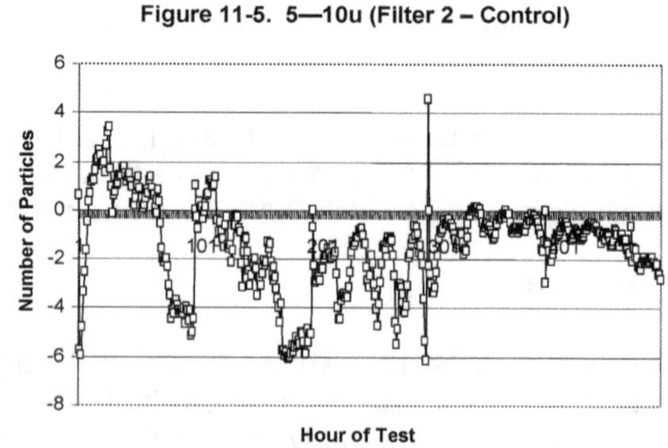

For both filters at Mohawk, during the first couple of weeks the data suggest an improvement in particle counts in the 2—5 μ range. During this time, there was a significant change in raw water (see Settled Water Turbidities – Figure 11-10). The data suggests that application of even small quantities of ClO_2 just prior to a filter helps particle reduction during periods of significant raw water changes. Once the settled water turbidity became somewhat more stable, there was less improvement with ClO_2.

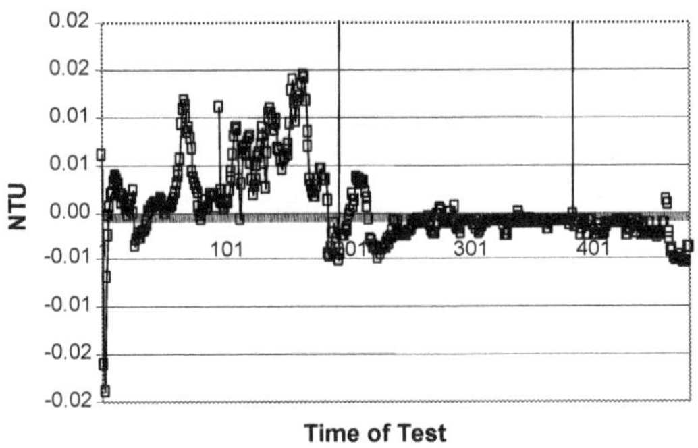

Figure 11-6. Turbidity (Filter 1 – Control)

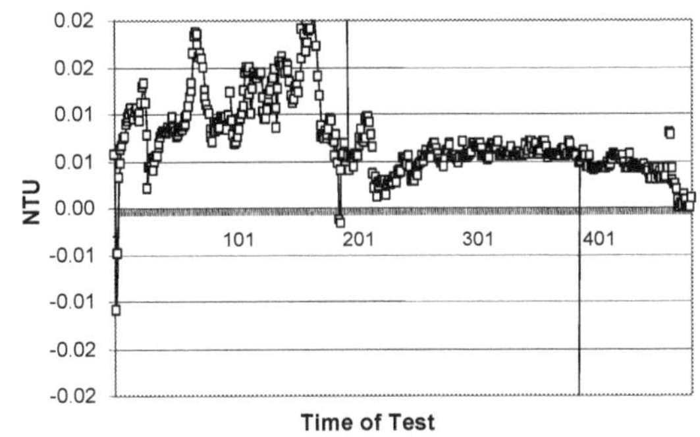

Figure 11-7. Turbidity (Filter 2 – Control)

Figures 11-4 and 11-5 show a similar Δ for particles in the 5—10 μ range (Filter – control). A similar trend is shown for the particles in the 5—10 μ range, although the effect in terms of

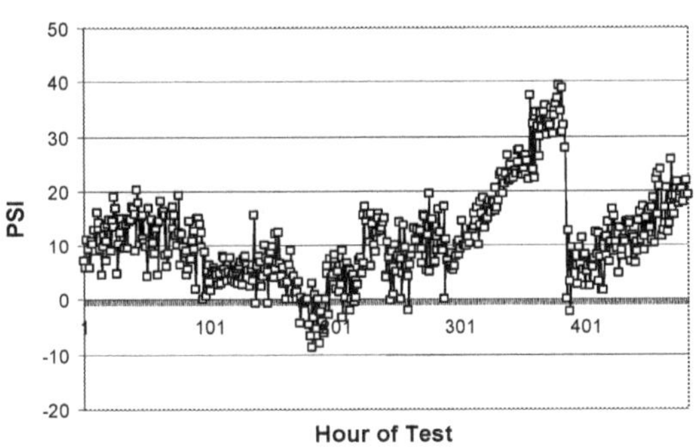

Figure 11-8. DP (Filter 1 - Control)

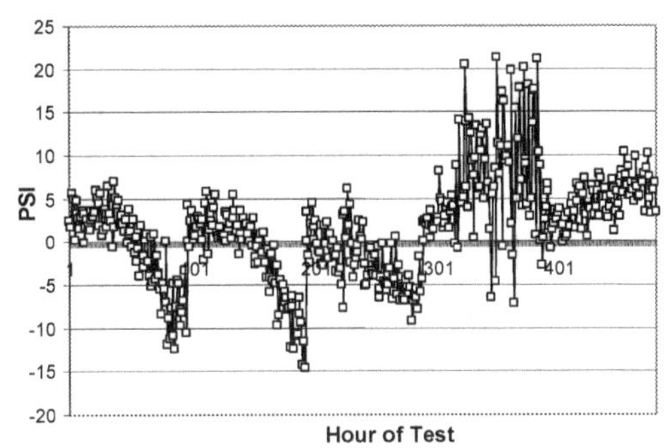

Figure 11-9. DP (Filter 2 - Control)

numbers of particles is smaller. This is expected, since the number of particles in the effluent for a particular range decreased with increasing size, as larger particles are removed by the filters.

Figures 11-6 and 11-7 show the Δ turbidity of each filter when compared with the control filter. For Filter 1, after the first week to 10 days, the turbidity actually shows some improvement. For Filter 2, the turbidity appears to have been degraded with application of ClO_2, and is higher for the entire test period than the control filter. This also suggests there may be a fundamental difference between Filter 1 and Filter 2.

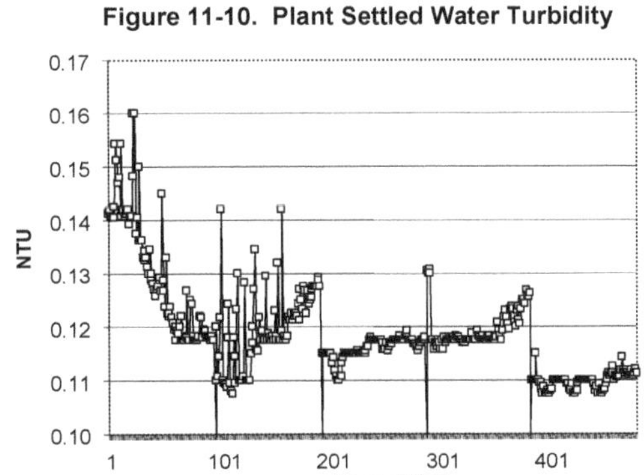

Figure 11-10. Plant Settled Water Turbidity

Figures 11-8 and 11-9 show the ΔDP across the filters, when compared to the control filter. Figure 11-8 shows that Filter 1 operated at a slightly elevated pressure during the entire

Figure 11-11. Particle Counts 2 - 5 Micron

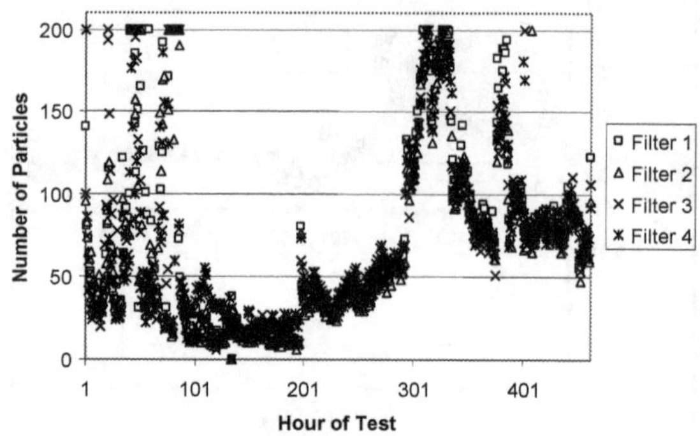

Figure 11-12. 2—5u (Filter1 - Control)

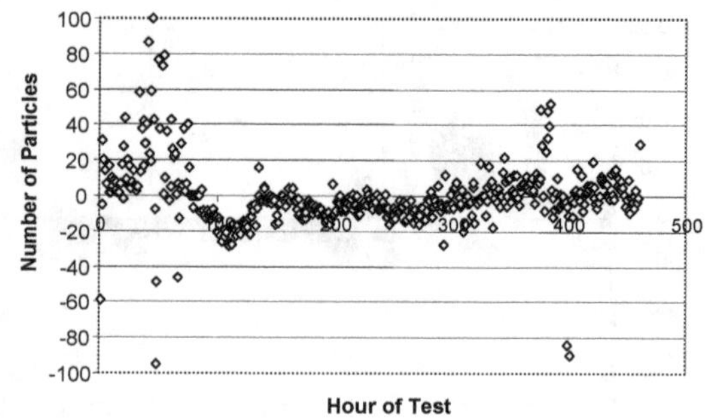

time, over control. This plot tends to support the theory proposed earlier that this filter is somewhat different than the other filters. Figure 11-9 shows the ΔDP for Filter 2. The trend is somewhat more like that expected, in that, over time, application of ClO_2 tended to keep the filter from fouling at the same rate as the untreated control. The data shown during the last week or two was more erratic and the reasons are not known.

Figure 11-10 shows the settled water turbidity of the plant, as this water was fed to the pilot filters. The data show a higher turbidity initially which settled down after the first week or so.

AB Jewell: Figure 11-11 shows the particle counts for all the filters at AB Jewell. As for the Mohawk data, there were some raw water changes at the beginning of this test. Again, due to the complexity in the data, deltas between data for each filter and for the control filter are plotted.

Figure 11-13. 2—5u (Filter 2 – Control)

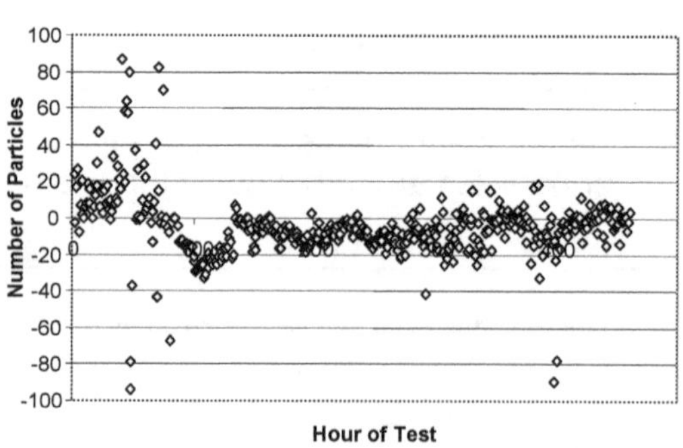

Figure 11-14. 2—5u (Filter 3 – Control)

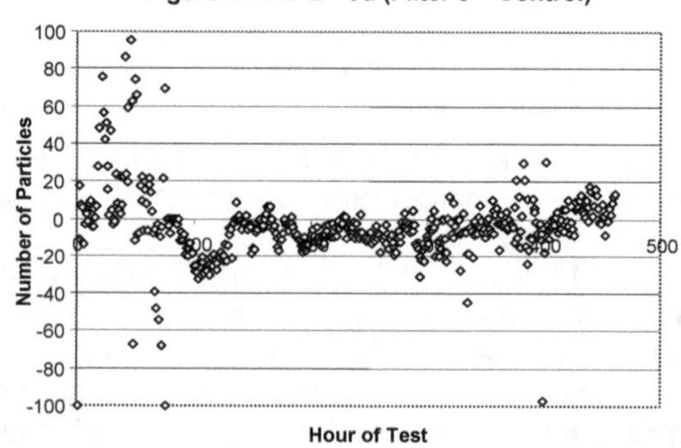

Figures 11-12 to 11-14 show the Δ for 2—5 μ particles for Filters 1-3 respectively. Generally, the data appears to fall slightly below the zero line, indicating a somewhat improved particle reduction with use of ClO_2, except for the initial period of the test, when the raw water was changing significantly. The particle reduction for Filter 1 (dosage ~ 0.25 mg/L ClO_2) is slightly less than for

Filter 2 (dosage ~ 0.5 mg/L ClO_2). The Δ for 2—5 μ particles for Filter 3 (target dosage ~ 0.75 mg/L ClO_2) also fell slightly below the zero line, indicating better particle removal.

Figure 11-15. 5—10u (Filter 1 – Control)

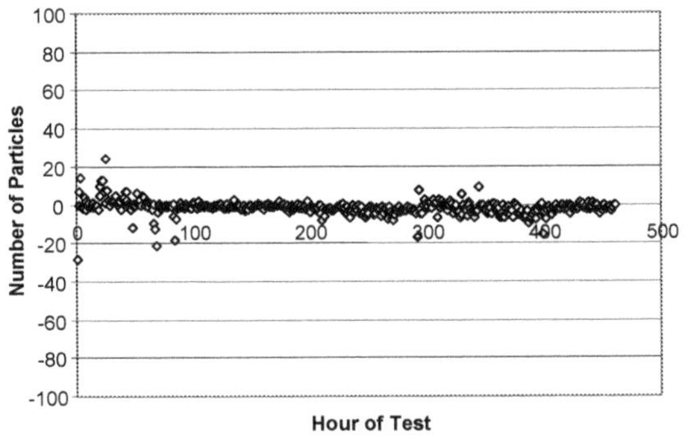

Figure 11-16. 5—10u (Filter 2 – Control)

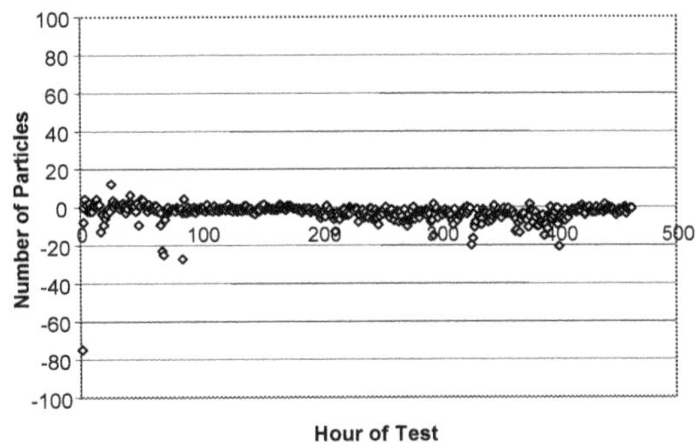

Figures 11-15 to 11-17 show the Δ for 5—10 μ particles for Filters 1 – 3, respectively. Again, the data falls slightly below the zero line, indicating improvement in particle reduction.

For the 10—15 μ particle range, the Δ for Filter 1 (~ 0.25 mg/L ClO_2), shown in Figure 11-18 also falls very close to, but slightly lower than the zero line. Surprisingly, for the filters with [ClO_2] > ~ 0.25 mg/L, the data indicated a significant particle reduction, as shown in Figures 11-19 and 11- 20.

Figure 11-17. 5—10u (Filter 3 – Control)

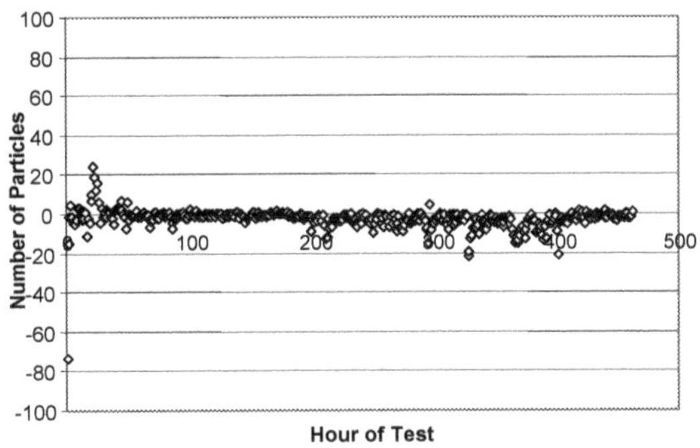

Figure 11-18. 10—15u (Filter 1 – Control)

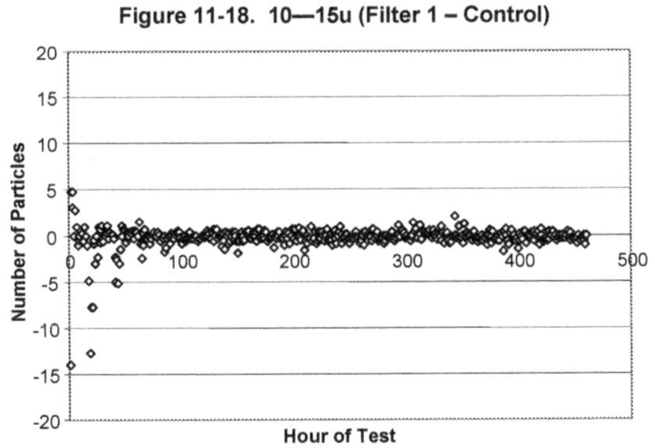

Figure 11-19. 10—15u (Filter 2 – Control)

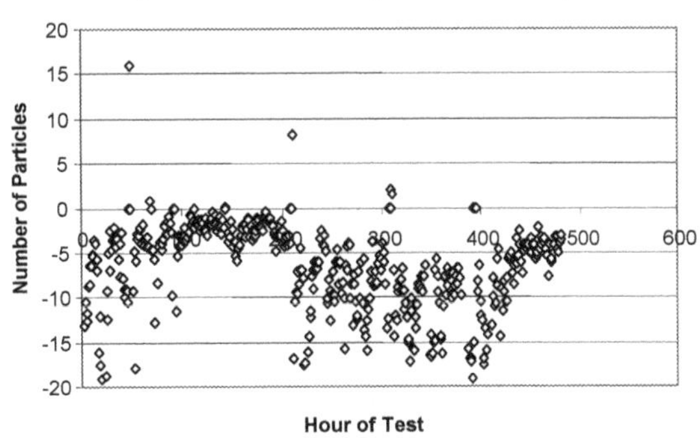

D. Discussion

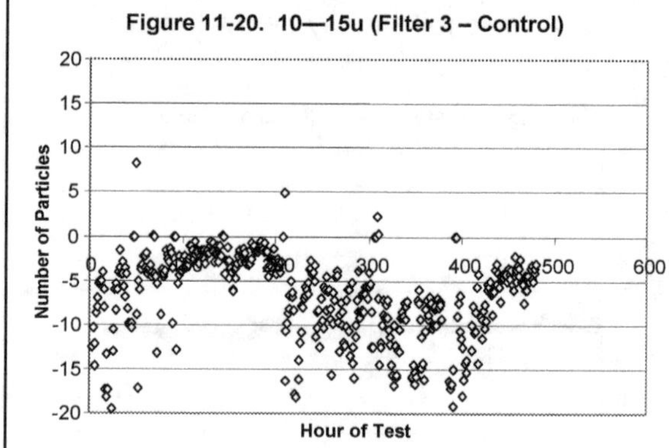

The goal of this work was to investigate how parameters change relative to each other over time. In terms of looking at both gross effects and the relative magnitude of any changes, the use of averages provides some insight. Table 11-2 shows averages over the entire test period for Mohawk plant.

In general, for ClO_2-treated filters at Mohawk, the data suggests that there was a general reduction in particle counts in all ranges, a slight increase in turbidities, and a modest to significant differential pressure increase when compared to control.

Table 11-2. Averages Over Test – Mohawk Plant					
	Filter			Δ	Δ
	1	2	3	F1-F3	F2-F3
[ClO_2]	0.47	0.41			
2—5 µ	22.2	24.7	34.9	-12.4	-10.1
5—10 µ	2.1	2.8	4.3	-2.2	-1.5
Total counts	64.3	64.9	93.1	-28.7	-28.1
Turbidities	0.051	0.057	0.050	0.001	0.007

In general, a negative number shows an *improvement* over the control. For example, in the 2—5 µ particle size range, particle counts for Filter 1 effluent (~ 0.47 mg/L ClO_2) were reduced by an average of just over 12 particles compared to that observed from Filter 3 (control) effluent.

Although the chlorine dioxide concentration was varied somewhat over the initial two weeks of the test (~ 0.2—0.8 mg/L), an analysis of [ClO_2] vs. other measured parameters showed no correlation. Thus, there does not appear to be any clear indication that any of the parameters measured varied as a direct result of making small changes in [ClO_2]. Therefore, for the purposes of looking at larger effects, average [ClO_2] values are considered.

The spread in ΔDP of the two filters at Mohawk supports the thought that Filter 1 is fundamentally different in some way from Filter 2.

Table 11-3 shows similar results for AB Jewell. Variations in [ClO_2] were also less for each filter than for the filters at Mohawk in that these dosages were set and remained unchanged over the test period. In general, the data appear to indicate a slight overall reduction in 5—10 and 10—15 µ particle counts. In the 2—5 µ range, at the lower ClO_2 dosage level, a slight increase in particle counts was observed. However, at the higher ClO_2 dosages, a reduction in particle counts was observed. This was especially true for the largest particles.

Table 11-3. Averages Over Test – AB Jewell Plant							
	Filter No				Δ		
	1	2	3	4	F1-F4	F2-F4	F3-F4
[ClO_2]	0.27	0.45	0.64				
2—5 μ	70.6	66.0	67.2	70.3	+0.3	-4.3	-3.0
5—10 μ	7.1	5.9	6.1	8.8	-1.6	-2.9	-2.6
10—15 μ	1.0	0.7	0.6	1.2	-0.3	-0.6	-0.6
DP	25.6	25.5	23.7	31.2	-6.4	-5.7	-7.5
Turbidities	0.063	0.059	0.059	0.062	0.001	-0.003	-0.003

For comparison purposes, it is instructive to look at the magnitude of decrease in terms of percentage reduction. For each data point, the control data is subtracted from that of the filter being investigated, and that value divided by the control data to give a percent change. These values were summed over the entire test period to give an average percent reduction. Table 11-4 shows a comparison of average percent reductions for each filter and plant.

Table 11-4. Average % Reductions					
	Mohawk		AB Jewell		
	F1-F3	F2-F3	F1-F4	F2-F4	F3-F4
2 – 5 μ	16.7	30.4	1.7	10.0	7.7
5 – 10 μ	35.6	35.6	17.4	27.6	24.0
10 – 15 μ			-9.1	13.8	9.1
Total counts	17.1	35.0			
DP	-61.4	-14.5	11.8	9.4	14.5
Turbidities	0.1	-14.0	-3.3	3.4	2.4

There are several differences between Mohawk and AB Jewell. The first difference is the nature and characteristic of the influent water. Differences between these raw water sources are listed in Table 11-5.

Table 11-5. Raw Water Characteristics		
	AB Jewell	Mohawk
Alkalinity (mg/L as $CaCO_3$)	100—120	75—105
Dissolved Solids (mg/L)	200—280	110—140
Total Hardness (mg/L as $CaCO_3$)	130—190	80—110
Turbidity (avg) (NTU)	9	5
TOC (mg/L)	4—5.5	<2

Secondly, the filter bed at Mohawk is considerably deeper (60" GAC) than AB Jewell (36" GAC), and, the type of GAC differs (Mohawk has lignite-based GAC, AB Jewell has bituminous-based GAC).

Thirdly, Mohawk influent water is treated with chlorine gas, while AB Jewell influent water is treated with chlorine dioxide.

Questions about the difference at Mohawk between Filter 1 and the other two filters have been raised. However, these differences might be attributable to the relative biological loading of these filters at the start of the trial. Subsequent testing of TOC reductions at the pilot plant filters at AB Jewell have revealed that Filter 3 (untreated with ClO_2) is providing slightly better TOC reduction and is operating at slightly higher pressures than either of the other two filters. This may be attributable to a lower biological loading on the ClO_2-treated filters than that of the filter which received no ClO_2.

E. Conclusions

Several conclusions can be made regarding the use of ClO_2 as an intermediate oxidant just prior to GAC filters in potable water facilities.

1. Use of ClO_2 can result in reduced particle counts in all ranges and in turn help with removal of Giardia and Cryptosporidium. The optimum [ClO_2] to accomplish desired particle count reductions in each range will likely be plant dependent and will have to be optimized for each facility.

2. Use of ClO_2 may result in decreased turbidities, but this will likely be dosage dependent and have to be optimized for each facility.

F. References

[Amy et al. 1987] Amy et al., "The Effects of Humic Substances on Particle Growth and Removal During Coagulation," ACS Division Spring Conference, Denver, CO, 1987.

[Becker et al. 2003] Becker, W., O'Melia, C., Kwod-Keung, A., and Young, J., "The Use of Oxidants to Minimize the Passage of Particles Through Granular Media Filters," American Water Works Association Research Foundation Report, 2003.

[Chandrakanth and Amy 1998] Chandrakanth, M. and Amy, G., "Effects of NOM Source Variations and Calcium Complexation Capacity on Ozone-Induced Particle Destabilization," *Water Research,* 32(1), 115(1998).

[Chang and Singer 1991] Chang, S. and Singer, P., "The Impact of Ozonation on Particle Stability and the Removal of TOC and THM Precursors," *Journal of the American Water Works Association*, 71(March 1991).

[Chowdhury et al. 2000] Chowdhury, Z., VanGelder, A., Lawler, D., and Moran, M., "Particle Count Method Development for Concentration Standards and Sample Stabilization," American Water Works Association Research Foundation Report, 2000.

[Dore et al. 1978] Dore, M., et al., "Influence of Oxidizing Treatments on the Formation and The Degradation of Haloform Reaction Precursors," *Progress in Water Technology,* 10(5), 853(1978).

[Dowbriggin and Singer 1989] Dowbriggin, W. and Singer, P., "Effects of Natural Organic Matter and Calcium on Ozone-Induced Particle Destabilization," *Journal of the American Water Works Association*, 81(6), 77(June 1989).

[Edwards and Benjamin 1991] Edwards, M. and Benjamin, M., "A Mechanistic Study of Ozone-Induced Particle Destabilization," *Journal of the American Water Works Association,* 83(6), 96(June 1991).

[Felix-Filho 1985] Felix-Filho, J., "The Effect of Aquatic Humic Substances and Ozonation on the Stability of Particles," Ph.D. Dissertation, University of North Carolina at Chapel Hill, 1985.

[Felix-Filho et al 1987] Felix-Filho, J., Singer, P., and Dowbiggin, W., "Effect of Humic Substances on the Colloidal Stabilty of Particles," ACS Division, Spring Conference, Denver, CO, 1987.

[Gibbs 1983] Gibbs, R., "Effect of Natural Organic Coatings on the Coagulation of Particles," *Environmental Science & Technology*, 17, 237(1983).

[Grasso 1987] Grasso, D., "Ozonation Dynamics in Water Treatment: Autocatalytic Decomposition, Mass Transfer, and Impact on Particle Stability," PhD Dissertation, University of Michigan, Ann Arbor, MI, 1987.

[Grasso and Weber 1988] Grasso, D. and Weber, W., "Ozone-Induced Particle Destabilization," *Journal of the American Water Works Association*, 73(August 1988).

[Gurol and Pidatella 1983] Gurol, M. and Pidatella, M., "Ozone-Induced Coagulation," Proceedings National Conference on Environmental Engineering, Boulder CO, 1983.

[Hall and Croll 1997] Hall, T. and Croll, B., "Particle Counters as Tools for Managing Cryptosporidium Risk in Water Treatment," *Water Science and Technology*, 36(4), 143(1997).

[Hatukai et al. 1997] Hatukai, S., Ben-Tzur, Y., and Rebhun, M., "Particle Counts and Size Distribution in System Design for Removal of Turbidity by Granular Deep Bed Filtration," *Water Science and Technology*, 36(4), 225(1997).

[Huck et al. 2001] Huck, P., Emelko, M., Coffey, B., Maurizio, D., O'Melia, C., "Filter Operation Effects of Pathogen Passage," American Water Works Association Research Foundation, 2001.

[Jekel 1983] Jekel, M., "The Benefits of Ozone Treatment Prior to Flocculation Processes," *Ozone Science and Engineering*, 5, 21(1983).

[Jekel 1986] Jekel, M., "The Stabilization of Dispersed Mineral Particles by Adsorption of Humic Substances," *Water Research*, 12, 1543(1986).

[Johnson and Randtke 1983] Johnson, D. and Randtke, S., "Removing Nonvalatile Organic Chlorine and Its Precursors by Coagulation and Softening," *Journal of the American Water Works Association*, 75(5), 249(May 1983).

[Korich et al. 1990] Korich, D., Mead, J., Madore, M., Sinclair, N., and Sterling, C., "Effects of Ozone, Chlorine Dioxide, Chlorine, and Monochloramine on *Cryptosporidium parvum* Oocyst Viability," *Applied and Environmental Microbiology*, 56, 1423 (May, 1990).

[Lewis and Manz 1991] Lewis, C. and Manz, D., "Light-Scattering Particle Counting: Improving Filtered Water Quality," *Journal of Environmental Engineering*, 117(2), 209(March/April 1991).

[McTigue et al. 1998] McTigue, N, LeChavallier, M, Arora, H, and Clancy, J, "National Assessment of Particle Removal by Filtration," American Water Works Association Research Foundation Report, 1998.

[Nickels et al. 2000] Nickels, A., Harriman, C., Nickels, A., Freedman, A., and Harris, A., "A Novel Method of Applying Chlorine Dioxide in a Recirculating Cooling Water System," AWT Annual Convention and Exposition, Water Technologies 2000, Honolulu, HI, October 31-November 4, 2000.

[Pickrell et al. 1999] Pickrell, W., Germer, M., Miller, J., and Simpson, G., "Plants Increase Production Through Use of Chlorine Dioxide," Practical Industrial Water Treatment Technology for the New Millenium, Houston, TX, January 27-29, 1999.

[Pitochelli 1999] Pitochelli, A., "Two-Component ClO_2 Generation Packets for Demand Studies and Small Pilot Studies," unpublished results, 1999.

[Reckhow and Singer 1984] Reckhow, D. and Singer, P., "The Removal of Organic Halide Precursors by Preozonation and Alum Coagulation," *Journal of the American Water Works Association*, 76(4), 151(April 1984).

[Reckhow et al. 1986] Reckhow, D., Legube, B., and Singer, P., "The Ozonation of Organic Halide Precursors: Effect of Bicarbonate," *Water Research,* 20(8), 987(1986).

[Reckhow et al. 1986] Reckhow, D., Singer, P., and Trussell, R., "Ozone as a Coagulant Aid," American Water Works Association Annual Conference, Denver, CO, 1986

[Reckhow et al. 1993] Reckhow, D., Edzwald, J., and Tobiason, J., "Ozone as an Aid to Coagulation and Filtration," American Water Works Association Research Foundation Report, 1993.

[Rice 1980] Rice, R., "The Use of Ozone to Control Trihalomethanes in Drinking Water Treatment," *Ozone Science & Engineering*, 2(1), 75(1980).

[Riley et al. 1978] Riley, L., Mancy, K., Boettner, E., "The Effect of Preozonation on Chloroform Production in the Chlorine Disinfection Process," Water Chlorination: Environmental Impact and Health Effects, 21, Jolley, R., et al., eds, Ann Arbor Sci, Publ., Ann Arbor, MI (1978).

[Saunier et al. 1983] Saunier, B., Selleck, R., and Trussell, R., "Preozonation as a Coagulant Aid in Drinking Water Treatment," *Journal of the American Water Works Association*, 75(5), 239(May 1983).

[Schalekamp 1979] Schalekamp, M., "Experience in Switzerland with Ozone, Particularly in Connection with the Change of Hygienically Undesirable Elements Present in Water," *Ozonews*, 7(5), 1(1979).

[Scrivner et al 1980] Scrivner, A., Singer, P., and Barry, J., "Control of Trihalomethanes in North Carolina Drinking Water by Oxidation/Coagulation," *Journal Water Supply Research and Technology – Aqua* (London), 7, 141(1980).

[Simpson 2001] Simpson, G., "Control of Biofilm in Cooling Towers with ClO_2," Fourth International Symposium on Chlorine Dioxide, Las Vegas, NV, February 15-16, 2001.

[Simpson and Miller 2000] Simpson, G., and Miller, J., "Control of Biofilm with Chlorine Dioxide," AWT Annual Convention and Exposition, Water Technologies 2000, Honolulu, HI, October 31-November 4, 2000.

[Simpson et al. 1993] Simpson, G., Laxton, G., Miller, R., and Clements, W., "A Focus on Chlorine Dioxide: Biocide of Choice for 'Stressed' Cooling Water Systems," WaterTech 93, Houston, TX, November 10-12, 1993.

[Singer and Chang 1989] Singer, P. and Chang, S., "Impact of Ozone on the Removal of Particles, TOC and THM Precursors," American Water Works Association Research Foundation Report, June, 1989.

[Sukenik et al. 1987] Sukenik, A., Teltch, B., Wachs, A., Shelef, G., Nir, I., and Levanon, D., "Effect of Oxidants on Microalgal Flocculation," *Water Research*, 21(5), 533(1987).

[Swietlik et al. 2002] Swietlik, J., Raczyk-Stanislawiak, U., Bilozor, S., Ilecki, W., and Nawrocki, J., "Adsorption of Natural Organic Matter Oxidized with ClO_2 on Granular Activated Carbon," *Water Research,* 36, 2328(2002).

[Tobiason et al. 1992] Tobiason, J., Edzwald, J., Schneider, O., Fox, M., and Dunn, H., "Pilot Study of the Effects of Ozone and Peroxone on In-Line Direct Filtration," *Journal of the American Water Works Association*, 72(December 1992).

[Trussell and Umphres 1978] Trussell, R. and Umphres, M., "The Formation of Trihalomethanes," *Journal of the American Water Works Association*, 70(11) 604(Nov 1978).

[Van Breeman et al. 1979] Van Breeman, A., Nieuwstad, T, and Van Der Meent-Olieman, G., "The Fate of Fulvic Acids During Water Treatment," *Water Research,* 13, 771(1979).

Oxidation of Fe with ClO_2

A. The Problem with Iron

Although iron exists in eight possible oxidation states, ranging from Fe^{-2} to Fe^{+6}, only three are common, Fe^{0}, Fe^{+2}, and Fe^{+3}. Iron causes different problems in different applications. In potable water, it causes one set of problems. In industrial cooling systems or in paper mill water, it may cause a different set of problems. Both are addressed below:

Potable Water: The problem of iron in potable water is one of aesthetics, as iron in potable water does not appear to cause significant health problems [Khoe and Waite 1989]. However, to minimize problems such as discoloration of water, laundry, or plumbing fixtures, or to minimize taste problems, a secondary maximum contaminant level (SMCL) of 0.3 mg/L has been established for iron.

Regulations impacting disinfection practices which are necessary to achieve goals of Cryptosporidium and Giardia cyst removal, THM reduction, and other goals of disinfection have resulted in processes with conditions that may not be optimal for removal of iron.

Industrial Water: Soluble iron, generally ferrous iron (Fe^{+2}), is present in makeup, particularly in some well waters. As the ferrous iron is exposed to oxygen that is present in open, recirculating cooling systems which are aerated by design, the ferrous iron is quickly oxidized to ferric iron (Fe^{+3}) by dissolved oxygen. Insoluble ferric oxide or ferric hydroxide forms, agglomerates, and forms deposits. Iron released by the corrosion process also contributes to overall deposits. Iron has been singled out as the largest contributor to deposit problems in cooling systems [Drew 1979].

Iron forms dense deposits in some water systems, particularly when the surfaces are hot, such as in industrial cooling systems and in boilers. The performance of most synthetic water treatment deposit control polymers is compromised, sometimes significantly, in the presence of iron. Thus the best way to deal with iron in an industrial water system is to remove iron before it enters the system and to make sure that the dispersants in systems plagued by iron are selected from the group which retains its performance characteristics in the presence of iron.

B. Common Methods for Iron Oxidation/Removal

Oxygen for Iron Oxidation – Early Work: The most commonly used approach for oxidation and removal of iron historically was the use of dissolved oxygen via aeration followed by filtration or sedimentation. The oxidation of iron by aeration has been studied extensively [Ghosh 1962; Robinson and Breland 1968; Robinson and Dixon 1968]. The oxidation of Fe^{+2} was found to be extremely pH dependent, with a 100-fold increase in reaction rate per unit increase in pH [Stumm and Lee 1961]. Many of the early studies were conducted using waters which contained minimal amounts of natural organic matter (NOM). The oxidation of iron with oxygen was found to follow the reaction pathway shown in Equation 12-1.

$$2Fe^{+2} + \tfrac{1}{2} O_2\,(aq) + 5H_2O \Rightarrow 2Fe(OH)_3\,(s) + 4H^{+} \qquad (12\text{-}1)$$

Chlorine for Iron Oxidation: In many systems, use of chlorine for disinfection has replaced aeration. For groundwater which contains little to no organics, chlorine has been used alone or with aeration, with minimal risk of THM formation. In waters with low organic content, Fe^{+2} can be oxidized readily by hypochlorous acid (HOCl) at a pH of 5.0 or above. Research showed that Fe^{+2} oxidation by free chlorine was much more rapid than oxidation by oxygen. Thus, free chlorine addition was found to be an effective (polishing) supplement to aeration in groundwater treatment facilities where Fe^{+2} removal efficiency was less than optimal. The reaction for the oxidation of iron by chlorine is shown in Equation 12-2.

$$2Fe^{+2} + HOCl + 5H_2O \Rightarrow 2Fe(OH)_3\,(s) + Cl^- + 5H^+ \qquad (12\text{-}2)$$

Other Oxidants for Iron Oxidation: Before 1975, iron was controlled by the addition of chlorine alone or chlorine in combination with potassium permanganate under alkaline conditions. Doses of free chlorine up to 5 mg/L were used. Free chlorine and potassium permanganate were shown to be more rapid than oxygen for oxidizing uncomplexed iron. Regulations requiring low THM water has forced many plants to consider alternatives to free chlorine for this purpose.

Experiments with alternative oxidants showed that uncomplexed Fe^{+2} was oxidized essentially instantaneously by potassium permanganate, chlorine dioxide, or ozone. Chlorine and hydrogen peroxide also produced rapid oxidation of uncomplexed Fe^{+2}.

Today, alternative oxidants are now used to a much greater extent, with Fe^{+2} removal from ground waters low in NOM being accomplished by oxidation with free chlorine, potassium permanganate, chlorine dioxide, ozone, and hydrogen peroxide. Due to their increased costs, many plants have significantly reduced their oxidant dosages.

Impact of Natural Organic Matter (NOM): Surface waters contain natural organic matter (NOM) which results from decay of vegetation. These vegetative decay products (humic and fulvic acids) can complex Fe^{+2} with a high efficiency, making the iron very difficult to oxidize.

Early workers noted the poor iron removal efficiency in several aerated systems [Ghosh et al. 1966]. Others, through infrared and polarographic analysis, showed that these naturally occurring organic acids had significant complexation capabilities for both ferrous and ferric iron, and that the rate of oxidation was significantly retarded [Oldham and Gloyna 1969]. Later work showed that water containing organic matter was able to hold iron in solution for indefinite periods of time after aeration [Jobin and Ghosh 1972].

In a manner similar to aeration, the ability of free chlorine to oxidize iron complexed by these acids was compromised. Further work showed that the organic matrix surrounding Fe^{+2} protect the Fe^{+2} from oxidation by any of the oxidants. The presence of significant concentrations of humic and fulvic acids (> 3 mg/L as NOM) may result in dissolved Fe^{+2} becoming so highly complexed that the required dosage for a given oxidant may increase to up to 800% of stoichiometric of that required to oxidize uncomplexed iron. Typically, less than 25% of the dissolved complexed iron was removed by chemical oxidation under these conditions.

Both the concentration and type of NOM present in surface water impact how iron removal is accomplished during treatment. The molecular weight of the NOM was found to be a very important factor, with success in iron oxidation and removal being inversely proportional to the molecular weight range of the NOM. Generally, organics that have higher molecular weights are more effective at shielding the iron from oxidation.

The worst iron removal was observed in systems where the dissolved iron was complexed by high molecular weight humic acids. More efficient iron removal was observed in systems where the iron was complexed by very low (< 1000 AMU) molecular weight fulvic acids. In systems with clarifiers, iron complexed by high molecular weigh organics may be removed when this organic matter is coagulated. Systems with clarifiers accomplish the most efficient removal by coagulation and sedimentation. Low temperatures have not been shown to adversely affect treatment.

Alum used for coagulation may also play an important role in the removal of iron, particularly iron complexed by organic material. The pH which is optimum for NOM removal with use of alum has also been shown to be optimal for iron removal [Knocke et al. 1987].

If Fe^{+2} is complexed by high molecular weight organic compounds, addition of oxidants such as potassium permanganate, chlorine dioxide or ozone may not produce more efficient iron removal than would be achieved by coagulation alone. If, however, the Fe^{+2} is complexed by lower molecular weight organics, then these oxidants may significantly enhance the efficiency of iron removal by coagulation. Characterization of the molecular weight distribution of organic compounds in the raw water would be helpful in making decisions on how best to treat surface waters for iron removal.

Iron Reduction by NOM: Some researchers showed that tannic and humic acids, gallic acid, pyrogallol, and other products of natural vegetative decay effectively retarded oxidation of Fe^{+2} for several days, even in water saturated with oxygen [Hem 1960; Theis and Singer 1973; Theis and Singer 1974]. These authors also showed that complexation could also promote reduction of Fe^{+3} to Fe^{+2} in a coupled redox reaction.

In parallel work, soil scientists, seeking to determine the mechanism of iron transport into plants from minerals in the soil, showed that ferric iron in minerals could be reduced to ferrous iron by interaction with humic acids under anaerobic conditions [Bloomfield 1952;Martell 1958; Schnitzer 1959; Coulson et al. 1960]. Iron-organic complexes were found to be extremely resistant to oxidation by oxygen [Weston 1909; Okura and Goto 1955; Oldham and Gloyna 1969].

Other Factors Important in the Oxidation and Removal of Iron: The important role other factors have played in the rate of oxidation of iron have also been reported, including the effect of buffering [Jobin and Ghosh 1972], of alkalinity [Applebaum 1956; Longley 1961; Ghosh 1962], of solids [Applebaum 1956], of ionic strength [Sung and Morgan 1980] and of pH [Ghosh 1962].

C. Literature Review – Oxidation of Iron with ClO_2

The oxidation of iron by ClO_2 was shown to follow the following reaction pathway [Gall 1976]:

$$ClO_2 + FeO + NaOH + H_2O \Rightarrow Fe(OH)_3 + NaClO_2 \qquad (12\text{-}3)$$

The oxidation of Fe^{+2} by ClO_2 was reported to produce insoluble hydrated ferric hydroxide by the following reaction pathway [Masschelein 1979].

$$ClO_2 + 5Fe^{+2} + 4H^{+} \Rightarrow 5Fe^{+3} + Cl^{-} + 4H_2O \qquad (12\text{-}4)$$

A later report described the very rapid oxidation of Fe^{+2} to Fe^{+3}, which forms insoluble iron hydroxide [Aieta and Berg 1986]. They showed the following reaction to dominate at neutral to alkaline pH:

$$ClO_2 + 5Fe(HCO_3)_2 + 3H_2O \Rightarrow 5Fe(OH)_3 + 10\,CO_2 + HCl \qquad (12\text{-}5)$$

One research report suggested that uncomplexed Fe^{+2} was oxidized almost instantaneously by ClO_2 [Knocke et al. 1990], but that NOM inhibited Fe^{+2} oxidation efficiency. They also showed that the most likely reaction pathway was that shown in Equation 12-6 [Knocke et al. 1990].

$$ClO_2 + Fe^{+2} + 3H_2O \Rightarrow Fe(OH)_3(s) + ClO_2^- + 3H^+ \quad (12\text{-}6)$$

Later workers found that Fe^{+2}, complexed with humic or fulvic acid, was strongly resistant to oxidation, with 800% stoichiometric excess required, and then only about 25% of the iron was oxidized [Knocke et al. 1990; Knocke et al. 1991].

In another report, researchers suggested that the oxidation of Fe^{+2} by ClO_2 was one of those few reactions for which ClO_2 could possibly utilize its full oxidation capacity, as shown in Equation 12-7 [Aieta and Berg 1986; Griese et al. 1991; Noack and Iacoviello 1992]:

$$ClO_2 \xrightarrow{+e^-} ClO_2^- \xrightarrow{+e^-} ClO \xrightarrow{+e^-} (OCl^- \text{ or } HOCl) \xrightarrow{+2e^-} Cl^- \quad (12\text{-}7)$$

Detailed kinetic studies of Fe^{+2} were done at low pH values [Hoigne and Bader 1994]. Rate constants could not be extrapolated to pH 8 because of the interference of the oxidation of uncomplexed iron(II) by oxygen at that pH.

One remaining question concerning the oxidation of Fe^{+2} by ClO_2 is whether the predominant reaction is that shown in Equation 12-3 or Equation 12-4. If indeed the oxidation of iron is one where ClO_2 makes use of its full oxidation capacity, then the use of ClO_2 may be economically favorable to other oxidants, particularly when gaseous chlorine is eliminated from consideration.

The following case history presents this issue in a real-world scenario.

D. Case History

One Midwest paper mill utilized water from a nearby well as the process water in the mill. This well contained almost 20 mg/L Fe, which was unusable in the mill due to the high potential for discoloration of the paper, formation of deposits, and staining of equipment. Gaseous chlorine was used to oxidize and disinfect the well water at the well. An older cooling tower and basin had been modified to provide a long residence time to aid the iron in settling (shown in Figure 12-1).

The large number of one-ton containers on site and new regulations resulted in the plant having to develop an emergency plan. This task was considered to be monumental and so the plant decided to replace gaseous chlorine with sodium hypochlorite. The reaction chemistry of chlorine was known with some degree of certainty, since gaseous chlorine had been used for this application historically. The possible reaction chemistry where ClO_2 would use all of its oxidation capacity led to an investigation of ClO_2. In addition, the use of ClO_2 would result in a much lower chemical requirement, and potentially, a lower cost. In addition, the ClO_2 generator

Figure 12-1. Schematic of Iron Treatment System

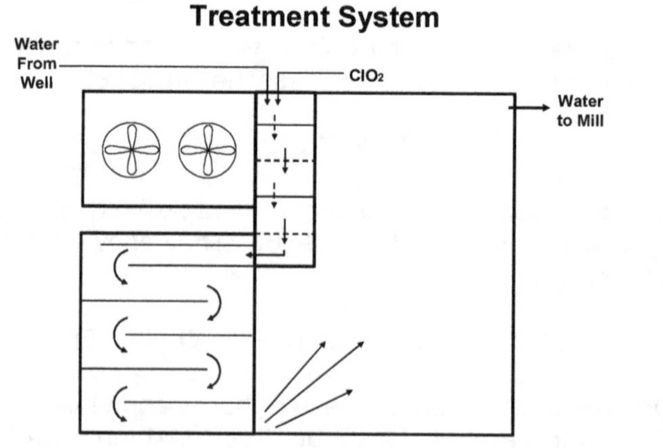

could be installed near the point of injection.

For a period of about 5 days, data was gathered. For the first day, iron was oxidized with chlorine. For the subsequent 4 days, ClO_2 was used to oxidize iron. This data included analysis of water samples from the well and from the effluent of the iron treatment system for both ferrous and ferric iron, for chlorine, and for ClO_2. The dosage of ClO_2 was set initially to be equal to that of chlorine, and was subsequently reduced slowly over the 4 day period. The results are plotted in Figure 12-2, along with iron test results for comparison.

Figure 12-2. Results – Cl_2 vs. ClO_2

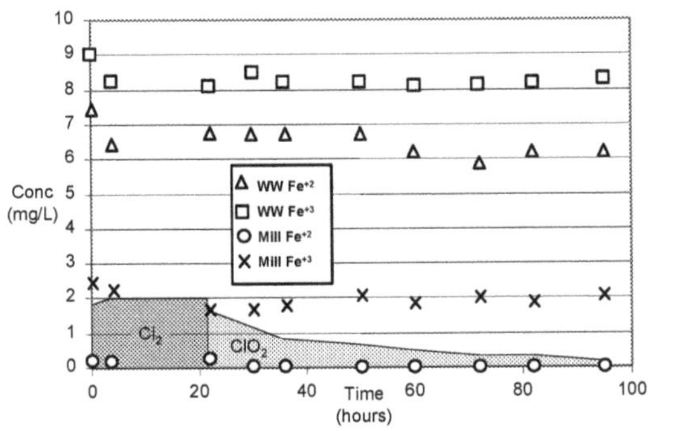

Figure 12-3. Comparison – Cl_2 vs. ClO_2

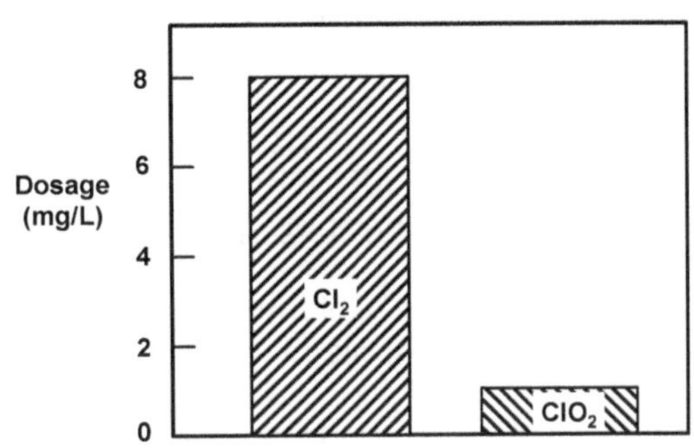

The data show that the dosage of ClO_2 was reduced substantially over the 4 day period, so that the measurable residual at the end of the trial was barely measurable, at a level of ~ 0.1 mg/L, without degradation of iron control.

After the trial had ended, the data was analyzed regarding the relative dosage of ClO_2 vs. that of chlorine and the results of this comparison are shown in Figure 12-3. This analysis showed that 7.5—8.5 mg/L chlorine was replaced by 1.0 mg/L ClO_2.

The field trial illustrated the ability of ClO_2 to control ferrous iron, as long as there was some mechanism to remove the Fe^{+3} formed.

E. Summary

The following important points have been obtained either from literature or from practical experience.

- ClO_2 can oxidize iron, and, by selecting the right reaction conditions the full oxidation capacity of ClO_2 can be utilized, resulting in a significant reduction in required dosage.
- Iron, complexed with organics, is much more difficult to oxidize, requiring up to 8 times the normal oxidant concentration to get reduction and even then a 100% oxidation may not be achieved. In such cases, some other method for iron removal may be required.
- Use of ClO_2 for iron oxidation can be cost effective, when compared with other chlorine alternatives, due to it utilization of its full oxidation capacity, for water with iron not complexed by natural organic matter.

F. References

[Aieta and Berg 1986] Aieta, E. and Berg, J., "A Review of Chlorine Dioxide in Drinking Water Treatment," *Journal of the American Water Works Association*, 78(6), 62(1986).

[Anderson et al. 1973] Andersen, D., Row, D., and Sindelar, G., "Iron and Manganese Studies of Nebraska Water Supplies," *Journal of the American Water Works Association*, 65(10), 635(1973).

[Applebaum 1956] Applebaum, S., "Iron and Manganese Removal," *Water and Sewage Works*, 103(6), 258(1956).

[Bloomfield 1952] Bloomfield, C., "Translocation of Iron in Podzol Formation," *Nature*, 170, 540(1952).

[Coulson et al. 1960] Coulson, C., Davies, R., and Lewis, D., "Polyphenols in Plants, Humus, and Soil: II Reduction and Transport by Polyphenols of Iron in Model Soil Columns," *Journal of Soil Science*, 11, 30(1960).

[Drew 1979] Drew, Principles of Industrial Water Treatment, Drew Chemical Corporation, publishers, 1979.

[Gall 1976] Gall, R., "Chlorine Dioxide: An Overview of its Preparation, Properties and Uses," Proceedings of IOI Meeting, Cincinnati, OH, Nov. 17-19, 1976.

[Ghosh 1962] Ghosh, M., "A Study of the Rate of Oxidation of Iron in Aerated Ground Waters," Sanitation Engineering Series No. 12, The University of Illinois, 1962.

[Ghosh et al. 1966] Ghosh, M., O'Connor, J., and Engelbrecht, R., "Precipitation of Iron in Aerated Ground Waters," *Journal Sanitary Engineering Division, American Society of Civil Engineers*, 92(SA1), 199(1966).

[Griese et al. 1991] Griese, M. H., Hauser, K., Berkemeier, M., and Gordon, G., "Using Reducing Agents to Eliminate Chlorine Dioxide and Chlorite Ion Residuals in Drinking Water," *Journal of the American Water Works Association*, 83(5), 56(May 1991).

[Hem 1960] Hem, J., "Complexes of Ferrous Iron with Tannic Acid," *Geological Survey Water-Supply,* Paper 1459-D, 1960.

[Hoigne and Bader 1994] Hoigne, J. and Bader, H., "Kinetics of Reactions of Chlorine Dioxide in Water - I. Rate Constants for Inorganic and Organic Compounds," *Water Research*, 28(1), 45(1994).

[Jobin and Ghosh 1972] Jobin, R. and Ghosh, M., "Effect of Buffer Intensity and Organic Matter on the Oxygenation of Ferrous Iron," *Journal of the American Water Works Association*, 64(9), 590(Sept 1972).

[Khoe and Waite 1989] Khoe, G. and Waite, T., "Manganese and Iron Related Problems in Australian Water Supplies," *Environmental Technology Letters*, 10, 479(1989).

[Knock et al. 1990] Knocke, W., Benschoten, J., Kearney, M., Soborski, A., and Reckhow, D., "Alternative Oxidants for the Removal of Soluble Iron and Manganese," American Water Works Association Research Foundation Report, March 1990.

[Knocke et al. 1991] Knocke, W., Benschoten, J., Kearney, M., Soborski, A., and Reckhow, D., "Kinetics of Manganese and Iron Oxidation by Potassium Permanganate and Chlorine Dioxide," *Journal of the American Water Works Association*, 80(June 1991).

[Longley 1961] Longley, J., "The Removal of Iron from Water by Aeration and Filtration," MS Thesis, The University of Illinois, 1961.

[Martell 1958] Martell, A., "The Chemistry of Metal Chlelates in Plant Nutrition," *Soil Science*, 84, 13(1958).

[Noack and Iacoviello 1992] Noack, M. G. and Iacoviello, S. A., "The Chemistry of Chlorine Dioxide in Industrial and Wastewater Treatment Applications," Chemical Oxidation, Technology for the Nineties, Second International Symposium, Vanderbilt University, Nashville, Feb 19-21 (1992).

[O'Connor 1971] O'Connor, J., "Iron and Manganese," Water Quality and Treatment, American Water Works Association, McGraw-Hill Publishers, New York, 1971.

[Okura and Goto 1955] Okura, T. and Goto, K., "Oxidation of Ferrous Iron in Natural Water," *Journal of the Chemical Society of Japan, Industrial Chemical Society*, 58, 239(1955).

[Oldham and Gloyna 1969] Oldham, W. and Gloyna, E., "Effect of Colored Organics on Iron Removal," *Journal of the American Water Works Association*, 61(11), 610(1969).

[Robinson and Breland 1968] Robinson, L. and Breland, E., "Removal of Iron and Manganese from Low Alkalinity Waters," *Public Works*, 99, 72(1968).

[Robinson and Dixon 1968] Robinson, L. and Dixon, R., "Iron and Manganese Precipitation in Low Alkalinity Ground Waters," *Water and Sewage Works*, 115, 514(1968).

[Schnitzer 1959] Schnitzer, M., "Interaction of Iron with Rainfall Leachates," *Journal of Soil Science*, 10, 300(1959).

[Stumm and Lee 1961] Stumm, W. and Lee, G., "Oxygenation of Ferrous Iron," *Industrial and Engineering Chemistry*, 53(2), 143(1961).

[Sung and Morgan 1980] Sung, W. and Morgan, J., "Kinetics and Product of Ferrous Iron Oxygenation in Aqueous Systems," *Environmental Science and Technology*, 14(5), 561(1980).

[Theis and Singer 1973] Theis, T. and Singer, P., "The Stabilization of Ferrous Iron by Organic Compounds in Natural Waters," Trace Metals and Metal-Organic Interactions in Natural Waters, Singer, P. ed., Ann Arbor Science Publishers, Ann Arbor, MI, 1973.

[Theis and Singer 1974] Theis, T. and Singer, P., "Complexation of Iron (II) by Organic Matter and its Effect on Iron(II) Oxygenation," *Environmental Science and Technology*, 8(6), 569(June 1974).

[Weston 1909] Weston, R., "The Purification of Ground Waters Containing Iron and Manganese," *Transactions ASCE*, 64, 112(1909).

Control of *Dressenia polymorpha* (Zebra Mussel) with ClO_2

A. Introduction – An Overview of Zebra Mussels and the Problems They Cause

History: The zebra mussel (henceforth ZM), or *Dreissena polymorpha*, is a freshwater bivalve or mollusk that is native to Europe. The ZM first appeared in the US in Lake St. Claire near Detroit in the mid 1980s. Due to their adaptability, the lack of natural predators in the US and their tremendous reproductive ability, they have spread along the Mississippi, and have resulted in the plugging of industrial water intakes of a variety of plants using Mississippi river water as cooling water. In fact, for the past several years, one refinery in southern Louisiana has treated their once-through cooling system annually for mitigation of ZM [Beckstrom 2005].

Morphology and Attachment Mechanism: The North American ZM is more thermally tolerant, has greater growth rates, and displays greater reproductive variation than its European cousin. The ZM is epifaunal, i.e., the ZM attaches to hard surfaces [Ram and McMahon 1996], forming dense communities, with the density of mussel fouling depending critically on the type of substrate [MacIsaac 1996]. The formation of dense communities results in great economic impact. Newly settled mussels reach such high densities (up to 700,000/m^2) and have such rapid post-settlement growth rates, that they rapidly form thick mats, occluding or blocking flow even in large diameter pipes.

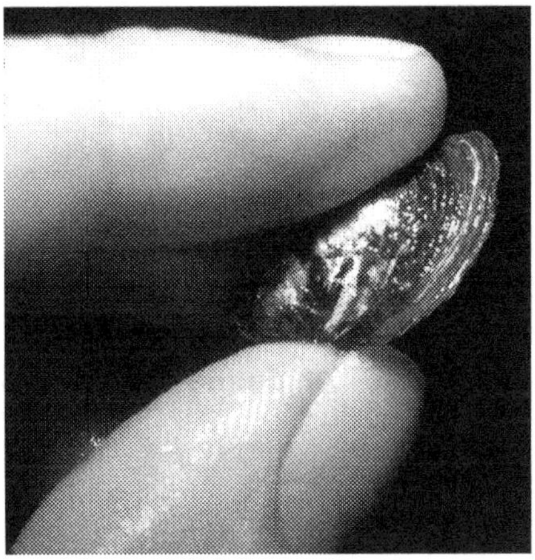

The ZM grows to about 1 inch in length, and a healthy female can produce up to 1.5 million veligers, or free-swimming larvae, in a single season.

Water temperature plays a very important role in the life-cycle of the ZM, specifically the growth rate of larvae [Mackie and Schloesser 1996]. In addition, the time from the development of a fertilized egg to a completely formed juvenile has been found to vary inversely with water temperature [Nichols 1996]. Finally, available data suggests that water temperature is a major factor in triggering the release of larvae, with larval production stopping once water temperatures drop below 10—12 °C. In warm water areas, such as thermally heated reservoirs or laboratory conditions, larval production has been documented year round [Nichols 1996], as long as the water does not drop below 12 °C.
Fertilization of ZM has been found to be impossible below 10 °C. Data suggests that spawning may start at 12 °C, but is maximized above 17—18 °C. Limited spawning may in fact occur below 17—18 °C [McMahon 1996]. Appearance of larvae during winter suggest that larvae were actually produced in the fall and that their development was delayed or very slow due to the colder temperatures [Nichols 1996].

The average life span of ZM is 3—5 years in most Polish lakes, 5 years in British waters, and 6—9 years in some Russian reservoirs. In North America, most ZM populations have a life span of 1.5—2 years [Mackie and Schloesser 1996].

Most studies indicate that ZM larvae must attach to hard substrate in order to complete metamorphosis and survive as juveniles, although some studies disagree [Nichols 1996].

The attachment mechanism has been the subject of much research. The ability to attach to a substratum exists by the pediveliger stage, although only a weak, single thread is formed during this time. The gland at the base of the foot that forms the byssus threads can also produce an enzyme that dissolves these threads [Nichols 1996].

The ZM attach to a surface by producing two types of threads: temporary and permanent. The threads differ in length, thickness, number, arrangement, and plaque morphology. Permanent threads make up the majority of a byssal mass. Permanent threads are formed in clumps or are arranged in rows and makeup the majority of a byssal mass. Temporary threads are few (1—6) arranged in tripod pattern, originate individually, and are separated spatially from the main byssal mass of permanent threads. Temporary threads are secreted before or after the initial byssal mass attachment. Adult mussels secrete about 23 threads/week. The threads are composed of DOPA (3,4-dihydroxyphenyl-alanine) [Mackie and Schloesser 1996].

The ZM is a very efficient filter feeder, being able to ingest particles that are sub micron in size. Any unpleasant material is eliminated through what is called pseudofeces production [MacIsaac 1996].

The ZM is a "leaky" organism, with a high paracellular permeability to solutes. That is, molecules present in the water being filtered rapidly move within the cell of the ZM. Work by Dietz has suggested that molecules as large as 5,000 Da (Daltons) may leak into a ZM [Dietz et al. 1996].

Freshwater bivalves, unlike the marine bivalves, have limited free amino acids in their body fluids and must rely on inorganic ions for osmotic regulation. The ZM has blood that is primarily a sodium chloride (NaCl) fluid with rather small amounts of bicarbonate ion (HCO_3^-) [Dietz et al. 1996].

Impact of Zebra Mussels: The impact of the ZM can be separated into three important areas. These include the impact on industry, the impact on other aquatic life, and the impact on the water.

Impact on Industry: Many industries are dependent on raw water from various lakes and rivers throughout the United States for cooling their process equipment. The industries most impacted by the ZM include electric power generation stations, public and private drinking water treatment plants, industrial facilities, navigation lock and dam structures, and other facilities throughout much of the eastern half of North America.

They have plugged inlet bays of once through cooling systems; they have caused pump failures; and they have blocked piping in a variety of facilities. Most notable are the electric power generating stations which rely on cooling to provide for a high degree of reliability.

Early experience in North America with infestation of these animals suggested that their fouling developed more rapidly and was more severe than that reported in Europe.

The economic impact of the zebra mussel invasion of North America in terms of the costs for repair, replacement and control has been estimated to be several billion dollars annually [Roberts 1990].

Impact on Other Aquatic Life: ZM infestation results in greatly diminished phytoplankton biomass production [MacIsaac 1996], and such infestations have resulted in wipeout of some populations of other bivalves in some locales [Schloesser et al. 1996].

Impact on Water Quality: Because the ZM is such an efficient water filter, one major effect of ZM infestation is enhanced water clarity [MacIsaac 1996].

Effect of Water Chemistry on Growth: Several water chemistry factors have been found to affect ZM growth. These include salinity, calcium hardness level, pH, and oxygen.

Salinity: The ZM can live in brackish water where the salinity does not exceed 8—12 ppt (parts per thousand). At higher temperatures (18—20 °C), the optimum salinity for adult mussels is about 1 ppt, but at lower temperatures (3—12 °C), the optimum salinity is 2—4 ppt [Mackie and Schloesser 1996].

pH: In one study, the minimum pHs at which ZM grow were found to be 6.5 for adults and 7.4 for veligers [McMahon 1996]. In other work, one researcher found 100% mortality in water having a pH less than 7.1, with mortality decreasing with increasing pH above 7.1, resulting in one group suggesting that infestation levels of abundance probably will not occur until pH exceeds 7.5—8.0 [Mackie and Schloesser 1996].

Calcium: Work suggested that the successful growth of ZM requires a water with a calcium concentration greater than 12 mg/L [McMahon 1996]. Others suggested that infestations will not occur unless the calcium levels exceed 15—20 mg/L [Mackie and Schloesser 1996].

Oxygen: Studies indicate that ZM are relatively intolerant of a lack of oxygen [McMahon 1996].

Dispersal Mechanisms: Dispersal mechanisms are not well understood. It appears that the planktonic veliger stage would be the most effective natural dispersal phase, but the planktonic veliger is not usually the first life stage found in a newly established mussel population. Juvenile and adult mussels can possibly be dispersed by extending mucous threads up to 1.5 cm long from the surface of the water by capillary action and drifting with surface water currents. Smaller mussels translocate at higher rates than larger mussels. Human mediated dispersal mechanisms are believed to be the most common means of distributing ZM, especially between unconnected water bodies. Live mussels attached to boats on trailers have been found as far west as Los Angeles, CA [Mackie and Schloesser 1996].

There appears to be a diurnal movement of veligers, with maximum densities occurring near the surface during early morning and at 5—7 meter depth during the day [Mackie and Schloesser 1996].

The largest problem with infestation by ZM does not appear to be settlement by veligers, but translocation by juveniles. Thus chemical treatments which target spawns is largely ineffective. A year round intermittent approach should be used. [Burton 1996; McMahon 1996]

Reproduction – Spawning: In North America, ZM become sexually mature in their first year of life, usually by 8—10 mm in shell length, but maybe as low as 5 mm. ZM are very prolific, generating from 30,000 to 1,610,000 eggs/female [Mackie and Schloesser 1996]. Some work suggests that slower growing ZM take longer to become sexually mature. ZM are sequential spawners, with each mussel releasing larvae over a period of 6—8 weeks [Nichols 1996].

Zebra mussels are dioecious, i.e., a zebra mussel has both male and female organs, and fertilization occurs externally. Therefore, coordinated maturation and spawning processes are important for successful reproduction [Ram et al. 1996].

Most of the data available for southern populations indicate that spawning does not begin until the water temperature exceeds 20 °C [Nichols 1996]. Mussels carry ripe larvae for long periods of time even though production may not be occurring. Larval production has been correlated to seasonal changes in water temperatures [Nichols 1996].

Researchers have found that spawning can be initiated by the addition of certain chemicals. In one work, the hypothesis was promoted that algae contain chemicals that can stimulate ZM spawns [Ram et al. 1996]. One chemical that was found to trigger spawning of ZM is sereotonin. Serotonin is a biogenic amine which is implicated in spawning behavior and can reliably trigger spawning [Ram et al. 1996].

Serotonin stimulates spawning in both male and female ZM by injection or by external application in surrounding aquarium water, at levels as low as 10^{-4} M and above. As little as a 5 minute application of serotonin is enough to trigger a complete spawning episode. In addition, spawning has been characterized by several serotonin-like compounds [Ram et al. 1996]. These include the chemicals shown in Table 13-1.

Table 13-1. Chemicals Which Trigger Spawns in ZM	
8-OH-DPAT	2-methylserotonin
TFMP	alpha-methylserotonin
ergotamine - most effective initiator of spawning; in females 10^{-5}M and in males 10^{-6}M	

Adaptability: Several books have been written about the ZM, but by the time they were published, much of the material was found to be dated and erroneous, primarily due to the ability of the ZM to adapt to their environment.

Recent work suggests that larvae are capable of acclimatizing to local conditions to a much greater extent than previously considered possible or that there has been natural selection for larvae with increased tolerances of elevated ambient temperatures or lower calcium levels [Nichols 1996].

In a warmer region where water temperatures are not below 10—12 °C for much of the year (when ZM eggs stop developing) and adult mussels seem to be adapting to warmer temperatures, larval production could easily continue throughout most of the year [Nichols 1996].

ZM have been shown to be able to adapt to higher salinities [Ram et al. 1996].

Because of this adaptability, many expect the ZM to eventually colonize the majority of permanent inland freshwater in the US and southern Canada [McMahon 1996].

B. Conventional Treatment Strategies for Control of the Zebra Mussel

A variety of approaches have been developed to control infestation by the ZM. These approaches include the following:

> **Mechanical Cleanings:** This approach is to let the mussels accumulate, and clean the system mechanically periodically. The downside of this approach is that it is expensive, infestations may hurt plant operations, odors are created as the ZM decompose, and the shells may be declared as a hazardous waste. The advantages are that the plant can be assured of the effectiveness of treatment, no permits are required, and no chemicals are required.

Coatings - This approach is to apply a ZM resistant or toxic coating to the surface of the system. The downside of this approach is that normally a clean, dry surface is required, this approach is expensive, and the toxic ingredient may leach into the water.

Heat - This approach involves periodic application of steam to select locations in the inlet bay or other parts of the plant to raise the temperature and kill the ZM throughout the system. The downside of this approach is the capital required to make plant modifications to allow this kind of treatment and that thermal discharge is regulated in many places, and so it may not be viable. The advantages are that it is less costly than some approaches.

Chemical Treatments - Generally these can be classed into either oxidizing or non-oxidizing.

Oxidizing - These include chlorine gas, bleach, chlorine (or bleach) with bromide ion, chlorine dioxide, and permanganate.

Non-Oxidizing - These include various poisons, i.e., quaternary ammonium compounds or surface active agents to poison the ZM. A few naturally occurring materials taken from plant extracts have also been used. Table 13-2 shows most of the commonly used non-oxidizers used, along with the companies which promote them.

Treatment Using ClO_2: There are two strategies required for the treatment of Zebra Mussels with ClO_2. These two strategies, along with the types of treatments belonging to each, are listed as follows:

1. **Killing and Removal of ZM after Attachment.** This method is important because the majority of the zebra mussels found in a given system are not the result of veligers settling after spawns but are in fact translocators. These translocators are typically juveniles which have settled elsewhere and because of rains, floods, upset conditions, or other reasons, have been moved by nature or have themselves cut their byssal threads to move to another location. This activity is not based on any specific event such as a spawn, and so timing a mitigation-type treatment has not been shown to be effective. An annual or twice-yearly mitigation treatment is required for removal of these animals.

2. **Prevention of ZM Settlement.** This routine type of treatment is necessary to minimize problems with ZM settlement during the year. It also makes the mitigation type treatment more reasonable with less impact on the plant and the environment. This settlement prevention type of treatment will not effectively eliminate mitigation treatment, unless this treatment is done year round, and even then, a mitigation treatment will likely be required.

The dosages required for mitigation, i.e., killing and removal after attachment, have been generally higher than those which would be required for prevention of settlement. Dosages required to prevent settlement are just high enough to irritate the organisms and cause them to move on. This type of treatment would be more environmentally friendly. This type of treatment would provide a means of keeping the surfaces free of ZM attachment, without any perceived negative environmental effects.

Table 13-2. Common Chemicals Used to Treat Zebra Mussels		
Name	**Supplier**	**Description**
Antimycin	Aquabiotics	isovaleric acid, 8-ester with 3-formamide-N-(7hexyl-3-yl) salicylamide
Bayluscicide	Mobay Corp.	2,5'-dichloro-4'-nitro-salicylanilide
Baythroid	Mobay Corp.	cyano(4-fluoro-e-phenoxy phenyl)methyl 3-(2,2-dichloroetheny)-2,2'-di-methyl-cyclopropanecar-carboxylate (pyrethrum)
Bulab 6002	Buckman	poly[oxyethylene-(dimethyl-imino) ethylene-(dimethyl-imino) ethylene dichloride] (polyquat ammon)
Bulab 6009	Buckman	2-(thiocyanomethylthio) benzothiazole
Clamtrol CT-1	GE Betz	n-alkyl dimethyl benzyl ammonium chloride and dodecylguanidine hydrochloride (polyquat ammon)
CuSO4	J.T. Baker	cupric sulfate, 5-hydrate
Endod	H.Lee, U Toledo	mixture of polycyclic algycone (sapogenin isolated from *Phytolacca dodecandra*
Juglone	Aldrich	5-hydroxy-1,4-naphthoquinone
KML V2	KML	chitosan
KML V54	KML	chitosan
Noxfish	Penick Corp	1,2,12,12a-tetrahydro-2-isopropenyl 8,9-dimethoxy-{1} benzopyrano-[3,4]furo[2,3-b][1] benzo pyran-6 (6aH) one (rotenone)
KCl	Fisher	potassium chloride
Salicylanilide	Aldrich	2-hydroxy-N-phenylbenzamide (salicylanilide)
TFM	Hoechts	3-trifluoro-4-nitrophenol
ClO2	Ashland, Vulcan	

C. Treatment of the Zebra Mussel with ClO_2 – A Literature Review

In general, a number of reports have found their way into the literature. The first report investigated the ability of oxidants, including ClO_2, to prevent settlement and to inactivate veligers [Fraleigh et al. 1991]. These researchers suggested that intermittent treatment would not likely be effective due to the ability of ZM to close up during treatment.

ClO_2 was also evaluated along with four other methods commonly used to control ZM [Khalanski 1993]. The report described the effectiveness of ClO_2, in that adult ZM were killed by a low-level treatment with ClO_2 (mean residual oxidant concentration of 0.15 mg/L) for 7 days with a water temperature of 15 °C.

In other work, ClO_2 was added at varying dosages and the time to achieve 100% mortality was recorded [Rusznak et al. 1994]. They found that feeding ClO_2 at dosages varying from 0.25—5.0 mg/L above oxidant demand produced 100% mortality ranging from 2.9—8.8 days at a temperature averaging 14.3 °C. In addition, ClO_2 at 0.25—0.50 mg/L gave a 100% mortality ranging from 1.6—3.7 days at an average temperature of 25.8 °C. Thus temperature was found to be a significant factor relating to ZM mortality. Once exposed to ClO_2, the ZM closed tightly and remained closed for as long as possible. This resulted in a lag time between introduction of ClO_2 and mortality.

Later work was done where ClO_2 was fed continuously at 0.15—0.5 mg/L above the oxidant demand [Smolik et al. 1995a]. When fed for 2-4 days at these dosages, ClO_2 effectively eradicated adult zebra mussels. These researchers also found that temperature was an important factor. On exposure to ClO_2, or any toxic chemical, ZM close up tightly for as long as possible. When forced to open to respire, they are exposed to the chemical and inactivated. The ability of the ZM to remain closed is a direct function of temperature, i.e., the higher the temperature, the shorter time they can remain closed.

In a later report, researchers found that treatment of veligers with 0.75—2.0 mg/L resulted in 16—42% mortality, based on 30 or 60 minute exposure times [Smolik et al. 1995b]. Also, results of an on site study indicated that an intermittent treatment of 0.25 mg/L ClO_2 produced a 95% reduction in post-veliger settlements.

In an application in a nuclear power plant on the eastern shore of Lake Michigan, settlement densities of < 500/m^2 were found to cause minimal problems [Freymark and Hale 1996]. They found that ClO_2, injected four times daily for 15 minutes at 0.25 mg/L, was able to reduce settlement of veligers by 95% and control settlement densities to an acceptable level.

In a later report, researchers found that a single 30 minute exposure to 20 mg/L or higher ClO_2 induces $\geq$ 50% mortality [Matisoff et al 1996]. Also, intermittent exposure for 30 minutes each day for 3—7 days at 10—20 mg/L ClO_2 was found to successfully control ZM. Intermittent exposures for 30 minutes each day for 28 consecutive days with 1—5 mg/L ClO_2 was ineffective (< 30 % mortality). And, continuous exposure to ClO_2 for 4 days produced > 50% mortality at > 0.5 mg/L and 100% mortality at > 1.0 mg/L.

In later work, an investigation was made as to the effect of ClO_2 on ZM veliger settlement [Mallen et al. 1997]. The results showed that feeding ClO_2 at 0.25 mg/L for 15 minutes, 4 times per day reduced veliger settlement by 95%, versus an untreated system and successfully controlled post-veliger mean densities to < 600 individuals / square meter .

D. Conclusions

Chlorine dioxide can be used effectively to control zebra mussels, both to eradicate adult and prevent settlement of veliger stages. One key appears to be the water temperature at which the treatment is applied. It is clear from the literature that the colder the water, the longer the mussels can remain closed. Increasing the water temperature causes mussels to open and close more frequently, thus forcing them to be exposed to the treatment.

E. Factors Limiting Use of ClO_2 for Zebra Mussel Treatment

Currently the factor most responsible for the lack of use of ClO_2 in the ZM market is the discharge limits placed on the chlorite ion. Although ClO_2 is very effective at controlling ZM, when ClO_2 reacts, from 50—80% of the ClO_2 reacts to form chlorite ion. In order for the ClO_2 to be effective, a residual of ClO_2 above the demand of the water (typically 2 mg/L or less) must be used. For a 2 mg/L feedrate, 1.0—1.6 mg/L chlorite ion would be formed and remain in the discharge water.

Work on the toxicity of chlorite to aquatic organisms has shown it to be quite toxic to Ceriodaphnia, a freshwater invertebrate commonly used in toxicity testing. As a result of this work, in some regions of the country a discharge limit of 0.005 mg/L has been placed on chlorite ion. Chlorite removal strategies have been relatively ineffective.

Other parts of the country have adopted a somewhat different approach. Total Residual Oxidant

(TRO) as measured by the colorimetric DPD method as an indication as to whether any oxidant is present or not. DPD has a number of interfering species. (DPD of Mississippi River water has shown ~ 2 mg/L without any oxidant added.) Some states have adopted their own policy, having been given primacy in this area by the EPA.

Until the past few years, only a few reports have been made regarding the investigation of the relative toxicity of the chlorite ion to freshwater aquatic species [Fisher and Burton 1993; Fisher and Burton 1995; Fisher and Burton 2000]. Because of the much fewer number of species of aquatic life that have been tested against the chlorite ion, and because the data has such a wide spread, the mathematical formula used to compute a "safe" level results in placing an extremely low value for the chlorite ion. A more detailed discussion of this topic is given elsewhere [Simpson and Phillips 1998; Simpson 2005].

Many scientists believe that the currently imposed limits on chlorite do not make sense from a scientific perspective [Burton 1996]. For example, in every toxicity study, chlorite ion has been shown to be less toxic towards aquatic life, in most cases far less (i.e. as much as three orders of magnitude less toxic) than chlorine. Yet the discharge limit placed on chlorite is the same as that imposed on chlorine, which exhibits relatively the same high toxicity towards aquatic life, irregardless of species.

As a result of this unrealistic constraint on the discharge of the chlorite ion, a toxicity assessment study was done [Fisher et al. 2003]. In this test, a large number of organisms were tested for toxicity and general guidelines were recommended based on the overall toxicity to aquatic life and not to just the toxicity towards one species. The results of this type of test have result in a more reasonable discharge limit for the chlorite ion.

F. Novel Approach [Dietz et al. 1995]

One possible approach to solving the zebra mussel problem, at a significantly reduced cost with less impact on the environment would be to couple chemical treatments.

Earlier in this chapter, the use of serotonin or one of its analogues was found to induce spawning. In fact, the zebra mussel, when exposed to serotonin, exhibited "exotic" behavior. When placed on a dry table top, the serotonin-exposed zebra mussel would open, the foot would come out, and it would push itself around the table. When placed in a 5% glutaraldehyde solution, the serotonin-exposed ZM would open and close, being totally oblivious that it was being exposed to a toxic chemical [Dietz 1996].

The leaky nature of the ZM plays an important role in this approach. Any chemical present in the bulk water to which an open ZM is exposed rapidly permeates the cell of a ZM. Thus, an open ZM, exposed to a sub mg/L concentration of serotonin or one of its analogues would, within a few minutes, begin to exhibit this exotic behavior and spawning would be induced.

Current treatment of the ZM with ClO_2 generally takes some time, because once the ZM has been exposed to a toxic chemical, the ZM can close up for some time depending upon water temperature. However, if a short dose of serotonin or one of its analogues were added for a few minutes, then those ZM which were open at a given time could be treated with a fairly high dose of ClO_2 for a very short period of time. The ZM would remain open and would be inactivated by the ClO_2.

While a significant amount of work remains to be done, the approach appears to be viable.

Several questions have to be answered.

1. What concentration of serotonin or its analogues would be required?
2. What is the environmental impact of serotonin or its analogues?
3. What concentration of ClO_2 would be required and for what duration?

Other questions would likely to be addressed. Still, treatment of a once through power plant at slightly above demand for 72 hours or so would require significantly greater time and manpower than treatment for 15 minutes to several hours at a higher dosage.

G. References

[Beckstrom 2005] Beckstrom, G., private communication, August 2005.

[Brooks et al. 1996] Brooks, G., Matisoff, G., and Bourland, B., "Toxicity of Chlorine Dioxide to Adult Zebra Mussels," *Journal of the American Water Works Association*, 88(8), 93(August, 1996).

[Brooks 1993] Brooks, G., "Treatment of Fresh Water for Zebra Mussel Infestation," US Patent 5,256,310, October 26, 1993.

[Burton 1996] Burton, D., private communication, University of Maryland, December, 1996.

[Claudi and Mackie 1994] Claudi, R. and Mackie, G. L., "Practical Manual for Zebra Mussel Monitoring and Control, Lewis Publishers, Boca Raton, FL, 227(1994).

[Dietz 1996] Dietz, T., private communication, Louisiana State University, October, 1996.

[Dietz et al. 1995] Dietz, T., Lynn, J., and Silverman, H., "Method for Controlling Bivalves such as Zebra Mussels," US Patent 5,417,987, May 23, 1995.

[Dietz et al. 1996] Dietz, T., Wilcox, S., Byrne, R., Lynn, J., and Silverman, H., "Osmotic and Ionic Regulation of North American Zebra Mussels (*Dreissena polymorpha*)," *American Zoologist*, 36(3), 364(June, 1996).

[Downing et al. 1995a] Downing, S., Smolik, N., Rusznak, L., Anderson, J., and Hale, L., "The Use of Chlorine Dioxide for Zebra Mussel Control - a Perspective of Treatment Histories," Third International Chlorine Dioxide Symposium, New Orleans, LA., September 15, 1995.

[Downing et al. 1995b] Downing, S., Rusznak, L., and Smolik, N., "Chlorine Dioxide - an Oxidizing Agent for Water Disinfection and Zebra Mussel Control," Third International Chlorine Dioxide Symposium, New Orleans, LA., September 15, 1995.

[Fisher and Burton 1993] Fisher, D., and Burton, D., "The Acute Effects of Continuous and Intermittent Application of Chlorine Dioxide and Chlorite on *Daphnia magna, Pimphales promelas,* and *Oncorhynchus mykiss,*" Report No. WREC-93-B4, University of Maryland System, Maryland Institute of Agriculture and Natural Resources, Wye Research and Education Center, Queenstown, MD 21658, August, 1993.

[Fisher and Burton 1995] Fisher, D. and Burton, D., "Determination of Acute Water Quality Criteria for Continuous and Intermittent Exposure of Chlorite for Freshwater Organisms," Report No. WREC-95-O3, University of Maryland System, Maryland Institute of Agriculture and Natural Resources, Wye Research and Education Center, Queenstown, MD 21658, June 1995.

[Fisher and Burton 2000] Fisher, D. and Burton, D., "Completion of the Acute Toxicity DataBase for Chlorite and Freshwater Organisms," WREC Report No 00-03, October 2000.

[Fisher et al. 2003] Fisher, D., Burton, D., Yonkos, L., Turley, S., Ziegler, G., and Turley, B., "Derivation of Acute Ecological Risk Criteria for Chlorite in Freshwater Ecosystems," *Water Research,* 37(18), 4359(Nov 2003).

[Fraleigh et al. 1991] Fraleigh, P., Matisoff, G., Gubanich, G., Hoffman, G., Klerks, P., McCall, P., and Wenning, M., "AWWA Zebra Mussel Research Project Results," Proceedings Annual Conference, American Water Works Association, 103(1991).

[Fraleigh et al. 1993] Fraleigh, P., Van Cott, W., Wenning, M., and DeKam, J., "Effects of Hypochlorite, Permanganate, Chlorine Dioxide and Chloramines on Zebra Mussel Settling," Proceedings: Third International Zebra Mussel Conference, Toronto, Ontario, Canada, February 23-26, 1993.

[Freymark and Hale 1996] Freymark, S. and Hale, L., "Tame Zebra Mussels at Lower Cost Using ClO2," *Power*, 140(1), 21(1996).

[Freymark and Hale 2000] Freymark, S. and Hale, L., "Process for Inhibiting the Settlement of Post-Veliger Zebra Mussels," US Patent 6,106,730, August 22, 2000.

[Garrett and Laylor 1995] Garrett, W. and Laylor, M., "The Effects of Low Level Chlorination and Chlorine Dioxide on Biofouling in a Once Through Service Water System." Proceedings Fifth International Zebra Mussel and Other Aquatic Nuisance Organisms Conference, Toronto, Canada, 135(February 1995).

[Germer 1994] Germer, M., "Zebra Mussel Control Using Chlorine Dioxide: Dairyland Power Coop, Alma WI," unpublished report, September 25, 1994.

[Germer 1995a] Germer, M., "Zebra Mussel Control Using Chlorine Dioxide: Dairyland Power Coop, Alma WI," unpublished report, October 3, 1995.

[Germer 1995b] Germer, M., "ClO_2 for Zebra Mussel Control: Dairyland Power, J.P. Madgett Plant," unpublished report, WI, September 10, 1995.

[Germer 1995c] Germer, M., "Zebra Mussel Control with ClO_2; Wisconsin Electric, Oak Creek Power Plant," unpublished report, 1995.

[Germer 1995d] Germer, M., "Three Year Study – Zebra Mussel Control with ClO_2; Wisconsin Electric, Point Beach Nuclear Power Plant," unpublished report, September 1, 1995.

[Germer 1996] Germer, M., "Two Year Study of Zebra Mussel Control Using Chlorine Dioxide: Dairyland Power Coop, Genoa, WI," unpublished report, October 10, 1996.

[Griese and Rosenblatt 1996] Griese, M. and Rosenblatt, A., "Chlorine Dioxide's New Phase: A Utility Case Study," 1996 Annual Conference Proceedings; American Water Works Association; Water Quality Conference, Toronto, Ontario, Canada, June 23-27, 1996.

[Holt and Ryan 1997] Holt, D. and Ryan, E., "Chlorine Dioxide--Adult Zebra Mussel Eradication at the Dofasco Steel Mill," Proceedings Seventh International Zebra Mussel and Aquatic Nuisance Species Conference, New Orleans, January 28-31, 1997.

[Khalanski 1993] Khalanski, M., "Testing of Five Methods for the Control of Zebra Mussels in Cooling Circuits of Power Plants Located on the Moselle River," Proceedings: Third International Zebra Mussel Conference, Toronto, Ontario, Canada, February 23-26, 1993.

[Klerks et al. 1993] Klerks, P., Fraleigh, P., and Stevenson, R., "Controlling Zebra Mussel (*Dreissena polymorpha*) Veligers with Three Oxidizing Chemicals: Chlorine, Permanganate, and Peroxide + Iron," Zebra Mussels: Biology, Impacts, and Control, T. F. Nalepa and D. W. Schloesser, eds., Lewis Publishers, Boca Raton, FL, 621(1993].

[MacIsaac 1996] MacIsaac, H., "Potential Abiotic and Biotic Impacts of Zebra Mussels on the Inland Waters of North America," *American Zoologist*, 36(3), 287(June, 1996).

[Mackie and Schloesser 1996] Mackie, G. and Schloesser, D., "Comparative Biology of Zebra Mussels in Europe and North America: An Overview," *American Zoologist*, 36(3), 244(June, 1996).

[Mallen et al. 1997] Mallen, E., Freymark, S., Hale, L., and Lippincott, B., "Use of Chlorine Dioxide to Prevent Zebra Mussel Settlement," Zebra Mussels and Aquatic Nuisance Species, F. M. D'Itri, ed., Ann Arbor Press, Chelsea, MI, 489(1997).

[Matisoff et al. 1991] Matisoff, G., Greenberg, A., Gubanich, G., and Ciaccia, J., "Effects of Potassium, Chloramine, and Chlorine Dioxide on Control of Adult Zebra Mussels," *Journal of Shellfish Research*, 11(1), 232(1992).

[Matisoff et al. 1996] Matisoff, G., Brooks, G., and Bourland, B., "Toxicity of Chlorine Dioxide to Adult Zebra Mussels," *Journal of the American Water Works Association*, 93(August 1996).

[McMahon 1996] McMahon, R., "The Physiological Ecology of the Zebra Mussel, *Dreissena polymorpha*, in North America and Europe," *American Zoologist*, 36(3), 339(June, 1996).

[McMahon 1996] McMahon, R., private communication, University of Texas at Arlington, December, 1996.

[Nichols 1996] Nichols, S., "Variations in the Reproductive Cycle of *Dreissena polymorpha* in Europe, Russia, and North America," *American Zoologist*, 36(3), 311(June, 1996).

[Nielson et al. 1993] Nielson, J., Greenburg, A. , Hoffman, G., and McNutt, R., "Zebra Mussel Abatement at Lake Erie Intake System for the City of Cleveland," Proceedings: Third International Zebra Mussel Conference, Toronto, Ontario, Canada, February 23-26, 1993.

[Ram and McMahon 1996] Ram, J. and McMahon, R., "Introduction: The Biology, Ecology, and Physiology of Zebra Mussels," *American Zoologist*, 36(3), 239(June, 1996).

[Ram et al. 1996] Ram, J., Fong, P., and Garton, D., "Psyiological Aspects of Zebra Mussel Reproduction: Maturation, Spawning, and Fertilization," *American Zoologist*, 36(3), 326(June, 1996).

[Roberts 1990] Roberts, L., "Zebra Mussel Invasion Threatens U.S. Waters," *Science*, 249, 1370(1990).

[Rusznak et al. 1994a] Rusznak, L., Mincar, G., and Smolik, N., "Chlorine Dioxide--a Molluscicidal Agent for Adult Zebra Mussel Eradication," Proceedings Fourth International Zebra Mussel Conference, Madison, WI, 127(March 1994).

[Rusznak et al. 1994b] Rusznak, L., Mincar, G., Smolik, N., "Chlorine Dioxide - A Molluscicidal Agent for Adult Zebra Mussel Eradication," 55th International Water Conference, 18(1994).

[Rusznak et al. 1994c] Rusznak, L., Mincar, G., Smolik, N., "Chlorine Dioxide - A Molluscicidal Agent for Adult Zebra Mussel Eradication," Ohio River Basin Consortium for Research and Education, Scientific Symposium, Huntington, WV, 1994

[Rusznak et al. 1995] Rusznak, L., Smolik, N., Hale, L., and Freymark, S. (1995). "Examination of the Potential of Chlorine Dioxide for Use in Zebra Mussel Veliger Control," Proceedings, Fifth International Zebra Mussel and Other Aquatic Nuisance Organisms Conference, Toronto, Canada, 371(February 1995).

[Rybarik et al. 1995] Rybarik, D., Byron, J., Germer, M., and Ackerman, J., "Chlorine Dioxide Treatment for Zebra Mussel Control," Proceedings Fifth International Zebra Mussel and Other Aquatic Nuisance Organisms Conference, Toronto, Canada, February 1995.

[Schalekamp, 1995] Schalekamp, M., "Zebra Mussels in Deep Lakewater- and River-Intakes. Actual Case Study on the Solution of the Problems," *Gas, Wasser, Abwasser*, 75(9), 744(1995).

[Schloesser et al. 1996] Schloesser, D., Nalepa, T., and Mackie, G., "Zebra Mussel Infestation of Unionid Byvalves (Unionidae) in North America," *American Zoologist*, 36(3), 300(June, 1996).

[Simpson 2005] Simpson, G., "The Environmental Impact of Chlorine Dioxide," Chapter 10 in <u>Practical Chlorine Dioxide: Volume I – Foundations</u>," Greg D. Simpson & Associates, Publishers, 2005.

[Simpson and Phillips 1998] Simpson, G. and Phillips, E., "The Environmental Impact of Chlorine Dioxide," Chemical Oxidation: Technologies for the Nineties, Eighth International Symposium, Nashville, TN, April 1-3, 1998.

[Smolik et al. 1995] Smolik, N., Rusznak, L., Anderson, J., and Hale, L. (1995). "The Use of Chlorine Dioxide for Zebra Mussel Control--A Perspective of Treatment Histories," Proceedings Fifth International Zebra Mussel and Other Aquatic Nuisance Organisms Conference, Toronto, Canada, February 1995.

[Smolik et al. 1995b] Smolik, N., Rusznak, L., Hale, L., and Freymark, S., "Examination of the Potential of Chlorine Dioxide for use in Zebra Mussel Veliger Control," 5th International Zebra Mussel Conference, Toronto, Ontario, Canada, 1995.

[Stein 1994] Stein, H., "Utilities Can Control Zebra Mussels ..., For Now," *Electric Light and Power*, 17(May 1994).

[Tsou et al. 1995] Tsou, J., Coyle, J., Pall, S., Mussalli, Y., and Martin, P., "Evaluation of the Use of Chlorine Dioxide to Control Zebra Mussels," Proceedings, Fifth International Zebra Mussel and Other Aquatic Nuisance Organisms Conference, Toronto, Canada, 1995.

[Van Benschoten et al. 1993] Van Benschoten, J., Jensen, J., Lewish, D., and Brady, T., "Chemical Oxidants for Controlling Zebra Mussels (*Dreissena polymorpha*): A Synthesis of Recent Laboratory and Field Studies." Zebra Mussels: Biology, Impacts, and Control, T. F. Nalepa and D. W. Schloesser, eds., Lewis Publishers, Boca Raton, FL, 599(1993).

Oxidation of Phenol with ClO_2

A. Introduction - The Problem with Phenol

Phenol is a benzene ring with an alcohol functional group attached (Figure 14-1). At high pH, the phenoxy anion forms, which is the chemically-reactive form.

Figure 14-1. Phenoxy Anion

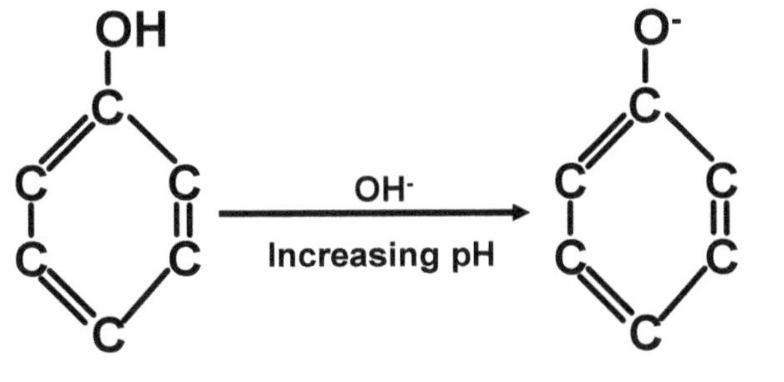

Phenol has a high toxicity to humans, with the symptoms of exposure including nausea, vomiting, circulatory collapse, paralysis, convulsions, coma, necrosis, and death from respiratory failure. Thought to be carcinogenic, the primary means of phenol exposure is by absorption through the skin.

At or above a concentration of 2 mg/L, phenol is also highly toxic to aquatic organisms, including fish. Phenol concentrations as low as 0.02 mg/L can impart an unpleasant taste to fish.

Because phenol has significant biocidal properties, it is difficult for bacteria to metabolize phenol (phenol is incorporated into several antibacterial formulations). Thus, high concentrations of phenol can be toxic to the bacterial populations in wastewater activated sludge plants. This means that there is an upper limit to the phenol concentration which can be present in the water going to the wastewater treatment plant (WWTP) without creating a highly toxic effluent and resulting in a massive kill of bacterial populations in the WWTP. Should this happen, the phenol-metabolizing capability of the WWTP bacteria can be lost and phenol along with other environmentally regulated compounds normally metabolized by bacteria can escape the WWTP and contaminate the receiving stream. Discharge violations can occur. This highly toxic nature of phenol towards bacteria means that phenol can persist in the environment for some time. For this reason, phenol is one the EPA's Priority Pollutants.

The term "phenol," as used in standard analytical procedures, refers not only to phenol, but also encompasses a range of compounds including substituted phenols, chlorophenols, nitrophenols, polyhydroxylate aromatics and phenoxyacids.

Phenol is a commonly found in the wastewater of many industrial plants. Because phenol is considered a priority pollutant, the levels which are tolerated in the effluent of industrial facilities are regulated by the EPA through a process called the National Pollutant Discharge Elimination System (NPDES). NPDES is a permitting program administered by the USEPA via state level environmental agencies. Through this program, point sources of pollution, defined as those pollutants which are discharged at discreet locations such as a pipe or channel overflow, are regulated.

The NPDES discharge guidelines established by the EPA for phenol are categorized by industry. Thus, established limits for phenol discharge can vary by industry. A list of some of the industries that discharge phenols is shown in Table 14-1. Although limits vary somewhat by industry, EPA effluent phenol guidelines generally establish phenol levels of 0.1 mg/L for the Best Practical Control Technology and 0.02 mg/L for the Best Available Control Technology.

Table 14-1. Industries That May Have Phenolic Discharges		
Plastics	Oil Refineries	Wood Preserving
Gas Works	Chemical Plants	Foundries
Coke Plants	Resin Manufacturers	Fiberboard
Explosives	Auto Plants	Sheep/Cattle Dip
Glass Production	Coal Gasification	Insecticides
Photography	Aircraft Maintenance	

B. Common Methods for Destruction or Removal of Phenol

The numerous methods that have been proposed for the destruction or removal of phenol can be separated into three broad categories: physical methods, biological methods, and chemical methods. The choice of method and the resulting economics will be dictated by a combination of factors, the most important of which are the phenol concentration, the wastewater flow rate, the presence of other chemical species present in the wastewater, and the goal of treatment, i.e, whether the treatment objective is gross reduction of high levels of phenol or removing the last traces of phenol. It is common to use a combination of methods to control phenol.

Physical Methods: Of the factors controlling the choice of the method used to remove or destroy phenol, the concentration of phenol is the most important. Physical methods may be used to recover high concentrations of phenol as a usable product.

In the case of a wastewater with a phenol concentration above about 500 mg/L, phenol recovery might be done by either solvent extraction or adsorption on granular activated carbon (GAC).

Solvent Extraction. One common method used to recover phenol is to extract the phenol into a recoverable solvent. In the solvent extraction process, the wastewater is acidified to convert the phenoxy anion to the neutral, undissociated phenol, which is then extracted into an organic solvent. The phenol is then separated from the solvent, and the solvent recovered, leaving several mg/L phenol in the water after extraction. Solvent extraction can recover 95—98% of the phenol.

Granular Activated Carbon (GAC). GAC adsorption can be used to remove phenol at concentrations of up to about 300 mg/L. A series of GAC units can reduce the phenol concentrations to 1—5 mg/L. The adsorbed phenol is removed from the exhausted GAC with caustic and the GAC regenerated thermally. The high capital and operating costs (up to several dollars per 1000 gallons of solution) limit this choice to streams with high phenol levels. GAC adsorption can exceed 98% recovery.

Bacterial Methods for Phenol Destruction: Bacterial destruction of phenol is another choice requiring a high capital investment. In spite of its toxicity, bacteria can be acclimated to handle phenol concentrations approaching 500 mg/L.

In some wastewater plants that receive periodic spikes of high phenol concentration, one successful approach is to supplementally dose phenol-acclimated bacteria to the system. Such bacteria might be fed to the wastewater to encourage growth of phenol-oxidizing bacteria in the WWTP and to provide some buffering capacity when slugs of phenol are received into the wastewater plant. Irregardless, spikes in phenol concentration exceeding 500 mg/L, can create a toxic environment to the activated sludge bacteria, and they become sick or die. When this happens, the plant can exceed its NPDES permit for a number of regulated species, as the bacterial population, which normally converts these species into innocuous by-products, has been killed or inhibited by the phenol.

Phenol concentrations of 200—300 mg/L can be reduced by bacterial action to 0.5—1.0 mg/L. Shock loads above 500 mg/L must be avoided. Conditions conducive to bacterial health must be promoted, such as a stable pH range of 6—9.5, a stable temperature between 65 °F and 95 °F, and the presence of the right concentration of nutrients such as nitrogen and phosphorous, which must be supplied if not naturally present in the wastewater. Shocks to the system, such as sudden pH swings, must be avoided. Oxygen is required, and is normally supplied by diffusion into the water, or by mechanical aeration. This process requires about 24 hours for the bacteria to destroy the phenol.

Chemical Oxidation of Phenol: In wastewater streams with a phenol concentration of less than about 5 mg/L, where other bacterial toxins are present or the phenol concentration must be reduced to below about 1.0 mg/L, biological treatment or GAC adsorption are not good choices. Chemical removal is the best option, and is usually accomplished by use of the chemical oxidants such as potassium permanganate ($KMnO_4$), hydrogen peroxide (H_2O_2), or chlorine dioxide.

The scenarios where chemical oxidation is appropriate are:

- **Batch treatment:** Chemical oxidation can be used to treat a small batch of wastewater high in phenols if it is cheaper than incineration or biological treatment.

- **Pretreatment before activated sludge plant:** Chemical oxidation can be used as a pretreatment before biological treatment to remove toxins from a high-phenol waste.

- **Polishing:** Chemical oxidation can be used as a final polishing step after other treatment or for dilute wastes low in other organics.

Potassium Permanganate: The equation for the oxidation of phenol by permanganate was reported to be that shown below:

$$3C_6H_5OH + 28KMnO_4 + 5H_2O \Rightarrow 18CO_2 + 28KOH + 28MnO_2 \qquad (14\text{-}1)$$

According to this balanced equation, the theoretical weight for the oxidation of phenol by $KMnO_4$ was stated to be 15.7 lb $KMnO_4$ per lb phenol, but 6—7 lb have been found to be effective [Vella and Munder 1991]. The reaction proceeds favorably in 7—10 pH range.

Potassium permanganate has some features which make it an attractive choice, including complete oxidation of phenols, simplicity of operation, minimal installation costs, and rapid, complete reaction. Thus $KMnO_4$ has been widely used for phenol oxidation.

While an attractive choice, the benefits of $KMnO_4$ come with a price. The final product of permanganate oxidation is manganese dioxide (MnO_2), which is a water-insoluble, brown, hydrous sludge. In some applications, the sludge aids in water clarification, but the use of $KMnO_4$ can also create a disposal and handling problem, in that increased color and suspended solids may adversely impact discharge of the treated water.

Hydrogen Peroxide: Although hydrogen peroxide will oxidize phenols, its action is too slow in most cases without the aid of a catalyst. Ferrous sulfate is often used as a catalyst with hydrogen peroxide, the iron-catalyzed peroxide blend being referred to as Fenton's chemistry or Fenton's reagent.

The ferrous iron catalyst, when applied with peroxide under carefully controlled conditions of pH and concentration, releases the hydroxyl radical, $OH^{-}\cdot$, which is an extremely reactive oxidant that will attack all but the most refractory organics.

The use of Fenton's reagent is limited to a narrow pH range, requiring tight pH control. The ideal pH is above 3 but lower than 5. Below 3 the reaction becomes inefficient. As the pH increases above about 4, the reaction slows. This is especially true with substituted phenolics, which react even more slowly than does pure phenol. For example, in the oxidation of 2,4 dichlorophenol, 1 mole equivalent of peroxide with 25 mole % iron catalyst oxidizes 26% of 2,4 dichlorophenol in 5 minutes. Five mole equivalents with 25 mole % iron catalyst achieved complete oxidation, but required 90 minutes to do so. By contrast, 1 mole equivalent of chlorine dioxide, at pH 4 (which is not the ideal pH for use of this oxidant) oxidizes 83% of 2,4 dichlorophenol in 5 minutes. A 5 mole equivalent ratio achieved complete oxidation in less than 1 minute.

One of the difficulties is the very low pH and extreme corrosivity of ferrous solutions to pumping equipment.

C. Literature Review – Chlorine Dioxide for Phenol Oxidation

Phenol destruction by chlorine dioxide is one of the most studied reactions in chlorine dioxide chemistry. In short, ClO_2 reacts with phenol to form a variety of organics, the composition of which and the speed of which depends upon both the pH under which the reaction is carried out and the ratio of ClO_2 to phenol.

In an early report, an overview of ways to treat phenolic type waste water materials was provided [Lanouette 1977]. In this work, phenolic compounds were defined to be compounds containing one or more hydroxyl groups attached to aromatic rings. Lanouette reported that common analytical methods for measuring phenol included some substituted phenols but not others.

The report indicated that when determining the oxygen demand of phenol, COD should be measured instead of BOD, as phenol is toxic to the bacteria added to the sample in the BOD test. If the BOD test is used however, care should be taken to insure that the bacteria used have been acclimated to phenol.

Chlorine dioxide oxidized phenol in a pH dependent fashion, with the requirements being 1.5 lb ClO_2 per lb of phenol at pH < about 8, 3.3 lb ClO_2 per lb of phenol at pH > about 10, and from 1.5—3.3 lbs ClO_2 per lb phenol in the intermediate pH range.

In another report, researchers used high-performance liquid chromatography (HPLC) to identify and quantify reaction products of dilute solutions of ClO_2 with phenols [Brueggemann et al. 1982]. Researchers prepared various standards of substituted phenols at concentrations of about 10^{-6} M. The concentration of phenol was 6.11×10^{-6} M (~ 0.6 mg/L).

With phenol in excess, chromatograms of unbuffered solutions (no pH given) showed a slow increase in the concentration of hydroquinone, and some chlorinated phenols. For reaction mixtures with a stoichiometric amount of phenol with ClO_2 or with ClO_2 in excess, no chlorinated products were found, and p-benzoquinone was the major product and was formed within a few seconds.

Several interesting observations were made by others [Wajoh et al. 1982]. First, when added to a phenol solution, ClO_2 disappeared very rapidly. In addition, chlorinated phenols formed slowly but no chlorinated quinones were formed, and then only in solutions containing phenol in stoichiometric excess. They also found that benzoquinone appeared within seconds, but not hydroquinone. Surprisingly, another undetermined oxidant species was found to be present for several minutes after mixing. Also 0.5 moles of chlorite per mole of ClO_2 was formed.

In another report [Hoigne and Bader 1984], researchers found phenolic compounds, when present as phenoxy anions, reacted with ClO_2 10^6 times faster than the protonated species. The apparent reaction rate constant increases for most phenolics by a factor of 10 per pH unit.

Russian workers [Belevzev and Maximenko 1985] found that at pH 5 and a ratio of 1.2:1 ClO_2 to phenol on a wt:wt basis, the main reaction product was p-benzoquinone. At pH 7.5 and 1.5:1 ClO_2 to phenol, the main reaction product was found to be again p-benzoquinone. However, in alkaline media, benzoquinone formed rapidly but was converted into carboxylic acids at a ratio of 5 mg ClO_2/mg phenol. At pH 10, the benzoquinone formed was further converted into carboxylic acids, at a ratio of 3 mg ClO_2/mg benzoquinone.

These workers noted that biochemically treated refinery plant effluent contained hydrocarbons in a range of COD from 100—150 mg/L. Effective treatment of phenol in waste of refinery are oxidized only with large excess of ClO_2, with the maximum effect in alkaline media.

It must be pointed out that much of the research prior to about 1990 utilized chlorine dioxide which may have been contaminated with chlorine, as the analytical methods available to separate chlorine dioxide from chlorine were not in common use at that time. In fact, Wajon, et al., remarked on this difficulty [Wajoh et al. 1982]. So there are at times contradictory reports.

Various oxidants, including permanganate, iron-catalyzed peroxide, and ClO_2 were compared for wastewater applications by another group [Vella and Munder 1991]. In pure water, ClO_2 was the most flexible oxidant studied. It was not affected by the pH or other species such as calcium or phosphate and worked well at all pH ranges studied. It also required the lowest dose on a wt:wt basis than the other oxidants studied.

For actual wastewaters, ClO_2 and $KMnO_4$ were effective, but iron-catalyzed peroxide was no better or was ineffective.

One group reported results of treatment of a wastewater sample from a specific organics manufacturer which contained a variety of organics included various chlorinated phenols, nitrates, xylenes, aniline, and formanilide [Vella et al. 1992].

Substituted phenols were reacted with $KMnO_4$, iron-catalyzed peroxide, and ClO_2. The effectiveness of treatment was determined along with a Microtox study. Microtox testing involves use of a special strain of fluorescing bacteria. This sensitive bacteria is added to the sample and the concentration of bacteria is measured photometrically. The results of exposure of the bacteria to a water sample provide a measure of the toxicity of the sample. They showed $KMnO_4$ and iron-catalyzed peroxide to have low toxicity, while ClO_2 actually increased the toxicity for this sample.

D. Effect of Substituted Phenols

Although benzene is fairly non-reactive, certain electron-donating functional groups, when attached to the ring, destabilize it and make the ring more reactive. These electron-donating functional include the alcohol group (-OH) or amide group ($-NH_2$), along with others shown in Figure 14-2.

Electron-withdrawing functional groups, i.e, those that withdraw electrons from the ring, make the phenol less reactive. Some of these functional groups and the relative impact of each are shown in Figure 14-2.

Figure 14-2. Effect of Groups on Electrophilic Aromatic Substitution

Strongly activating
$-NH_2$ ($-NHR$, $-NR_2$)
-OH

Moderately activating
$-OCH_3$ ($-OC_2H_5$)
$-NHCOCH_3$

Weakly activating
$-C_6H_5$
$-CH_3$ ($-C_2H_5$)

Deactivating
Meta directing
$-NO_2$
$-N(CH_3)_3$
-CN
-COOH (-COOR)
$-SO_3H$
-CHO, -COR

Deactivating
Ortho, para directing
-F, -Cl, -Br, -I

When more than one group is attached to benzene ring, the effect is less clear. Therefore, for applications where substituted phenols are expected to be present, demand testing is highly recommended.

E. Case History

In one field application, a tank of 8500 gallons of chemical plant waste ([phenol] = ~ 150 mg/L, TOC = ~ 2500 mg/L) was treated with ClO_2 [Simpson 1996]. The waste had a layer of emulsified hydrocarbon floating on the wastewater. Approximately 6—10 lbs of ClO_2 was generated. Equipment limitations prevented an exact determination of the amount of ClO_2 produced. At pH of slightly less than 7, within an hour, phenols were reduced to around 2 mg/L. After several hours, the ClO_2 residual had been reduced to below detectible levels, and the phenol had actually increased slightly, indicating either inadequate mixing, or some of the phenol was leaching out of the organic layer on top of the water. Further testing over the next several days showed no benzoquinone present, and the presence of maleic acids. Actual results are shown in Table 14-2.

Table 14-2. Field Results

Date	Time	Phenol	TOC	ClO_2
7/11	5:00	85	2619	
7/11	14:00	150	2695	570
7/11	14:30	5.10	2295	
7/11	15:00	2.96	1759	133
7/11	16:00	2.38	1855	98
7/12	8:00	11.0		ND

In addition to 11 mg/L phenol, the sample tested on 7/12 contained <10 µg/L p-benzoquinone, 22 mg/L hydroquinone, and 6.4 mg/L maleic acid.

F. Summary

A concise summary of the reaction of phenol with ClO_2 under various conditions is shown in Table 14-3.

Table 14-3. Products formed from ClO_2-Phenol Reaction Under a Variety of Conditions

	pH < 6	6 < pH < 8	pH > 8
excess phenol	chlorinated phenols p-benzoquinone	chlorinated phenols quinones	quinones
stoichiometric	p-benzoquinone	non-chlorinated products p-benzoquinone	maleic acid oxalic acid other organic acids
excess ClO_2	p-benzoquinone	non-chlorinated products p-benzoquinone	maleic acid oxalic acid other organic acids

According to published literature, treatment to destroy phenol (all the way to organic acids) can be done if the following guidelines are used:

- If possible, the pH of the solution to be treated should be adjusted to an alkaline pH, i.e., a pH > 8, and preferably ≥ 10.
- The concentration of ClO_2 to phenol should be ~ 3:1 on a weight:weight ratio. If the solution includes organics or sulfides, the level of ClO_2 required to destroy phenol may be higher. A demand or treatability study should be done to determine the appropriate dosage.
- Sufficient time should be allowed for the reaction to go to completion (likely minutes). If there is a significant concentration of other ClO_2-reactive species in the sample, the time for the reaction to go to completion may be longer.

G. References

[Belevzev and Maximenko 1985] Belevzev, A. and Maximenko, Y., "Russian Study: ClO_2 Oxidation of Cyanides and Phenols," *Effluent and Water Treatment Journal*, 25(5), 171(1985).

[Brueggemann et al. 1982] Brueggemann, E., Wajoh, J., Wade, C., and Burrows, E., "Analysis of Unquenched Reaction Mixtures of Chlorine Dioxide and Phenols by Reversed Phase High Performance Liquid Chromatography," *Journal of Chromatography*, 237, 484(1982).

[Glabisz 1967] Glabisz, U., "Uber die Wirkung von Chlorioxide auf einwertige Phenole," *ChemischeTechik*, 19(6), 352(1967).

[Hoigne and Bader 1984] Hoigne, J. and Bader, H., "Kinetics of Reactions of Chlorine Dioxide in Water - I. Rate Constants for Inorganic and Organic Compounds," *Water Research*, 28(1), 45(1994).

[Lanouette 1977] Lanouette, K., "Treatment of Phenolic Wastes," *Chemical Engineering*, 99(October 17, 1977).

[Simpson 1996] Simpson, G., unpublished report, July 24, 1996.

[Throop 1975/77] Throop, W., "Alternative Methods of Phenol Wastewater Control," *Journal of Hazardous Materials*, 1, 319(1975/77).

[Vella and Munder 1991] Vella, P. and Munder, J., "Uses of $KMnO_4$, H_2O_2, and ClO_2 in

Wastewater Applications," Chemical Oxidation, Technology for the Nineties, First International Symposium, Vanderbilt University, Nashville, Tenn, February 20-22, 1991.

[Vella et al 1992] Vella, P., Munder, J., Patel, B., Veronda, B., "Chemical Oxidation: A Tool for Toxicity Reduction," Proceedings, 47th Purdue Industrial Waste Conference, Lewis Publishers, 73, 1992.

[Wajoh et al. 1982] Wajon, J., Rosenblatt, D., and Burrows, E., "Oxidation of Phenol and Hydroquinone by Chlorine Dioxide," *Environmental Science and Technology*, 16(7), 396(1982).

Cyanide Removal with ClO_2

A. Cyanide – What It is and What Problems Does It Cause

What is Cyanide: Cyanide ion, $C{\equiv}N^-$, is a very stable ion with the carbon atom triple-bonded to the nitrogen atom. The cyanide is an extremely good complexing agent, due to the pair of electrons attached to the nitrogen, which makes it readily available to form a coordination compound with metal ions. Therefore, it is frequently found complexed to metals. In this chapter, the term cyanide refers to both simple cyanide (free cyanide) and complexed cyanides (cyanides bound to metals).

In aqueous solution, simple cyanides exist as a mixture of the cyanide ion and hydrocyanic acid, the relative abundance of each form depending upon pH. As hydrocyanic acid has a pKa of about 9.2, in most natural waters the HCN form predominates.

Cyanides complexed with metal ions display considerable variation in bond strength. For example, iron-cyanide complexes are so stable that oxidation alone cannot break the bonds. Cyanides complexed to other metal compounds such as silver, copper and others can be oxidized. The following shows the relative strength of the cyanide-metal bond, and thus the relative ease of oxidizing the cyanide.

Co > Fe > Mn, Cr, Ni > Au, Ag > Cu, Zn, Cd
Relative ease of oxidation ⇨

The cyanide ion is most commonly known as a poison. Potassium cyanide (KCN) or sodium cyanide (NaCN), when added to water, releases the poisonous hydrogen cyanide gas (HCN). The poison used in Nazi gas chambers during the Holocaust, Zyklon B, was prussic acid (hydrogen cyanide or hydrocyanic acid). Cyanide is also the compound used in U.S. gas chambers in those states which utilize this method of execution.

The Problem with Cyanide: In addition to hydrogen cyanide being a highly poisonous gas, both simple and complexed cyanide are very toxic to aquatic organisms in the aqueous phase [Doudoroff and Katz 1950; Doudoroff 1956; Doudoroff et al. 1966; Doudoroff 1976]. The toxicity of most complex cyanides results from the dissociation of the complexes and the HCN that is formed, as HCN is more toxic than the cyanide ion [Doudoroff 1956; Doudoroff 1976].

Its mechanism of inactivation is to shut down cellular respiration. It interrupts the electron transport chain in the inner membrane of the mitochondrion because it is a much stronger metal chelating agent than oxygen, i.e., it binds more strongly than oxygen to the Fe^{3+} (ferric iron ion) in cytochrome a_3, preventing this cytochrome from combining electrons with oxygen. Cyanide does not bind well to ferrous hemoproteins, such as hemoglobin, the mechanism that makes carbon monoxide toxic, contrary to popular belief.

Because it is so toxic, cyanide is highly regulated in the outfall of industrial plants. In addition, because various metal complexes of cyanide can decompose as a result of photolysis, even those that are generally very stable and resistant to oxidation, such as cyanide complexes of iron, are included [Burdick and Lipschuetz 1948].

Cyanide is also regulated in drinking water. The Safe Drinking Water Act, passed by Congress in 1974, requires the EPA to determine safe levels of various chemical compounds in potable water which have the potential to be harmful. These non-enforceable levels, based solely on possible health risks and exposure, are called Maximum Contaminant Level Goals. The MCLG for cyanide was set at 0.2 mg/L.

Sources of Cyanide: Cyanides are produced in some refining and petrochemical processes. It is desirable from an environmental perspective to destroy cyanides, as they are quite toxic to aquatic organisms. As a result, cyanide limits in the outfall of industrial plants are quite low. Cyanide is normally taken care of in biological treatment systems. In plants that have biological treatment systems, the cyanide levels are toxic to the bacteria above a certain level. Thus even in facilities which have biological treatment systems, cyanide can be a problem.

Due to the high solubilities of cyanides of gold and silver, cyanide is used in mining, electroplating, metallurgy and jewelry for chemical gilding, buffing, and extraction of gold [Noack and Iacoviello 1992].

Cyanides can be produced by certain bacteria, fungi, and algae, and are found in a number of foods and plants. Hydrogen cyanide is a component of vehicle exhaust gas and tobacco smoke. Some burning plastics release hydrogen cyanide, sometimes resulting in cyanide poisonings of victims in house fires.

Cyanides are used in the chemical industry, particularly in production of plastics. Cyanides are used in photographic development and in the fumigation of ships.

Cyanide production in the late 1980s and early 1990s was roughly 5 billion pounds per year. The major environmental sources of cyanide are discharges from metal finishing industries, iron and steel mills, and organic chemical industries.

Health Effects of Aqueous Cyanide Exposure: People exposed to cyanide in potable water above the 0.2 mg/L concentration for relatively short periods of time, cyanide can experience rapid breathing, tremors and other neurological effects. People exposed to cyanide in potable water over long periods of time may experience weight loss, thyroid effects, and nerve damage.

B. Common Methods for Cyanide Removal

The most commonly used method of cyanide oxidation is by use of alkaline chlorination. It is the most economical and easily controllable method for cyanide removal. The alkaline oxidation of cyanide by chlorine can be thought of as occurring in two distinct steps. In the first step, chlorine reacts with cyanide to form cyanogen chloride (CNCl) as an intermediate which forms cyanate ion (CNO^-) at the elevated pH. During the second step, cyanate ion is further oxidized to carbon dioxide and nitrogen.

Step 1 (Equation 15-1) is slow below a pH of about 8, with the reaction kinetics increasing as a function of pH. At a pH of about 8.5, the reaction goes to completion in fewer than 30 minutes. As the pH increases, the time for this reaction to go to completion decreases.

Cyanogen chloride (CNCl) is a toxic, volatile, and odorous lachrymator (an irritating gas that causes tearing, like tear gas). Since the solubility of cyanogen chloride increases with pH and the rate of reaction dramatically increases with the pH, the first step is usually conducted at pH 11.

$$NaCN + NaOCl \Rightarrow NaCNO + NaCl \qquad (15\text{-}1)$$

Step 2 involves the oxidation of cyanate ion (CNO-) to carbon dioxide and nitrogen. This step shows a reverse pH dependency, with decreasing pH resulting in a more rapid reaction. That is, at a pH of about 10, the reaction usually requires about 40 minutes, while at pH of about 8.5, the reaction is usually complete in about 10 minutes.

$$2NaCNO + 4NaOH + 3Cl_2 \Rightarrow 2H_2O + 6NaCl + 2CO_2 + N_2 \qquad (15\text{-}2)$$

The theoretical dosage requirements for the destruction of cyanide by alkaline chlorination is shown in Table 15-1.

Table 15-1. Theoretical Dosages Required for Destruction of Cyanide Ion by Alkaline Chlorination

Reaction Step	Cl_2 (mg/L)	NaOH (mg/L)	typical pH
Step 1. CN^- to CNO^-	2.73	1.13	~11
Step 2. CNO^- to CO_2 and N_2	4.09	3.08	~8.5
Total	6.82	5.21	

While the reaction may be thought of as a process with two distinct steps, and with the rate of Step 2 being slow conditions which are optimum for Step 1, in practice the reaction does not appear to occur in two distinct steps. For example, some cyanate is converted to carbon dioxide and nitrogen at the elevated pH of Step 1. The stoichiometry shows that 40% of the total chlorine used would be consumed in Step 1 and 60% in Step 2. In practical applications the amount of chlorine consumed in Step 1 is 70—80% of the total chlorine demand. Therefore, a significant portion of the cyanate has been oxidized to carbon dioxide and nitrogen in Step 1. In addition, the actual amount of chlorine consumed by the reaction may be twice the theoretical amount.

While alkaline chlorination is by far the most commonly used method for destruction of cyanide, other methods have been used. Such methods include alternative oxidants such as ozone, and peroxygens such as hydrogen peroxide, peroxymonosulfuric acid (H_2SO_5, made from the reaction of hydrogen peroxide and sulfuric acid and commonly called Caro's acid), and sodium percarbonate. Other non-chemical methods include electrochemical oxidation, and thermal oxidation. In addition, some methods employ cyanide removal by precipitation with ferrous sulfate.

C. Literature Review – Use of ClO_2 for Oxidation of Cyanide

Only a few studies have been reported that describe both laboratory and field studies for the destruction of cyanide by ClO_2.

In one early report, Russian researchers discussed the results of the use of ClO_2 to oxidize simple cyanides [Lur'e and Belevtsev 1969]. They reported the reaction to proceed instantaneously according to Equation 15-3.

$$CN^- + 2ClO_2 + 2H_2O \Rightarrow CNO^- + H^+ + 2ClO_2^- \qquad (15\text{-}3)$$

In another report, the author suggested that the reaction of ClO_2 with cyanide proceeded according to Equations 15-4, with the production of cyanate ion [Hall 1981].

$$CN^- + 2ClO_2 + 2OH^- \Rightarrow CNO^- + H_2O + 2ClO_2^- \qquad (15\text{-}4)$$

Complex cyanide of copper was reported to react according to Equation 15-5.

$$5[Cu(CN)_3]^{-2} + 7ClO_2 + 12OH^- \Rightarrow 15CNO^- + 5Cu(OH)_2 + 7Cl^- + H_2O \qquad (15\text{-}5)$$

Cyanide Removal with ClO_2

The reaction of the complexed copper cyanide complex was reported to occur in two steps. In the first step, the ClO_2 reacted to form the chlorite ion, which then proceeded to react catalytically. In this reaction the ClO_2 was found to make use of its full oxidation capacity, accepting five electrons and forming chloride ion.

Another researcher suggested that the oxidation stoichiometry of simple cyanide by ClO_2 showed considerable variations with pH (Equations 15-6 to 15-8) [Wood 1983].

$$CN^- + 2ClO_2 + 2OH^- \Rightarrow NCO^- + 2ClO_2^- + H_2O \quad \text{Basic} \quad (15\text{-}6)$$
$$5CN^- + 2ClO_2 + 2OH^- \Rightarrow 5NCO^- + 2Cl^- + H_2O \quad \text{Neutral} \quad (15\text{-}7)$$
$$2CN^- + 2ClO_2 \Rightarrow 2CO_2 + N_2 + 2Cl^- \quad \text{Acidic} \quad (15\text{-}8)$$

Other researchers suggested that the oxidation of cyanide ion by any oxidant was a 2-step reaction, illustrated by Equations 15-9 and 15-10 [Belevzev and Maximenko 1985]. In Step 1, the cyanide ion was oxidized to cyanate ion. The cyanate ion was then reported to hydrolyze forming carbonate and ammonium ions in Step 2.

$$\text{Step 1} \quad CN^- - 2e^- + 2OH^- \Rightarrow CNO^- + 2H_2O \quad (15\text{-}9)$$
$$\text{Step 2} \quad CNO^- + 2H_2O \Rightarrow CO_3^{-2} + NH_4^+ \quad (15\text{-}10)$$

They showed that the reaction of both simple and complex cyanides of zinc and cadmium with ClO_2 were described by Equations 15-11 and 15-12.

$$CN^- + 2ClO_2 + 2OH^- \Rightarrow CNO^- + 2ClO_2^- + H_2O \quad (15\text{-}11)$$
$$[Me(CN)_4]^{-2} + 8ClO_2 + 10OH^- \Rightarrow 4CNO^- + Me(OH)_2 + 8ClO_2^- \quad (15\text{-}12)$$

Cyanide bound to copper was found to follow a different reaction pathway (Equation 15-13). In addition, although the copper cyanide complex is more stable that that of zinc and cadmium, the reaction rate was almost five times that of the reaction of ClO_2 with zinc and cadmium cyanide complexes.

$$[Cu(CN)_3]^{-2} + 7ClO_2 + 8OH^- \Rightarrow 3CNO^- + Cu(OH)_2 + 7ClO_2^- + 3H_2O \quad (15\text{-}13)$$

In this reaction scheme, chlorine dioxide is reduced to chlorite ion stoichiometrically. The chlorite continues to react catalytically (Equation 15-14).

$$5[Cu(CN)_3]^{-2} + 7ClO_2^- + 12OH^- \Rightarrow 15CNO^- + 5Cu(OH)_2 + 7Cl^- + 3H_2O \quad (15\text{-}14)$$

In this case, ClO_2 makes use of all of its oxidation capacity, accepting 5 electrons overall in the reaction. These results are in good agreement with previous results [Hall 1981].

Not all possible complexes of copper with cyanide showed the catalytic effect, although the presence of copper in the zinc and cadmium solutions did show the catalytic effect of copper.

Cyanide complexed to iron could not be oxidized by ClO_2, and complexes with nickel and cobalt were very resistant to oxidation by ClO_2. In fact, the cobalt complex is so stable that it is impossible to liberate CN- quantitatively [Leschber 1968].

In a more recent report, workers described the novel application of the oxidation of cyanide in a hydrocyclone reactor by use of gaseous ClO_2 [Parga and Cocke 2001]. They found that efficiencies of simple cyanide oxidation approached 99% at all pH values in 5 minutes. The data suggests that gaseous ClO_2 may be significantly more potent as an oxidant in the gas phase when compared to aqueous phase ClO_2.

D. Summary

Simple cyanide is easily oxidized by ClO_2. However, cyanide is an extremely strong chelating agent, reacting readily with any available transition metal ions, forming very stable complexes, of which those containing iron and cobalt are extremely resistant to oxidation [Noack and Iacoviello 1992].

The reactions of chlorine dioxide with various cyanide complexes depend upon the relative ease of dissociating the cyanide ion from the complex. Several of the complexes of cyanide, including the complexes of cyanide with Ag^+, Cu^+, Zn^{+2}, Ni^{+2}, Pb^{+2}, and Cd^{+2}, dissociate relatively easily and so are easily destroyed by chlorine dioxide. The cobalt complex is not very important as it is not found in the effluent of petrochemical or electroplating plants. The iron complex, unfortunately, is ubiquitous in the wastewater of petrochemical plants. Cyanide comes from the processing units, and iron comes from the corrosion of the units, lines, and other places where iron is used in piping or process equipment. Despite the extreme stability of the iron complex, it is a regulated species because ultraviolet light can cause the degradation of the complex and cyanide can be released.

In general, if the cyanide is present as free cyanide or as one of the weakly bound complexes then ClO_2 may have applicability. Otherwise, if the cyanide is complexed to iron or cobalt, it is best to find another alternative.

E. References

[Belevzev and Maximenko 1985] Belevzev, A. and Maximenko, Y., "Russian Study: ClO_2 Oxidation of Cyanides and Phenols," *Effluent and Water Treatment Journal,* 25(5), 171(1985).

[Burdick and Lipschuetz 1948] Burdick, G. and Lipschuetz, M., "Toxicity of Ferro and Ferricyanide to Fish," *Transactions of the American Fisheries Society,* 78, 192(1948).

[Doudoroff 1956] Doudoroff, P., "Some Experiments on the Toxicity of Complex Cyanides to Fish," *Sewage and Industrial Wastes*, 28, 1020(1956).

[Doudoroff 1976] Doudoroff, P., "Toxicity of Fish to Cyanides and Related Compounds. A Review," EPA 600/3-76-038, U.S. EPA, 1976.

[Doudoroff and Katz 1950] Doudoroff, P. and Katz, M., "Critical Review of the Literature on the Toxicity of Industrial Wastes and their Components to Fish," *Sewage and Industrial Wastes*, 22, 1432(1950).

[Doudoroff et al. 1966] Doudoroff, P., Leduc, G., and Schneider, C., "Acute Toxicity to Fish of Solutions Containing Complex Metal Cyanides, in Relation to Concentrations of Molecular Hydrocyanic Acid," *Transactions of the American Fisheries Society*, 95, 6(1966).

[Hall 1981] Hall, M., "Cyanide Destruction by Chlorine Dioxide," unpublished report, December 9, 1981.

[Leschber 1968] Leschber, R., *Galvanotechnik*, 59(10), 823(1968).

[Lur'e and Belevsev 1969] Lur'e, Y. and Belevsev, A., *Ochistka Proizvodstvennylh Stochnvkh Vod*, 4, 56(1969).

[Noack and Iacoviello 1992] Noack, M. and Iacoviello, S., "The Chemistry of Chlorine Dioxide in Industrial and Wastewater Treatment Applications," Chemical Oxidation, Technology for the Nineties, Second International Symposium, Vanderbilt University, Nashville, Feb 19-21, 1992.

[Parga and Cocke 2001] Parga, J. and Cocke, D., "Oxidation of Cyanide in a Hydrocyclone Reactor by Chlorine Dioxide," Desalination, 140, 289(2001).

[Wood 1983] Wood, R., "*Industrial Waste Water Treatment Technology*," unpublished report, August 19, 1983.

More Frequently Asked Questions

A. Can ClO_2 Oxidize Bromide Ion?

The Problem with Bromide: Bromide in potable water can be oxidized by several disinfectants commonly used in potable water. If the water contains organics from decaying vegetation, the formation of brominated hydrocarbons can result, specifically bromine-containing trihalomethanes (THMs). THMs have been linked to increased incidences of stomach cancer [Page 1976; Kuzuma et al. 1977; Alvanja et al. 1977], and there has been some work to determine whether THMs have any relationship to spontaneous abortion [Waller et al. 1998]. Bromate, one of the disinfection by-products, is considered as a possible carcinogenic agent [World Health Organization 1986; World Health Organization 1987; World Health Organization 1993; Richardson 1993]. The importance of this topic deals with the health effects that the choice of oxidant makes in potable water. To explore this topic further requires an understanding of oxidation-reduction chemistry.

Oxidation-Reduction Chemistry of Chlorine and Bromine Species: Oxidation-reduction reactions are common in chemistry. Equation 16-A-1 shows the oxidation of ferrous ion (Fe^{+2}) to ferric ion (Fe^{+3}). In this reaction an oxidant (OX) pulls an electron from ferrous ion, which is oxidized to ferric ion, i.e., its charge increases (i.e., $Fe^{+2} \Rightarrow Fe^{+3}$). OX is reduced in the process, i.e., the charge on OX decreases (i.e., $OX \Rightarrow OX^-$).

$$Fe^{+2} + OX \Rightarrow Fe^{+3} + OX^- \qquad (16\text{-}A\text{-}1)$$

The standard half cell reactions and corresponding oxidation potential for some of the relevant chlorine and bromine species is shown in Table 16-A-1. The oxidation potential for many, if not all, of these species exhibits a strong pH dependency.

Table 16-A-1. Standard Half-Cell Potentials (E°) for Common Oxidants

Reaction	E°	Eq.
$HClO_2 + 3H^+ + 4e^- \Rightarrow Cl^- + 2H_2O$	1.57 V	(16-A-4)
$HOCl + H^+ + 2e^- \Rightarrow Cl^- + H_2O$	1.49 V	(16-A-5)
$HOBr + H^+ + 2e^- \Rightarrow Br^- + H_2O$	1.34 V	(16-A-6)
$ClO_2 + e^- \Rightarrow ClO_2^-$	0.95 V	(16-A-7)
$OCl^- + 2H_2O + 2e^- \Rightarrow Cl^- + 2OH^-$	0.9 V	(16-A-8)
$ClO_2^- + 2H_2O + 4e^- \Rightarrow Cl^- + 4OH^-$	0.78 V	(16-A-9)
$OBr- + H_2O + 2e^- \Rightarrow Br^- + 2OH^-$	0.77 V	(16-A-10)

Common oxidation-reduction reactions in water treatment include the oxidation of the bromide ion by hypochlorous acid, to form hypobromous acid (HOBr) as shown in Equation 16-A-2.

$$HOCl + Br^- \Rightarrow HOBr + Cl^- \qquad (16\text{-}A\text{-}2)$$

Oxidation reactions of chlorine are complicated by the pH dependency of the forms of chlorine (as the pH changes, the relative predominance of the oxochlorine species changes), because each of the species have markedly different oxidation potentials.

When chlorine dissolves in water, hypochlorous acid (HOCl) is formed (Equation 16-A-3). (Compounds formed that are not relevant to this discussion are not shown.) The hypochlorous

$$\mathbf{Cl_2} + H_2O \Leftrightarrow \mathbf{HOCl} \Leftrightarrow \mathbf{OCl^-} \qquad (16\text{-}A\text{-}3)$$

increasing pH ⇨

acid dissociates into hypochlorite ion (OCl^-) as the pH increases (Figure 16-A-1). At a pH of 7.5, approximately 50% of the chlorine is in the hypochlorous acid form, and 50% is in the hypochlorite ion form.

Bromine exhibits a similar reaction chemistry with water and a similar pH dependency, with the bromine species vs. pH curve shifted upwards by about 1 pH unit. Thus, at a pH of 8.5, about half of the bromine is in the form of hypobromous acid (HOBr), and about half is in the form of hypobromite ion (OBr^-).

Figure 16-A-1. Chlorine Species vs. pH

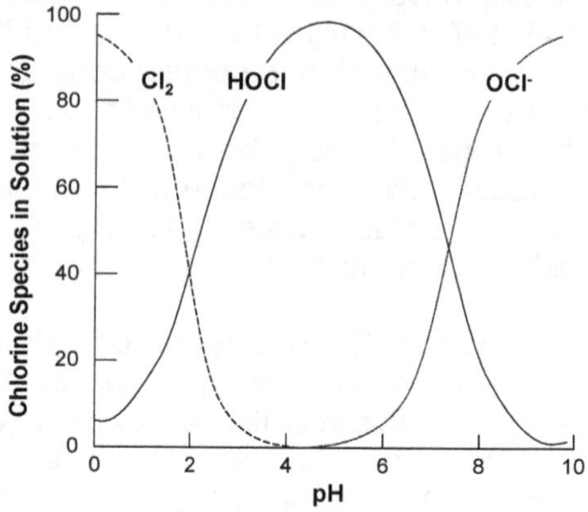

Oxidation of Bromide by Chlorine: Aqueous chlorine oxidizes bromide ion to HOBr. HOBr can react with naturally occurring organics to form significant levels of bromoform and other bromine-containing trihalomethanes. These naturally occurring organics include humic acid [Rook 1974; Bellar et al. 1974; Bean et al. 1978; Lange and Kawczynski 1978; Brauch et al. 1981; Cooper et al. 1985; Guttman-Bass et al. 1987; Grossman et al. 1999]. In addition, water containing bromide treated with chlorine in the presence of sunlight can form other brominated organics, specifically bromate ion (BrO_3^-) [Macalady et al. 1977]. Many of these brominated organics are known to be more toxic than their chlorinated counterparts [Cooper et al. 1985; Aieta and Berg 1986; Simmon and Tardiff 1978; Richardson et al. 2001].

Oxidation of Bromide by Ozone: Another common oxidation-reduction reaction is the oxidation of bromide ion by ozone. Ozone can oxidize the bromide ion to hypobromous acid (Equation 16-A-11). In a manner similar to that of hypochlorous acid, HOBr can dissociate in solution to give hypobromite ion, OBr^-.

$$O_3 + Br^- + H_2O \Rightarrow HOBr + Cl^- \qquad (16\text{-}A\text{-}11)$$

Ozone has also been found to oxidize Br^- to bromate ion (BrO_3^-) [Haag and Hoigne 1983; von Gunten and Hoigne 1992; von Gunten and Hoigne 1994].

Although ozone has not been shown to form THMs in water that does not contain bromide ion, if bromide is present, use of ozone in water containing naturally occurring organics results in THMs [Greenberg 1981; Rice and Gomez-Taylor 1986; Guttman-Bass et al. 1987]. The ability of chlorine and ozone to oxidize bromide and produce brominated THMs has raised the question as to whether ClO_2 will also oxidize bromide ion.

Oxidation of Bromide by ClO_2: Calculations and experimental data have suggested that the oxidation of bromide by ClO_2 is thermodynamically unfavored [Latimer 1952]. ClO_2 has not been shown, from theoretical considerations, laboratory studies [Hoigne and Bader 1994] or field experience, to oxidize bromide ion to hypobromous acid, under conditions of water treatment with bromide levels at even the highest bromide level found in natural waters (2—3 mg/L in the Sea of Galilee) [Richardson et al. 2001]. Neither has the production of brominated THMs been observed under similar conditions [Vilagines et al. 1977; Lange and Kawczynski 1978; Brauch et al. 1981; Hoigne and Bader 1982; Hoigne 1985; Cooper et al. 1983; Hoigne and Bader 1994]. Work has also shown that bromate ion is not produced by use of ClO_2 under conditions typically found in potable water treatment [Moll and Krasner 2003]. Since the pH range of natural waters generally falls into the pH range of abut 6 to about 8.5, the question arises as to whether ClO_2 might oxidize bromide ion at pH outside that range.

From consideration of the pertinent half-cell reactions, it does not appear that ClO_2 is a sufficiently strong oxidant to oxidize bromide ion to HOBr, as the oxidation potential for HOBr is greater than that of ClO_2. However, the oxidation potential of ClO_2 shows an interesting pH dependency (Figure 16-A-2). As the pH increases, the oxidation potential of ClO_2 decreases linearly as a function of the pH with a decrease of 0.062 per pH unit [Masschelein 1979]. This curve raises the question as to whether there might be a fairly narrow pH range where the oxidation potential for ClO_2 is greater than that of OBr^-.

Figure 16-A-2. Oxidation Potential of ClO_2 vs. Amount of OBr^-

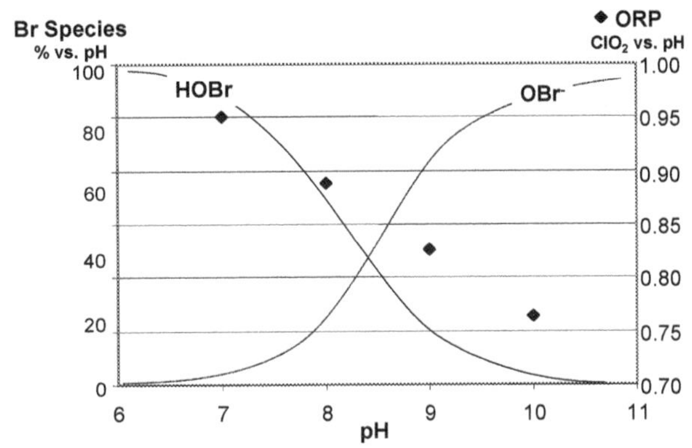

Two factors combine to raise the question as to the possibility of this reaction. These adverse effects include the decrease in the oxidation potential of ClO_2 with pH, and the relative predominance of hypobromite ion with increasing pH.

Given the pH dependency of the dissociation curve for HOBr, as the pH increases above about 8.5, hypobromite becomes the predominant form. But examination of the oxidation potential curve in Figure 16-A-2 shows that the oxidation potential for ClO_2 declines, so that at pH of about 9.5, the oxidation potential of ClO_2 declines below the oxidation potential for hypobromite ion, so at this pH the oxidation of bromide ion by ClO_2 becomes unfavorable. Still, between a pH of about 8, where the concentration of hypobromite form is very small as compared to HOBr, and a pH of 9.5, when hypobromite ion is the predominant form, might it be possible for ClO_2 to oxidize bromide to hypobromite ion? As this pH falls outside the range for all but the most extreme waters, this has not been observed.

From practical considerations, there are three possible circumstances under which ClO_2 may oxidize bromide ion.

1. The ClO_2 used to treat the water is contaminated by significant amounts of free chlorine. This happens with some generators that require excess chlorine to drive the oxidation of chlorite reaction [Vogt 1980] or some poorly tuned generators.

2. There is the possibility that photochemical reactions of water containing ClO_2 or ClO_2^- that also contains Br- may result in the formation of brominated THMs [Cooper et al. 1985; Zika et al. 1985; Gordon et al. 1995; Gordon et al. 1996].

3. The third possible exception is that if an acid-chlorite generator is used, chlorous acid is

formed which disproportionates to ClO_2. The disproportionation reaction can take a fairly long time (minutes to days) [Pitochelli 2005]. Chlorous acid has a standard oxidation potential (E°) of 1.56V. An oxidation potential of this magnitude would likely make the oxidation of Br^- to HOBr thermodynamically possible.

Oxidation of Chloride Ion by ClO_2: A similar argument could be made for chloride ion, raising the possibility that under some conditions, ClO_2 may oxidize chloride ion to hypochlorite ion, especially above pH of about 7. Although this has never been borne out under experimental work, even if it had been observed, the hypochlorite would likely have been attributed to a byproduct of ClO_2 generation.

Conclusions: No report has been found which shows experimentally that ClO_2 can oxidize bromide ion to hypobromite ion under conditions of water typically found in water treatment applications.

References:

[Aieta and Berg 1986] Aieta, E. and Berg, J., "A Review of Chlorine Dioxide in Drinking Water Treatment," *Journal of the American Water Works Association*, 78(6), 62(1986).

[Alvanja et al. 1977] Alvanja, M., Goldstein, I., and Susser, M., "Case Control Study of Gastrointestinal Cancer Mortality in Seven Selected New York Counties in Relation to Drinking Water Chlorination," Water Chlorination, Environmental Impact and Health Effects Conference, Gatlinburg, TN, 1977.

[Bean et al. 1978] Bean, R., Riley, R., and Ryan, P., "Investigations of Halogenated Components Formed From Chlorination of Marine Water," Water Chlorination: Environmental Impact and Health Effects, Volume 2, Jolley et al., eds., Ann Arbor Science, Ann Arbor, MI, 1978.

[Bellar et al. 1974] Bellar, T., Lichtenberg, J., and Kroner, R., "The Occurrence of Organohalides in Chlorinated Drinking Water," *Journal of the American Water Works Association*, 66(11), 703(1974).

[Brauch et al. 1981] Brauch, H., Baldauf, G., and Sontheimer, H., "Studies on the Formation of Organic Chlorine Compounds When Treating Humic Acid Waters with Chlorine Dioxide," *Sonderdruck aus Vom Wasser*, 56, 25(1981).

[Bunn 1975] Bunn, W., Haas, B., Deane, R., and Kleopfer, R., "Formation of Trihalomethanes by Chlorination of Surface Water," *Environmental Letters*, 10, 205(1975).

[Cooper et al. 1983] Cooper, W., Zika, R., and Steinhauer, M., "The Influence of Bromide Ion on the Formation of Trihalomethanes During Water Treatment: Evaluation of Alternative Oxidants and Research Needs," Third Quarterly Report of Cooperative Agreement CR 810277-01, USEPA, Cincinnati, OH, June 6, 1983.

[Cooper et al. 1984] Cooper, W., Zika, R., and Steinhauer, M., "The Effect of Bromide in Water Treatment: A Literature Review of Ozone and Bromide Interactions and the Formation of Organic Bromine," Proceedings International Ozone Association, Pretoria, South Africa, March 26-29, 1984.

[Cooper et al. 1985] Cooper, W., Zika, R., and Steinhauer, M., "Bromide-Oxidant Interactions and THM Formation: A Literature Review," *Journal of the American Water Works Association*, 77(4), 116(April 1985).

[Gordon and Bubnis 1995] Gordon, G. and Bubnis, B. "Bromate Ion Formation in Water When Chlorine Dioxide is Photolyzed in the Presence of Bromide Ion." Proceedings of the American Water Works Association, Water Quality Technology Conference, New Orleans, Louisiana, November 12-16, 1995.

[Greenberg, 1981] Greenberg, A., "Public Health Aspects of Alternative Water Disinfectants," *Journal of the American Water Works Association*, 31(January 1981).

[Grossman et al. 1999] Grossman, L, Idin, A., Relis, B, and Rebhun, M., "Formation of Cyanogen

Bromide and Other Volatile DBPs in the Disinfection of Bromide-Rich Lake Water," *Environmental Science and Technology*, 33, 932(1999).

[Guttman-Bass et al. 1987] Guttman-Bass, N., Balrey-Albuquerque, M., Ulltzur, S., Chartrand, A., and Rav-Acha, C., "Effects of Chlorine and Chlorine Dioxide on Mutagenic Activity of Lake Kinnereth Water," *Environmental Science and Technology*. 21(3), 252(1987).

[Haag and Hoigne 1983] Haag, W. and Hoigne, J., "Ozonation of Bromide-Containing Waters: Kinetics of Formation of Hypobromous Acid and Bromate," *Environmental Science and Technology*, 17, 261(1983).

[Hoigne 1985] Hoigne, J., "Organic Micropollutants and Treatment Processes: Kinetics and Final Effects of Ozone and Chlorine Dioxide," *Science of the Total Environment*, 47, 169(1985).

[Hoigne and Bader 1982] Hoigne, J. and Bader, H., "Kinetic Typischer Reaktionen von Chlorodioxid mit Wasserinhaltsstoffen," (Kinetics of Typical Reactions of Chlorine Dioxide with Substances in Water), *Vorn Wasser*, 59: 253(1982).

[Hoigne and Bader 1994] Hoigne, J. and Bader, H., "Kinetics of Reactions of Chlorine Dioxide in Water - I. Rate Constants for Inorganic and Organic Compounds," Water Research, 28(1), 45(1994).

[Kuzuma et al. 1977] Kuzuma, R., Kuzuma, D., and Buncher, C., "Ohio Drinking Water Source and Cancer Rates," *American Journal of Public Health*, 67(8), 725(1977).

[Lange and Kawczynski 1978] Lange, A. and Kawczynski, E., "Controlling Organics – The Contra Costa Water District Experience," *Journal of the American Water Works Association*, 70(11), 653(November 1978).

[Latimer 1952] Latimer, W., Oxidation Potentials, Prentice-Hall Inc., Englewood Cliffs, N.J., 1952.

[Masschelein 1979] Masschelein, W. J., Chlorine Dioxide: Chemistry and Environmental Impact of Oxychlorine Compounds, Ann Arbor Science, Ann Arbor, Michigan, 1979.

[Macalady et al. 1977] Macalady, D., Carpenter, J., and Moore, C., "Sunlight-induced Bromate Formation in Chlorinated Seawater," *Science*, 195, 1335(1977).

[Moll and Krasner 2003] Moll, D. and Krasner, S., "Information Collection Rule Data Analysis," Chapter 9, Bromate Occurrence in Disinfected Water," AWWRF, 225(March 2003).

[Nawrocki and Bilozor 1997] Nawrocki, J. and Bilozor, S., "Brominated Oxidation By-Products in Drinking Water Treatment," *Journal Water Supply Research and Technology – Aqua*, 46(6), 304(1997).

[Page et al. 1976] Page, T., Harris, R., and Epstein, S., "Drinking Water and Cancer Mortality in Louisiana," *Science*, 193, 55(1976).

[Pitochelli 2005] Pitochelli, A., private communication, October 2005.

[Rice and Gomez-Taylor 1986] Rice, R. and Gomez-Taylor, M., "Occurrence of By-Products of Strong Oxidants Reacting with Drinking Water Contaminants - Scope of the Problem," *Environmental Health Perspectives*, 69, 31(1986).

[Richardson 1993] Richardson, M., "Some Chemistry/Hazards of Halogen Biocides," *Environmental Management and Health*, 4(3), 6(1993).

[Richardson et al. 2001] Richardson, S., Thruston, A., Rav-Acha, C., Groisman, L., Popilevsky, I., Glezer, V., "Chlorine Dioxide DBPS: Overview and Results of Recent Work, " Proceedings of the Fourth International Symposium: Chlorine Dioxide: The State of Science, Regulatory, Environmental Issues, and Case Histories, Las Vegas, NV, 47(February 15-16, 2001).

[Rook 1974] Rook, J., "Formation of Haloforms During Chlorination of Natural Waters," *Water Treatment and Examination*, 23, 234(1974).

[Simmon and Tardiff 1978] Simmon, V. and Tardiff, R., "The Mutagenic Activity of Halogenated Compounds Found in Chlorinated Drinking Water," Water Chlorination, Environmental Impact and Health Effects, Jolley, R, Gorchev, H., Hamilton, D., eds, Ann Arbor Science, Ann Arbor, MI, 2, 417(1978).

[Vilagines et al. 1977] Vilagines, R., Monteil, A., Derreumaux, A., and Lambert, M, "A Comparative Study of Halomethanes Formation during Drinking Water Treatment by Chlorine or Its Derivatives in a Slow Sand Filtration Treatment Plant and in Waste Water Treatment Plants," Proceedings American Water Works Association Annual Conference, Anaheim, CA, May 1977.

[Vogt 1980] Vogt, C., "Controlling Trihalomethanes While Achieving Disinfection," American Water Works Annual Conference Annual Conference, Atlanta, GA, June 1980.

[von Gunten and Hoigne 1992] von Gunten, U. and Hoigne, J., "Factors Controlling the Formation of Bromate During Ozonation of Bromide-Containing Waters," *Journal Water Supply Research and Technology – Aqua*,, 41(5), 299(October 1992).

[von Gunten and Hoigne 1994] von Gunten, W. and Hoigne, J., "Bromate Formation During Ozonation of Bromide-Containing Waters: Interaction of Ozone and Hydroxyl Radical Reactions," *Environmental Science and Technology*, 28, 1234(1994).

[Waller et al. 1998] Waller, K., Swan, S., DeLorenze, G., and Hopkins, B., "Trihalomethanes in Drinking Water and Spontaneous Abortion," *Epidemiology*, 9, 134(1998).

[World Health Organization 1986] World Health Organization, International Agency for Research on Cancer, "Potassium Bromate," Lyon, 40, 207(1986).

[World Health Organization 1987] World Health Organization, International Agency for Research on Cancer, Lyon, 1987, 7, 70(1987).

[World Health Organization 1993] World Health Organization, Guidelines for Drinking Water Quality, 2nd Edition, Vol. 1. Recommendations, WHO, Geneva, 1993.

[Zika et al. 1985] Zika, R., et al., "Sunlight-Induced Photodecomposition of Chlorine Dioxide," Water Chlorination: Environmental Impact and Health Effects, Vol 5, Jolley, et al., eds. Ann Arbor Science, Ann Arbor, MI, 1985.

B. How is Chlorite Made?

A common objection to use of ClO_2 is the cost to produce it on site. The reason for this relatively high cost per pound is directly related to the cost of the precursor sodium chlorite. Sodium chlorite is significantly more expensive than the other precursors, for the simple reason that the cost to produce sodium chlorite is relatively high. The reason for this high cost is that in order to make sodium chlorite, ClO_2 has to be produced first, by some other method, and the ClO_2 is then converted to sodium chlorite by addition of a reducing agent, generally under alkaline conditions.

In the early manufacturing history of sodium chlorite, several methods were employed. In one published account [Holst 1950], ClO_2 was reduced by carbon powder which was dispersed in a solution of caustic soda and lime [Equation 16-B-1].

$$4ClO_2 + 4NaOH + Ca(OH)_2 + C \Rightarrow 4NaClO_2 + CaCO_3 + 3H_2O \qquad (16\text{-}B\text{-}1)$$

The $CaCO_3$ and any remaining carbon were filtered out, and the solution was evaporated to leave dry sodium chlorite (~ 80—88% on a dry weight basis).

In another process developed in the 40s, ClO_2 gas was reduced with zinc dust dispersed in water to produce zinc chlorite (Equation 16-B-2) [Gardiner and Karr 1946].

$$2ClO_2 + Zn \Rightarrow Zn(ClO_2)_2 \qquad (16\text{-}B\text{-}2)$$

The zinc chlorite was converted to sodium chlorite by reaction with caustic soda. The resulting solution was filtered, and dry sodium chlorite was made by evaporation of the water. Yield with this process was not acceptable.

A number of processes were patented over the years for production of sodium chlorite. Almost all involved adding reducing agents in alkaline media to ClO_2. Reducing agents included hydrogen peroxide [Levi 1922a; Levi 1922b], ammonia [Koninklijke 1947a], and metals, including silver, nickel, manganese, copper, cobalt, iron, and gold [Koninklijke 1947b].

The common method today for manufacture of sodium chlorite is to produce ClO_2 gas from a large-scale chlorate-based generator, dissolving the ClO_2 into a very cold aqueous solution, reducing the pH to allow removal of molecular chlorine, then reducing the ClO_2 using hydrogen peroxide in alkaline solution [Kaczur and Cawlfield 1992; Aieta and Roberts 1984].

The reaction chemistry is given by Equation 16-B-3.

$$2ClO_2 + H_2O_2 + 2NaOH \Rightarrow 2NaClO_2 + 2H_2O + O_2 \qquad (16\text{-}B\text{-}3)$$

References

[Aieta and Roberts 1984] Aieta, E. and Roberts, P., "The Chemistry of Oxo-Chlorine Compounds Relevant to Chlorine Dioxide Generation," Fifth Conference on Water Chlorination: Environmental Impact and Health Effects, Williamsburg, VA, June 3-8, 1984.

[Gardiner and Karr 1946] Gardiner, W. and Karr, E., Department of Commerce, OTS Report, PB 33218 (1946).

[Holst 1950] Holst, G., "Production of Sodium Chlorite: Survey of Prevalent Manufacturing Methods," *Industrial and Engineering Chemistry*, 42(11), 2359(November 1950).

[Kaczur and Cawlfield 1992] Kaczur, J. and Cawlfield, D., "Chlorous Acid, Chlorites, and Chlorine Dioxide," Kirk-Othmer Encyclopedia of Chemical Technology, 4th edition, 5, 968(1992).

[Koninklijke 1947a] Koninklijke, N., Dutch Patent 60,345, December 15, 1947.

[Koninklijke 1947b] Koninklijke, N., Dutch Patent 60,346, December 15, 1947.

[Levi 1922a] Levi, G., Atti accad. Lincei, 31(I), 371(1922).

[Levi 1922b] Levi, G., Gazz. Chim. Ital., 52(II), 56(1922).

[White 1949] White, J., "Chlorites and Chlorine Dioxide," Kirk-Othmer Encyclopedia of Chemical Technology, 3, 696(1949).

C. How Can ClO_2 Be Inactivated?

ClO_2 can be inactivated in a number of ways. Most of these will result in ClO_2 forming the chlorite ion, which itself may require removal. Reduced sulfur compounds are known to react readily with ClO_2. However, the reaction chemistry is not too straightforward, in that, because of the many stable oxidation states for both sulfur and chlorine, many side reactions are possible. As a result, the overall reaction may not be a single reaction, but a sum of several possible reactions.

In addition, the reaction of many reduced sulfur compounds results in a moderate amount of acid production, and some alkalinity reduction may result. This factor may be important in the selection of the reducing agent to be used.

Sodium Sulfite, Na_2SO_3 (formula weight = 126) is commonly used to remove oxidants, including ClO_2. Because sodium sulfite is a white hygroscopic powder, sodium sulfite is normally used as an aqueous solution. The predominant reaction appears to be:

$$5Na_2SO_3 + 2ClO_2 + H_2O \Rightarrow 5Na_2SO_4 + 2HCl \qquad (16\text{-}C\text{-}1)$$

Potassium Sulfite, K_2SO_3 (formula weight = 158) is commonly used in water treatment formulations. It is more commonly used in aqueous form and the chemistry is analogous to that shown in Equation 16-C-1.

Sodium Thiosulfate, $Na_2S_2O_3$. It is not widely used as a dechlorinating agent because its reaction with chlorine is slow. Several reaction pathways are possible with one commonly used reaction shown in Equation 16-C-2.

$$5Na_2S_2O_3 + 8ClO_2 + 9H_2O \Rightarrow 10NaHSO_4 + 8HCl \qquad (16\text{-}C\text{-}2)$$

Sodium Bisulfite, $NaHSO_3$. Available as a powder or solution (up to 44%), this acidic chemical is compatible with stainless, PVC, or Fiberglass. As the odor is significant and harsh, use of this chemical is not recommended. The reaction is:

$$5NaHSO_3 + 2ClO_2 + H_2O \Rightarrow 5NaHSO_4 + 2HCl \qquad (16\text{-}C\text{-}3)$$

Sulfur Dioxide, SO_2 (formula weight 64) is a gas which can be injected into a stream. When SO_2 is dissolved in water, it reacts with water to form sulfurous acid, as shown in Equation 16-C-4. All forms of chlorine oxidant are removed. Use of sulfur dioxide is common in very large applications due to its low cost.

$$SO_2 + H_2O \Rightarrow H_2SO_3 \qquad (16\text{-}C\text{-}4$$

The H_2SO_3 can then react with ClO_2, as shown in Equation 16-C-5.

$$5H_2SO_3 + 2ClO_2 + H_2O \Rightarrow 5H_2SO_4 + 2HCl \qquad (16\text{-}C\text{-}5)$$

The sum of reactions 16-C-4 and 16-C-5 is shown in Equation 16-C-6.

$$5SO_2 + 2ClO_2 + 6H_2O \Rightarrow 5H_2SO_4 + 2HCl \qquad (16\text{-}C\text{-}6)$$

How Can ClO_2 Be Inactivated?

Hydrogen Peroxide, H_2O_2 (formula weight 34), is used in the production of chlorite from ClO_2 (Equation 16-C-7).

$$2ClO_2 + H_2O_2 + 2OH^- \Rightarrow 2ClO_2^- + O_2 + 2H_2O \qquad (16\text{-}C\text{-}7)$$

The reduction in alkalinity by various reducing agents producing acidic reaction products when reacted with ClO_2 is shown in Table 16-C-1. It should be clear that in low alkalinity waters, the choice of reducing agent may have a dramatic impact on pH.

Table 16-C-1. Summary: Dosages Required to Inactivate ClO_2 and the Alkalinity Reduction Which Results

Reducing Agent	Formula	~ mg/L per mg/L ClO_2	~mg/L alkalinity reduction per mg/L ClO_2
Sodium sulfite	Na_2SO_3	4.7	2
Potassium sulfite	K_2SO_3	5.9	2
Sodium thiosulfate	$Na_2S_2O_3$	1.5	5
Sodium bisulfite	$NaHSO_3$	3.9	8.3
Sulfur dioxide	SO_2	2.4	9
Hydrogen peroxide	H_2O_2	0.3	-

D. What Is the Difference Between Bulk Water Demand and System Demand?

What is Demand? Demand is the dosage requirement for a given oxidant, chlorine, bromine, ClO_2, ozone, etc., which must be met in order for a measurable residual to be achieved after a specified time. This type of demand is called Bulk Water Demand.

Figure 16-D-1. Demand

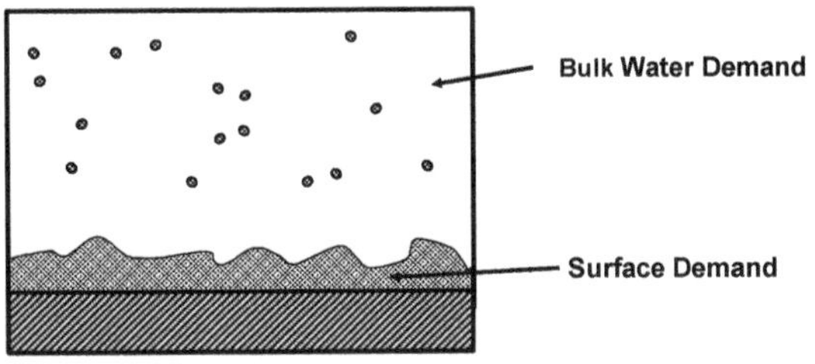

Total demand = Bulk Water Demand + Surface Demand

There is another type of demand in a system, Surface Demand, which is a measure of the demand for oxidant that lies on system surfaces. Total Demand, then, is the sum of Bulk Water Demand and System Demand.

For most oxidants, which are not very effective on biofilm, this point is not very significant. For ClO_2, this point is very significant, as it demonstrates one major difference between ClO_2 and other oxidants.

Bulk Water Demand: Bulk water demand can be measured. An oxidant is dosed into a water sample, and, after some period of time, the residual is measured [Simpson 2005].

System Demand: System demand can be thought of primarily as a measure of biofilm. This cannot easily be measured. To illustrate the importance of this factor, the following anecdotal stories are provided.

1: A Texas municipality used ClO_2. Prior to implementation of ClO_2, when chlorine was dosed into the water, as soon as that water would get to the end of their 26-mile distribution system, a chlorine residual could be measured. When chlorine was replaced by ClO_2 for disinfection, 4—6 months passed before a residual could be measured at the end of the distribution system. Those involved attributed this to the demand for ClO_2 created by biofilm on distribution pipe walls.

2: In a large Gulf Coast olefin plant cooling system, continuous chlorination was used as the primary method for microbiological control. A good residual of chlorine was measurable at every point in the system. Still, several exchangers required backwashing daily to control process water temperatures. In addition, water had to be sprayed on the outside of one of these exchangers to help cool the process.

ClO_2 feed was implemented in place of chlorine for microbiological control. Even though chlorine residuals were present throughout the plant, once ClO_2 was implemented, two weeks passed before a ClO_2 residual could be measured at the tower return. By this time, required exchanger backwashing was virtually eliminated, as was the water spray on the outside of the one exchanger.

What is the Difference Between Bulk Water Demand and System Demand?

These two anecdotal stories are somewhat extreme examples of phenomenon that is observed in virtually every cooling water field trial. In most cases, the phenomenon is not as extreme, but it is there nevertheless.

Discussion: Chlorine is known to be relatively ineffective at controlling slime attached to surfaces [LeChevallier et. al. 1990; de Beer et. al. 1994]. Chlorine tends to react with biofilm constituents; it does not penetrate the slime layer, but works only at the surface.

A continuous feed of chlorine is common in large industrial cooling towers. For fairly clean cooling towers, continuous feed of chlorine is sufficient to provide adequate control of bacterial biofilm. Contaminated towers generally have a significant growth of biofilm, depending upon the type and nature of bacterial nutrient. In such a case it is possible to have a measurable concentration of chlorine in the water, and yet significant amounts of biofilm may be present. In extreme cases, it may not be possible to achieve a free chlorine residual [Pickrell et al. 1999]. This phenomenon is well supported in the literature [Seidler et al. 1977; Nagy et al. 1982; Kozelski 1983; LeChevallier et al. 1988; Boivin et al. 1992; Chen et al. 1993; de Beer et al. 1994; Stewart et al. 1994; Hang et al. 1995; Chen and Stewart 1996].

Chlorine dioxide has been proven to control biofilm very well, even when fed intermittently [Mayack et. al. 1984]. When applied correctly, ClO_2 controls biofilm effectively under a variety of conditions [Simpson et. al. 1993]. Because it is relatively non-reactive with most biofilm constituents, it readily penetrates the biofilm and inactivates biofilm bacteria. It has been shown in numerous cases to have superior biofilm-removing capability.

References

[Boivin et al. 1992] Boivin, J., Shapka, R., Khoury, A., Blenkinsopp, S., and Costerton, J., "An Old and a New Method of Control for Biofilm Bacteria," Paper No 171, Corrosion 92, 1992.

[Chen and Stewart 1996] Chen, X. and Stewart, P., "Chlorine Penetration into Artificial Biofilm is Limited by a Reaction-Diffusion Interaction," *Environmental Science & Technology* 30(6), 2078(1996).

[Chen et al. 1993] Chen, C., Griebe, T., and Characklis, W., "Biocide Action of Monochloramine on Biofilm Systems of *Pseudomonas aeruginosa*," *Biofouling*, 7, 1(1993).

[de Beer et al. 1994] de Beer, D., Srinivasan, R., and Stewart, P., "Direct Measurement of Chlorine Penetration into Biofilms during Disinfection," *Applied and Environmental Microbiology*, 60(12), 4339(1994).

[Hang et al. 1995] Huang, C., Yu, F., McFeters, G., and Stewart, P., "Nonuniform Spatial Patterns of Respiratory Activity within Biofilms during Disinfection," *Applied and Environmental Microbiology*, 61(6), 2252(June 1995).

[Kozelski 1983] Kozelski, K. J., "Field Experience with a Simple Cooling Water Biofilm Monitoring Device," TP IWC-83-46, International Water Conference, Pittsburgh, PA, October 24-26, 1983.

[LeChevallier et al. 1988] LeChevallier, M., Cawthon, C., and Lee, R., "Inaction of Biofilm Bacteria," *Applied and Environmental Microbiology*, 54(10), 2492(October, 1988).

[LeChevallier et al. 1990] LeChevallier, M., Lowry, C., and Lee, R., "Disinfecting Biofilms in a Model Distribution System", *Journal of the American Water Works Association*, 87, July 1990.

[Nagy et al. 1982] Nagy, L. A. Kelly, A.,Thun, M., Olson, B., "Biofilm Composition, Formation and Control in the Los Angeles Aqueduct System," Proceedings American Water Works Association, WQTC, Nashville, TN, 1982.

[Pickrell et al. 1999] Pickrell, W., Germer, M., Miller, J., and Simpson, G., "Plants Increase Production Through Use of Chlorine Dioxide," First International Conference: Practical

Industrial Water Treatment Technology for the New Millenium, Houston, Texas, January 27-29, 1999.

[Seidler et al. 1977] Seidler, R., Morrow, J., and Bagley, S., "Klebsielleae in Drinking Water Emanating From Redwood Tanks," *Applied and Environmental Microbiology*, 33, 893(1977).

[Simpson 2005] Simpson, G., Chapter 8, *Practical Chlorine Dioxide: Volume I – Foundations*, Greg D. Simpson & Associates, Publishers, 157(2005).

[Stewart et al. 1994] Stewart, P., Griebe,T., Srinivasan, R., Chen, C.-I. Yu, F., DeBeer, D., and McFeters, G., "Comparison of Respiratory Activity and Culturability during Monochloramine Disinfection of Binary Population Biofilms," *Applied and Environmental Microbiology*, 60(5), 1690(May1994).

What is the Difference Between Bulk Water Demand and System Demand?

E. What Concentrations of Sodium Chlorite Are Available Commercially?

Table 16-E-1. North American Chlorite Producers

Manufacturer	**ERCO (formerly Sterling)**
Address	**ERCO Worldwide** 302 The East Mall, Suite 200 Toronto, Ontario, Canada M9B 6C7
Web site	**www.clo2.com**
Telephone	416-239-7111
% $NaClO_2$	**Trade Name**
37%	Solution 37%
31.2	Ercopure 31%
25.2	Ercopure 25%

Manufacturer	**Oxy (formerly Vulcan)**
Address	**Basic Chemicals Co., LLC** 5005 LBJ Freeway Dallas, Texas 75244-6119
Web site	**www.basicchemicalsllc.com**
Telephone	800-633-8280
% $NaClO_2$	**Trade Name**
37%	Technical Sodium Chlorite Solution 50
31	31% Active Sodium Chlorite Solution
25.2	Technical Sodium Chlorite Solution 31.25
15	Technical Sodium Chlorite Solution 18.75
7.5	AktaKlor 7.5
2.8	CD-2

Table 16-E-2. European Chlorite Producers

Manufacturer	**Caffaro S.r.l** Industrie Chimiche Caffaro SpA (Italy)
Addresses	Via F. Nullo, 8 25126 Brescia Piazzale Marinotti, 1 33050 TORVISCOSA (Ud) C.so Garibaldi, 20/22 00034 COLLEFERRO (Roma)
Web site	**www.caffaro.it**
Telephone	+39 030 3192457
% $NaClO_2$	**Trade Name**
31%	Biocaf® Stericlar®
25	
8	

Manufacturer	**Ercros** **(formerly Aragonesas)**
Address	Avda. Diagonal, 595 08014 Barcelona
Web site	**www.ercros.es**
Telephone	+34 934 393 009
% $NaClO_2$	**Trade Name**
25%	
30-31%	
80% solid material	

Manufacturer	**Arkema Inc.** **(formerly Atofina)**
Address	2000 Market Street Philadelphia, PA 19103-3222
Web site	**www.arkema-inc.com**
Telephone	800-346-5757
% $NaClO_2$	**Trade Name**
31%	Sodium Chlorite Solution

Table 16-E-3. Far East Chlorite Producers

Manufacturer	Shree Chlorates
Address	101-102 Block B Radha Krishna Complex 10/1 Manoramagen A.B. Road Indore, 452-001 India
Web site	www.shreechlorates.com
Telephone	+91-731-249-1495
% $NaClO_2$	**Trade Name**
31	
5 – 25%	

Manufacturer	Yencheng City Huaou Industry Ltd.
Address	No. 3 People Road Yancheng Jiangsu Chine 224011
Web site	huaouchem.en.ec21.com
Telephone	+86-515-870-6020
% $NaClO_2$	**Trade Name**
variable	

What Concentrations of Sodium Chlorite Are Available Commercially?

F. How Can the Chlorite Ion be Removed?

The Problem with the Chlorite Ion: Chlorite has been found to be toxic to several invertebrates which are important in the food chain [Masschelein 1991; Fisher and Burton 1993; Fisher and Burton 1995; Fisher and Burton 2000; Fisher et al. 2003]. Thus, in applications such as zebra mussel treatment, chlorite levels are commonly removed or reduced to a very low level.

From a human consumption perspective, there has been no evidence that there is any long term negative effect for the vast majority of the population [Michael et al. 1981; Bianchine et al. 1981; Lubbers et al. 1982; Lubbers et al. 1984a; Lubbers et al. 1984b; Lubbers and Bianchine 1984]. Some tests showed that disinfectants fed at very high concentrations caused some problems in rats [Daniel et al. 1990]. It has also been suggested that chlorite ion may cause hemolytic anemia in certain groups of people [Condie 1986; Condie 1988], particularly those lacking the enzyme glucose-6-phosphate dehydrogenase (G6PD) [Greenberg 1981].

While an exhaustive review of toxicological data has not been presented above, the brief summary above illustrates the problems associated with the chlorite ion, which has resulted in chlorite being a regulated species in potable water. For a more comprehensive discussion of toxicology associated with chlorite ion, the reader is referred to the monograph by Gates [Gates 1998]

Review of Existing Chlorite Reduction Methodologies: There are a number of methodologies which have been evaluated or considered for removal or reduction of the chlorite ion. These are described briefly below.

Granular Activated Carbon, Reverse Osmosis, and Softening: Although partial or complete removal of chlorite is possible with these technologies, none of these technologies are practical on a large scale.

Sulfur Dioxide and Sulfite Ion: The addition of sulfur dioxide (SO_2) and sulfite ion (SO_3^{2-}) have been shown to remove ClO_2^- [Gordon et al. 1990]. The chemistry of these species is essentially equivalent because when SO2 is added to water, it forms sulfurous acid, which dissociates to form sulfite ion.

$$SO_2 + H_2O \Rightarrow H_2SO_3 \qquad (16\text{-}M\text{-}1)$$

In the pH range 4.0 to 7.5, the reaction of chlorite ion with sulfite ion is shown below:

$$2SO_3^{2-} + ClO_2^- \Rightarrow 2SO_4^{2-} + Cl^- \qquad (16\text{-}M\text{-}2)$$

In the presence of oxygen and at elevated pHs, the reaction chemistry was shown to deviate from that shown in Equation 16-M-2. For example, other researchers [Dixon and Lee 1991; Griese et al. 1991] observed that the presence of oxygen in the reduction of chlorite with SO_2 or SO_3^{2-} resulted in the formation of chlorate. As a result, one group suggested that use of sulfur-based reducing agents for removal of chlorite in potable water is not a viable option [Dixon and Lee 1991].

The reaction of chlorite with SO_2 or SO_3^{2-} is complete in less than a minute below a pH of about 5.0.

How Can the Chlorite Ion be Removed?

However, for large industrial once through applications, where 100,000 gpm throughput is common, it is not feasible nor environmentally acceptable in many cases to add sufficient acid to depress the pH of the water to < 5.0 to get rapid chlorite destruction.

Ferrous Iron: Chlorite ion can be reduced to Cl^- by ferrous iron (Fe^{2+}), as shown in the following equation.

$$4Fe^{2+} + ClO_2^- + 10H_2O \Rightarrow 4Fe(OH)_{3(s)} + Cl^- + 8H^+ \quad (16\text{-M-}3)$$

In this reaction, chlorite ion is reduced to chloride ion, and the iron forms a ferric hydroxide floc which ultimately settles out in the water.

One group investigated the kinetics of this reaction and found that at pH < 2.0 and high ionic strength condition (2.00 M), the reaction proceeded at a rate that would be acceptable for potable water plants [Ondrus and Gordon 1972].

Another group indicated that total removal of chlorite was usually achieved in less than 30 minutes with no significant chlorate formation [Griese et al. 1991].

The use of ferrous iron may be acceptable for potable water facilities, because they have the capability of handling the sludge produced. In addition, their holding time permits a somewhat slower reaction to proceed. Still some difficulties have been observed [Henderson et al. 2001]. For large industrial facilities, particularly once-through facilities, ferrous iron is much too slow and the sludge produced is generally environmentally unacceptable for its use to be practical.

The use of ferrous iron may be acceptable for potable water facilities, because they have the capability of handling the sludge produced. In addition, their holding time permits a somewhat slower reaction to proceed. Thus, the use of ferrous iron (Fe^{2+}) has been used with good results by several potable water facilities [Tarquin et al. 1995].

Ascorbic or Erythorbic Acid for Chlorite Reduction: A serendipitous observation in a paper mill where ClO_2 was being used for iron oxidation and removal led to the discovery that under some circumstances, ascorbic acid and its anion, along with its isomer erythorbic acid, reacts stoichiometrically with chlorite ion to effect its removal.

Ascorbic acid, more commonly known as Vitamin C, was originally thought to consume chlorite ion independent of pH [Simpson 2002]. Subsequent work, particularly field trials, showed that the reaction was not pH independent, as originally thought. Instead, the reaction shows a marked pH dependence the details of which have not been completely worked out.

Field Trial 1 – Ascorbic Acid (AA) for Chlorite Reduction: This Southeastern Wastewater facility discharged 6 MGD. A change in state regulations involving reductions in THM limits resulted in the plant having to move away from chlorine gas to meet the new regulations. An engineering company, after a thorough review of available technologies, recommended two possibilities.

1. The first option involved the use of UV light to achieve disinfection without exceeding THM limits. Total projected cost of this approach was ~$15 Million over 5 years.

2. The second possibility was to use chlorine dioxide with chlorine. Total projected costs for this option were estimated to be from 15—30% of the costs for the UV approach over the same time period.

A chlorine dioxide trial resulted. THM results were well below the new limits. However, the effluent began to fail toxicity tests. Bench scale work was done to investigate the toxic effect of ClO_2. The wastewater was tested using ion chromatography and chlorite ion was found, even though ascorbic acid was being fed.

Since normal reactions of ClO_2 involve a one electron gain to produce chlorite ion, somewhere between 50—80% of the applied ClO_2 typically ends up as chlorite ion. The sensitivity of the organism used in toxicity testing was quite sensitive to chlorite ion. Thus, the effluent began to fail aquatic toxicity tests.

After further internal discussion, it became clear that the rapid reduction of chlorite shown by the original work was an artifact of the amperometric titration. The test for chlorite involves adding acid and allowing the sample to sit in the dark for 5 minutes. The reaction of chlorite with most reducing agents, including ascorbic acid (AA), is very fast at low pHs. However, as was ultimately found for ferrous iron, the chlorite ion at pHs above 7 is not very reactive, even with very strong reducing agents. It is only as the pH is depressed that the reaction kinetics increases.

Therefore, a catalyst was added to enhance the reaction kinetics. Sub ppm levels of copper or iron are known to catalyze the reaction of ascorbic acid. Therefore, the experiment was repeated with copper and iron. Also, the conditions of the experiment were chosen so as to simulate addition of ascorbic acid earlier in the system. A dosage of AA and catalyst (0.3 mg/L iron) were sufficient to achieve both the desired THM reduction and pass toxicological tests.

Approximately 15 minutes were required for Fe-catalyzed ascorbic acid to reduce chlorite ion to below detectable limits. The dosage of ascorbic acid required to reduce chlorite ion varied from 3.76—3.96 mg/L, compared to the theoretical dosage of 5.22 mg/L. The reasons for this difference are not known.

Field Trial 2A – Sodium Erythorbate (NaET) for Chlorite Reduction*:* This South Texas Municipality was operating two parallel systems at ~ 10.8 MGD each. One side was selected for the field trial. This municipality had used ClO_2 for a number of years. Prior to the implementation of the limits on chlorite, this facility used ClO_2 and maintained a residual in the distribution system. Not only were there no taste and odor problems, numerous customers would mention that the water had a very pleasing taste. Since the implementation of the chlorite restrictions and the subsequent reduction in feedrate of ClO_2, such compliments have not occurred. In fact, during certain hot parts of the year, taste and odor can be a problem because of algae. To eliminate these complaints, and to try and recover the good tasting water, a higher level of ClO_2 was desired.

To allow for a higher ClO_2 feedrate without busting the chlorite limit, a field trial of sodium erythorbate (NaET), an isomer of AA, was implemented. The goal of this trial was complete elimination of the chlorite residual, just to establish the technical feasibility of the approach.

This municipality uses ClO_2 for primary disinfection, aluminum salts for clarification, and chloramines to provide a measurable disinfectant residual in the distribution system. Since the engineering work showed that a measurable ClO_2 residual must be found at the end of the flocculation basin to achieve the appropriate Ct credits, the NaET had to be added after this point. This was acceptable, since there was approximately a 3 hour residence time in the clarifier at high flows, and a 6 hour residence time in the clarifier at low flows.

Chloramine was added just downstream of the erythorbate injection point. Therefore, the chloramines residual was projected to disappear due to reaction with NaET (NaET was projected to react quite a bit faster with chloramines than with chlorite). Therefore, the injection point of the chloramines was moved to after the clarifier.

NaET, a powder commonly used as an anti-oxidant in food, was a powder available in 50 lb bags. These bags were emptied into a Merrick dry feeder. The feeder was modified to provide approximately 3 times the projected NaET required to reduce all of the chlorite ion, since some demand for NaET was expected. The unit was set at this maximum feed rate and the trial initiated.

After a dozen or so hours of product addition, no change in the chloramines residual was observed. As this was unexpected, discussions were held as to the reasons for this observation. One possibility was that the NaET was just not working. This was at odds with our earlier laboratory work, but to determine whether this was indeed the fact, additional laboratory work was done on site. NaET effectively inactivated both chlorine and chloramines.

Another remote possibility was that the NaET was not adequately dissolving, and that most of the undissolved powder was ending up at the bottom of the flocculation basin. Therefore, the trial was interrupted, and a different approach was taken.

Field Trial 2B: To insure that the NaET was being thoroughly mixed, a small bulk tank was set up, with a dynamic mixer. An appropriate number of bags were emptied into a certain amount of water. No chemical was fed until the solution was completely dissolved. By mixing chemicals in this way, a higher concentration was made, allowing an approximately 4.6 times the required stoichiometry.

Once again, the field trial began. After a dozen or so hours, even at this extremely high dosage rate, no reduction in chloramines was observed. The trial was ended.

Follow up: An intensive review of the literature and discussions with the supplier revealed that both AA and EA and their salts are extremely good metal complexing agents. Once this piece of information was factored in, it appears that the presence of aluminum salts provided a demand for erythorbate. This demand had not been considered in the pre-trial laboratory study.

Conclusions: These case histories established some very important parameters which must be taken into account if use of ascorbic or erythorbic acid or their salts are to be used for chlorite reduction. These parameters include the following:

1. pH is very important. The higher the pH, the longer the reaction time required for chlorite reduction. Use of a sub-mg/L iron or copper catalyst can significantly increase the rate of reaction.

2. Ascorbic acid, given sufficient time to react (time can be shortened if a copper or iron catalyst is used), can drop chlorite levels to any desired level. In the first field trial, use of catalyzed ascorbic acid resulted in a solution that was less toxic even than the blank.

3. AA or NaET must be added at a point where no significant contact with iron or aluminum coagulants can occur. Given the need for some reaction time (system specific), AA or EA or their salts may not be appropriate for some systems, even with the small amount of iron or copper catalyst.

4. Bench studies in the facility should be used to insure sufficient reaction time for chlorite reduction to occur.

Dosage Requirements: The approximate dosage requirements are illustrated in Table 16-F-1.

Table 16-F-1. Approximate Dosage Requirements

Reductant	mg / mg ClO_2^-	Conditions
SO_2	~ 10 mg SO_2	Reaction pH dependent, ClO_3^- forms, minutes at pH 5—7
SO_3^{-2}	~ 10 mg SO_3^{-2}	Reaction pH dependent, ClO_3^- forms, minutes at pH 5—7
HSO_3^-	~ 10 mg SO_3^{-2}	Reaction pH dependent, ClO_3^- forms, minutes at pH 5—7
$S_2O_3^{-2}$	~ 5 mg $S_2O_3^{-2}$	No production of ClO_3^-, dependent on pH and reaction time.
Fe^{+2}	~ 3 mg as Fe^{+2}	Reaction complete within minutes at pH < 7.
AA	~ 4 mg as AA	Reaction pH dependent

References

[Bianchine et al. 1981] Bianchine, J., Lubbers, J., Chauhan, S., and Miller, J., "Study of Chlorine Dioxide and Its Metabolites in Man," EPA-600/S1-81-068 December, 1981.

[Condie 1986] Condie, L., "Toxicological Problems Associated with Chlorine Dioxide," *Journal of the American Water Works Association*, 78(6), 73(June 1986).

[Condie 1988] Condie, L., "Toxicological Effects Associated with Drinking Water Disinfectants and Their By-Products," EPA/600/D88/043, March, 1988.

[Daniel et al. 1990] Daniel, F., Condie, L., Robinson, M., Stober, J., York, R., Olson, G., and Wang, S., "Comparative Subchronic Toxicity Studies of Three Disinfectants," *Journal of the American Water Works Association*, 61(October 1990).

[Dixon and Lee 1991] Dixon, K. and Lee, R., "The Effect of Sulfur-Based Reducing Agents and GAC Filtration on Chlorine Dioxide By Products," *Journal of the American Water Works Association*, 48(May 1991).

[Fisher and Burton 1993] Fisher, D. and Burton, D., "The Acute Effects of Continuous and Intermittent Application of Chlorine Dioxide and Chlorite on *Daphnia magna, Pimphales promelas*, and *Oncorhynchus mykiss*," Report No. WREC-93-B4, University of Maryland System, Maryland Institute of Agriculture and Natural Resources, Wye Research and Education Center, Queenstown, MD 21658, August, 1993.

[Fisher and Burton 1995] Fisher, D. and Burton, D., "Determination of Acute Water Quality Criteria for Continuous and Intermittent Exposure of Chlorite for Freshwater Organisms," Report No. WREC-95-O3, University of Maryland System, Maryland Institute of Agriculture and Natural Resources, Wye Research and Education Center, Queenstown, MD 21658, June 1995.

[Fisher and Burton 2000] Fisher, D. and Burton, D., "Completion of the Acute Toxicity Database for Chlorite and Freshwater Organisms," WREC Report No 00-03, October 2000.

[Fisher et al. 2003] Fisher, D., Burton, D., Yonkos, L., Turley, S., Ziegler, G., and Turley, B., "Derivation of Acute Ecological Risk Criteria for Chlorite in Freshwater Ecosystems," *Water Research*, 37(18), 4359(Nov 2003).

[Gates 1998] Gates, D., The Chlorine Dioxide Handbook, Water Disinfection Series, American Water Works Association, Denver CO, 1998.

[Gordon et al. 1990] Gordon, G., Slootmaekers, B., Tachiyashiki, S., and Wood, D., "Minimizing Chlorite Ion and Chlorate Ion in Water Treated with Chlorine Dioxide," *Journal of the American Water Works Association*, 160(April 1990).

[Greenberg 1981] Greenberg, A., "Public Health Aspects of Alternative Water Disinfectants," *Journal of the American Water Works Association*, 31(January 1981).

[Griese et al. 1991] Griese, M., Hauser, K., Berkemeier, M., and Gordon, G., "Using Reducing Agents to Elimnate Chlorine Dioxide and Chlorite Ion Residuals in Drinking Water," *Journal of the American Water Works Association*, 56(May 1991).

[Iatrou 1991] Iatrou, A., "Removal of Chlorite by Reaction with Ferrous Iron," MS Thesis, Virginia Polytechnic Institute and State University, Blacksburg, VA, July, 1991.

[Lubbers and Bianchine 1984] Lubbers, J. and Bianchine, J., "Effects of the Acute Rising Dose Administration of Chlorine Dioxide, Chlorate and Chlorite to Normal Healthy Adult Male Volunteers," *JEPTO* 5-4/5:215, 1984.

[Lubbers et al. 1982] Lubbers, J., Chauan, S., and Bianchine, J., "Controlled Clinical Evaluations of Chlorine Dioxide, Chlorite, and Chlorate in Man," *Environmental Health Perspectives*, 46, 57(1982).

[Lubbers et al. 1984a] Lubbers, J., Chauhan, S., Miller, J., and Bianchine, J., "The Effects of Chronic Administration of Chlorite to Glucose-6-Phosphate Dehydrogenase Deficient Healthy Adult Male Volunteers," *JEPTO* 5-4/5:239, 1984.

[Lubbers et al. 1984b] Lubbers, J., Chauhan, S., Miller, J., and Bianchine, J., "The Effects of Chronic Administration of Chlorine Dioxide, Chlorite and Chlorate to Normal Healthy Adult Male Volunteers," *JEPTO* 5-4/5:229, 1984.

[Masschelein 1991] Masschelein, W., "Chlorine Dioxide," Chemical Oxidation, Technology for the Nineties, First International Symposium, Vanderbilt University, Nashville, TN, February 20-22, 1991.

[Michael et al. 1981] Michael, G., Miday, R., Bercz, J., Miller, R., Greathouse, D., Kraemer, D., and Lucas, J., "Chlorine Dioxide Water Disinfection: A Prospective Epidemiology Study," *Archives of Environmental Health*, 36, 20(1981).

[Ondrus and Gordon 1972] Ondrus, M. and Gordon, G., "The Oxidation of Hexaaquoiron(II) by Chlorine(III) in Aqueous Solution," *Inorganic Chemistry*, 11(5), 985(1972).

[Simpson 2002] Simpson, G., "Method for Destroying Chlorite in Solution," US Patent 6,440,314, August 27, 2002.

[Tarquin et al. 1995] Tarquin, A., Hansel, G., and Rittman, D., "Reduction of Chlorite Concentrations in Potable Water with Ferrous Chloride," *Water Engineering and Management*, 36(February 1995).

G. The Concentration of Aqueous ClO_2 vs. That of Gaseous ClO_2

What happens if the water from a 1-Liter, 1 mg/L ClO_2 solution was removed, and the 1 mg ClO_2 was left behind? There would still be 1 mg/L, but in this case, the Liter would not be a volume of water, but of air, which weighs much less than water.

$$1\ m^3 = 10^6\ cc$$
$$1\ L = 1000\ cc$$

$$\text{Therefore, } 1\ m^3 = 1000\ L$$

$$\text{So, } 1\ mg/L\ ClO_2 \text{ would be } 1000\ mg/m^3$$

The original 1 mg/L ClO_2 would equate to 1000 mg/m^3.

As the 8-hour Permissible Exposure Limit is 0.1 ppm (or 0.3 mg/m^3), the original ClO_2 would be 3300 times higher than this.

This illustrates why the evolution of ClO_2 from solution, even a very dilute solution, can be a concern for people who work around this material.

H. Is There Equipment for Monitoring Gaseous ClO_2?

The following is a partial listing of vendors of various supplies. This list is not intended to be a complete listing of all monitor vendors, nor is inclusion in this listing an endorsement of these supplies or vendors over others that may not be listed.

Three types of monitors are listed. The first table (Table 16-H-1) includes electronic equipment which is used to monitor a gaseous ClO_2 application, with relatively high concentrations of ClO_2, such as mold remediation, to measure and provide confirmation that a minimum ClO_2 concentration has been achieved.

The second table (Table 16-H-2) includes low level electronic monitors that are used to provide continuous monitoring in the workplace.

The third table (Table 16-H-3) includes passive badges that a worker might wear when working around ClO_2 and which would change color, the color change being proportional to exposure.

Table 16-H-1. Partial Listing of Providers of High-Level ClO_2 Gas Monitors

Manufacturer	Model, Type, and Limits of Detection
ATI 6 Iron Bridge Drive Collegeville, PA 19426 (800) 959-0299 www.analyticaltechnology.com	**PortaSens II – Portable with three ranges, 0—1 ppm, 0—5 ppm, and 0—200 ppm ranges. ATI also has several models for wall-mount or fixed.**
Interscan Corporation P.O. Box 2496 Chatsworth, CA 91313 (800)458-6153 www.gasdetection.com	**Fixed: LD Series and RM Series** **Portable: 4000 Series** **All Series have several ranges, 0—1999 ppm, 0—199.9 ppm, 0—19.99 ppm, and 0—1999 ppb.**
Mil-Ram Technologies 4135 Business Center Drive Fremont, CA 94538 (510) 656-2001 www.mil-ram.com	**2-Wire Transmitter – 0—5 ppm, 0—200 ppm ranges** **3- or 4-Wire Transmitter – 0—5 ppm, 0—200 ppm ranges.**
Scott Instruments 251 Welsh Pool Rd. Exton, PA 19341 (800) 736-9118 www.scottinstruments.com	**Freedom 5600 Explosion Proof- Fixed – (0—1, 3, 5, 10, 15, 20, and 30 ppm)** **Mini-SA – Portable Single Gas Monitor – (0—25 ppm)**

Is There Equipment for Monitoring Gaseous ClO_2?

Table 16-H-2. Partial Listing of Providers of Low-Level ClO_2 Gas Monitors for the Workplace	
Manufacturer	**Model, Type, and Limits of Detection**
ATI 6 Iron Bridge Drive Collegeville, PA 19426 (800) 959-0299 www.analyticaltechnology.com	**PortaSens II – Portable with three ranges, 0—1 ppm 0—5 ppm, and 0—200 ppm ranges. ATI also ha several models for wall-mount or fixed.**
Brandt Instruments 18568 Oak Grove Parkway Prairieville, LA 70769 225-673-6776 www.brandtinst.com	**PhD Lite – Portable – 0—5 ppm**
BW Technologies 2840-2 Avenue SE Calgary, Alberta Canada T2A 7X9 888-749-8878 www.gasmonitors.com	**GasAlert Extreme – portable – 0—1 ppm**
CEA Instruments, Inc. 16 Chestnut Street Emerson, NJ 07630 (888) 893-9640 www.ceainstruments.com	**Triple Plus + - Portable with range of 0—1 ppm** **GasMaster TXGARD Series – Wall Mount – 0—1 ppm** **CEA 420-B Series – Wall Mount – 0—1 ppm** **electrochemical detection**
ENMET Corporation P.O. Box 979 Ann Arbor, MI 48106-0979 (734) 761-1270 www.enmet.com	**Target – Multi Gas Detector – SDS-97D – ppm level**
Industrial Scientific Corporation 1001 Oakdale Road Oakdale, PA 15071 (800) 338-3287 www.indsci.com	**GasBadge® Pro Personal Gas Alarm – range 0—1 ppm**
Interscan Corporation P.O. Box 2496 Chatsworth, CA 91313 (800)458-6153 www.gasdetection.com	**Fixed: LD Series and RM Series** **Portable: 4000 Series** **All Series have several ranges, 0—1999 ppm, 0—199.9 ppm, 0—19.99 ppm, and 0—1999 ppb.**
Mil-Ram Technologies 4135 Business Center Drive Fremont, CA 94538 (510) 656-2001 www.mil-ram.com	**2-Wire Transmitter – 0—5 ppm, 0—200 ppm ranges** **3- or 4-Wire Transmitter – 0—5 ppm, 0—200 ppm ranges.**
Scott Instruments 251 Welsh Pool Rd. Exton, PA 19341 (800) 736-9118 www.scottinstruments.com	**Freedom 5600 Explosion Proof- Fixed – (0—1, 3, 5, 10, 15, 20, and 30 ppm)** **Mini-SA – Portable Single Gas Monitor – (0—25 ppm)**

Table 16-H-3. Partial Listing of Passive Badges	
Manufacturer	**Model, Type, and Limits of Detection**
Morphix Technologies 2557 Production Road Virginia Beach, VA 23454 800-808-2234 www.kandmenvironmental.com	**Safeair System – passive colorimetric badge**

Is There Equipment for Monitoring Gaseous ClO_2?

I. What is the Freezing Point of Chemicals Used to Produce ClO_2?

Sodium Chlorite: The crystallization temperature of aqueous sodium chlorite is important, particularly when solutions of chlorite are stored outdoors.

The graph [Robson 1964; Noack and Doerr 1979] and Table 16-I-1 [Taylor et al. 1940] show the crystallization temperature of aqueous sodium chlorite. The graph clearly shows why 25% sodium chlorite has become the primary concentration of chlorite used.

Figure 16-I-1. Crystallization Temperature vs. Wt% $NaClO_2$

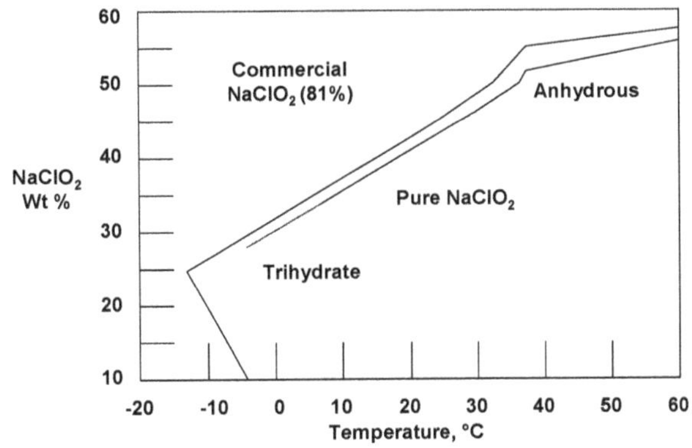

Table 16-I-1. Solubility of Sodium Chlorite in Water [Taylor et al. 1940]		
Temp °C	Moles H_2O per Mole of Total Solute	$NaClO_2$ Wt%
5	9.6	34
17	7.8	39
30	5.8	46
45	4.5	53
60	4.1	55

Hydrochloric Acid: HCl is frequently used in the 3-chemical generator. It is also used in the acid-chlorite generator.

Figure 16-I-2 shows the freezing point of various concentrations of HCl.

Figure 16-I-2. Wt% HCl vs. Freezing Point (°F)

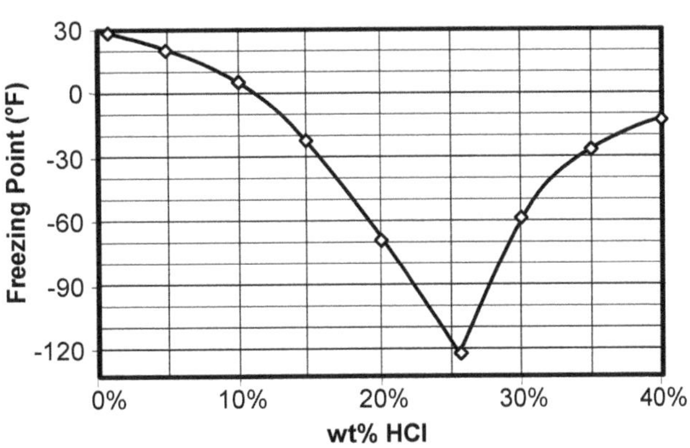

What is the Freezing Point of Chemicals Used to Produce ClO_2?

Sulfuric Acid: Though less frequently used, it can be used, as long as it has been diluted first. The heat of solvation would melt the plastic used in many ClO_2 generators.

Table 16-I-2. Wt% H_2SO_4 vs. Freezing Point

Bé	H_2SO_4 wt%	Sp. Gr.	Freezing Point (°F)
0	0	1.0000	+32.0
10	10.77	1.0741	+22.8
20	22.25	1.1600	+1.6
30	34.63	1.2609	-74
40	48.10	1.3810	-41
50	62.18	1.5263	-27

Bleach: Bleach, or sodium hypochlorite, is used in the 3-chemical generation process. One way bleach is made is by adding Cl_2 gas to a solution of water. In order to stabilize the bleach, somewhere around 0.5 wt% NaOH is added. Some manufacturers add more, some add less. The variability of the amount of excess caustic causes some variability in the freezing point.

The data shown in Figure 16-I-3 was from three different MSD sheets. There will likely be some variability in these numbers.

Figure 16-I-3. Wt% NaOCl vs. Freezing Point (°F)

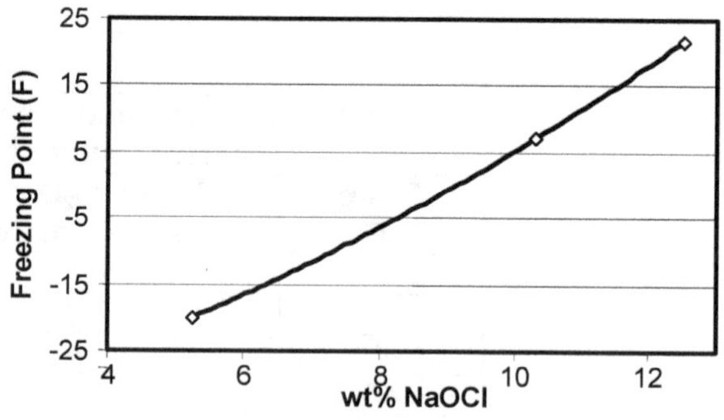

References

[Noack and Doerr 1979] Noack, M., and Doerr, R., "Chlorine Oxygen Acids and Salts - Chlorine Dioxide, Chlorous Acid, and Chlorites," Kirk-Othmer Encyclopedia of Chemical Technology, 3rd Ed., Volume 5, 612-32., 1979.

[Robson 1964] Robson, H., "Chlorous Acid, Chlorites, and Chlorine Dioxide," Kirk-Othmer Encyclopedia of Chemical Technology, 2nd Edition, 5, 27(1964).

[Taylor et al. 1940] Taylor, M., White, J., Vincent, G., and Cunningham, G., "Sodium Chlorite, Properties and Reactions," *Industrial and Engineering Chemistry*, 32(7), 899(1940).

J. How Were the Human Exposure Levels Determined?

There have been limited quantitative dose-effect studies, particularly with regard to Human subjects. Therefore, human exposure limits were established based on ozone numbers, and by extrapolation of data taken from laboratory studies on small animals [ACGIH 1991]. A summary of recommendations from various agencies are shown in Table 16-J-1.

Table 16-J-1. Exposure Limits Guidelines

Exposure Limit	Recommending Agency	Recommended Level	Reference
8-Hour Time-Weighted Averages (TWA)			
Permissible concentration	WCB of B.C.	0.1 ppm (0.3 mg/m^3)	WCB 1980
TLV	USA-ACGIH	0.1 ppm	ACGIH 1991
PEL	USA-OSHA	0.1 ppm	OSHA 1988
REL	USA-NIOSH	0.1 ppm	NIOSH 1988
MAK	Germany	0.1 ppm	DFG 1994
Various	Sweden, UK, Belgium, France, Finland, Australia, Mexico, Denmark, Italy, Indonesia, China	0.1 ppm	FIHRP 1988 ACGIH 1991 Fidino 1996
Short-Term Exposure Limits (STEL)†			
STEL (15 min)	WCB of B.C.	0.3 ppm (0.9 mg/m^3)	WCB 1980
STEL (15 min)	USA-ACGIH	0.3 ppm	ACGIH 1991
STEL (15 min)	As above for 8-h TWA	0.3 ppm	ACGIH 1991 Fidino 1996
STEL (10 min)	UK	0.3 ppm	ACGIH 1991
Short-term (5 min, 8 times/shift)	Germany	0.2 ppm (0.6 mg/m^3)	DFG 1994
Other Human Toxicity Limits			
IDLH*	NIOSH	5 ppm (15 mg/m^3)	NIOSH 1994
Reference Conc (R_fC*†)	EPA	0.0002 mg/m^3	IRIS 1995

† **STEL** – Short Term Exposure Limit - Maximum allowable concentration not to be exceeded at any time during a 15-minute excursion period. There must be no more than 4—15-minute excursions in an 8-hour period, with at least 60 minutes between excursion, and provided the 8-hour average limit is not exceeded (WCB 1980).

* IDLH – Immediately Dangerous to Life or Health. This is the concentration that is likely to cause immediate death, delayed permanent adverse health effects, or to prevent escape (NIOSH 1994).

*† Inhalation Reference Concentration is the estimated continuous lifetime exposure level that is unlikely to cause appreciable risk of harmful effects in the general population, including sensitive individuals (IRIS 1995).

How Were the Human Exposure Levels Determined?

As mentioned initially, no dose-response studies have been conducted on humans. One report described the fatality of a worker in a confined space in which 19 ppm of ClO_2 gas was present [Elkins 1959]. The author also mentioned that 5 ppm was definitely irritating.

Table 16-J-2 shows some symptoms of accidental exposures to ClO_2 of unknown concentrations.

Table 16-J-2. Physical Effects of ClO_2 Exposure		
Exposure/Duration	**Physical Effects**	**Reference**
Repeated exposure over several years	Asthmatic bronchitis, emphysema, dyspnoea	Petry 1954
Repeatedly over 5 years at a sulfite mill (SO_2 exposure also)	Cough, wheezing, shortness of breath, headache, nausea, halophenomenon*, conjunctivitis	Gloemme and Lundgren 1957
Above 0.1 ppm	Irritation of eyes, nose, throat; cough, wheezing, bronchitis (may become chronic), pulmonary edema	NIOSH 1994
Swimmers in ClO_2 disinfected pool, single exposure	Asthma-like symptoms – tight chest, sore throat, difficulty breathing, cough	FIHRP 1988
Bleached kraft pulp mill workers – one or more exposures (also chlorine)	Chest tightness, wheezing, cough, mild airflow obstruction, decreased lung function	Salisbury et al. 1991 Kennedy et al. 1991

* Rings seen around point sources of light

The author can attest to some of the physical effects noted above.

One other phenomenon not mentioned is the persistence of ClO_2 odor. After a day of being around noticeable ClO_2 odors, as many as two or three days later, one can open a car door, or go outside from being inside, and there will be the sudden impression of a noticeable ClO_2 odor, even if no ClO_2 is present. It is not clear if this is a physical phenomenon, where ClO_2 may be desorbing from mucous membranes or if this is some other phenomenon [Simpson 2005].

References

[ACGIH 1991] American Conference of Governmental Industrial Hygienists, "Documentation of Threshold Limit Values and Biological Exposure Indices," 6th edition, 1991.

[DFG 1994] Federal Republic of Germany, "Maximum Concentrations in the Workplace and Biotolerance Values for Working Materials, Report No. 30, Commission for Investigation of the Health Hazards of Chemicals in the Work Area," 1994.

[Fidino 1996] Fildino, V., cited in the British Columbia Council of Forest Industries, "Environmental and Technical Information Manual, 1997.

[FIHRP 1988] Forest Industry Health Research Program, "A Review of Exposure Limits for Chlorine Dioxide," 1998.

[IRIS 1995] - Integrated Risk Information System, Environmental Protection Agency, Washington, D.C., 1995.

[NIOSH 1988] National Institute for Occupational Safety and Health, "Testimony of NIOSH on the Occupational Safety and Health Administration's Proposed Rule on Air Contaminants, 29CFR Part 1910, Table N1, Appendix A., 1988.

[OSHA1988] Occupational Safety and Health Administration, U.S. Department of Labor, 29CFR Part 1910, Air Contaminants; Final Rule, Federal Register 54(12), 2508(1988).

[Simpson 2005] Simpson, G., unpublished results, 2005.
[WCB 1980] WCB, "Industrial Health and Safety Regulations," Workers Compensation Board of British Columbia, 1980.

How Were the Human Exposure Levels Determined?

Sulfide/H_2S Control with Chlorite

A. Problems with H_2S and Sulfide

One of the most difficult problems facing those involved in the production of crude oil or natural gas involves the removal of sulfur-containing compounds, particularly hydrogen sulfide (H_2S). In addition to the odor problems, H_2S poses a serious health risk. H_2S provides an environment which supports growth and proliferation of the highly corrosive and anaerobic sulfate-reducing bacteria, *Desulfovibrio desulfuricans* (SRB). Under the right conditions, these corrosive bacteria produce H_2S. This H_2S causes corrosion and then H_2S reacts with the iron generated from the corrosion cell, to produce iron sulfide. Iron sulfide and biofilm associated with SRB can foul piping, oil-producing formations, and collection equipment (A more detailed description of the properties of H_2S and some of the resulting problems were given in Chapter 10.).

A number of approaches have been employed to treat such problems, including mechanical methods, chemical methods, or both. A variety of different types of chemicals have been used to address the problem of H_2S. Chemicals used to treat the H_2S/sulfide-containing water include bleach (sodium hypochlorite) or hydrogen peroxide to oxidize the sulfide to colloidal sulfur and sulfate. Colloidal sulfur can be a problem, however, if it precipitates or forms in an uncontrolled manner.

B. Common Methods for Control of H_2S and Sulfide

Three basic chemistries have been employed for the control of sulfide. These include the use of chemicals that tie up or complex H_2S, the use of chemicals to prevent formation or consume biogenically generated H_2S, and the use of chemicals to oxidize H_2S.

Chemicals Which Complex the Sulfide: Chemicals used to treat the H_2S/sulfide in downhole oil production include various triazine compounds [Dillon 1990]. These triazine compounds can effectively complex sulfide, but in the case of very high levels of sulfide, these compounds can suffer from three significant disadvantages. First, use of high levels of the triazine solution can result in an increased pH of the produced water. Consequently, the formation potential for calcium carbonate scale/deposit or some other scale/deposit also increases with pH. This problem can be controlled to varying degrees by use of appropriately selected scale and deposit control agents. Secondly, the triazine compounds only serve to tie up the sulfides, forcing downstream refineries to re-process the crude for sulfur removal. Thirdly, the solubility limits of triazine compounds, the relatively small lines normally available to feed chemicals downhole, and the very high levels of H_2S which may be found in some producing wells, mean that in systems with high levels of H_2S, it may not be possible to feed sufficient chemical to complex all of the H_2S/sulfide. Thus, to prevent off-gassing of the H_2S when the crude is brought to the surface, production of crude is limited to that amount of crude in which all the H_2S has been complexed.

In municipal sewer systems and in sludge dewatering applications, iron salts are also sometimes used to form an insoluble complex with the sulfide, thus removing it from a water stream.

Acrolein, a highly unstable and toxic gas, has been used effectively as both an H_2S scavenger and a biocide. Some have suggested that bacteria can become acclimated to acrolein [McCoy 1974].

Preventing and Consuming Biogenic H_2S: Nitrate ion has been found to control sulfide odors in municipal sewer lines and is commonly used for that purpose [Fales 1929; Carpenter 1932; Heukelekian 1943; Allen 1949; Hunniford and Davis 1990; Holden and Simpson 2005]. One theory as to how nitrate interrupts sulfide production is that the nitrate ion provides a more suitable terminal electron acceptor than sulfate, which is the terminal electron acceptor normally utilized by anaerobic bacteria in its metabolic production of sulfide. (One way of thinking about this is that aerobic bacteria use oxygen to respire. Anaerobic bacteria do not use oxygen in this way; instead they use sulfate or nitrate or some other species which can act as a terminal electron acceptor.) In this theory, the sulfate reducer uses nitrate instead of sulfate, the nitrate ion being kinetically favored over sulfate ion [Jenneman et al. 1986].

Another theory is that the nitrate acts as a nutrient for bacteria which compete with SRB for predominance. As many bacteria produce toxins that are inhibitory to other bacteria in their competition for nutrients, the sulfate reducer could be inhibited by such toxins. This theory is supported by the fact that a reduction in existing sulfide levels has been observed when nitrate is added to the system [Goeres 1995; Davidova et al. 2001]. In this theory, it is other sulfide-consuming bacteria which are accomplishing the reduction of existing H_2S.

Oxidation of Sulfide/H_2S by ClO_2: The third approach is to utilize chemicals to oxidize the sulfide. A number of oxidants have been used, including bleach and hydrogen peroxide. All oxidants produce some elemental sulfur, and, under the right conditions, the potential exists for ClO_2 to also produce some elemental sulfur.

$$5H_2S + 8ClO_2 + 4NaOH \Rightarrow 5H_2SO_4 + 4NaCl + 4HCl \quad (17\text{-}1)$$

$$2H_2S + 2ClO_2 \Rightarrow 2HCl + H_2SO_4 + S^\circ \quad (17\text{-}2)$$

There are several concerns surrounding the use of ClO_2 to oxidize sulfide in downhole applications. These include the acids produced from the reaction and the associated corrosion which can result from use of ClO_2 [Prues et al. 1985; Edmondson and Holder 1991; Teeters 1992]. For example, Equation 17-1 shows the reaction stoichiometry most commonly associated with the oxidation of sulfide by ClO_2. For every 5 molecules of H_2S oxidized, 9 molecules of strong mineral acid are produced.

A second area of concern is the potential for production of elemental sulfur (Equation 17-2). While in practice little elemental sulfur production has been observed, it is possible to produce elemental sulfur under some conditions [Smeck 1978; Oakes 1984].

A third area of concern is the oxidation of iron sulfide which is generally present as a deposit on piping walls. In Equation 17-1, the sulfide is generally oxidized to sulfate, while the ferrous ion (Fe^{+2}) is generally oxidized to an insoluble ferric ion (Fe^{+3}) species. Concerns over the potential plugging of producing wells has resulted in implementation of an acidizing step after a ClO_2-based well-stimulation treatment, to solubilize the ferric ion floc that has been produced.

A recent novel and proprietary approach to control the iron produced without this additional acidizing step is to utilize citric acid as the acid in the 3-chemical generation of ClO_2 [Simpson 2005a]. The citric acid effectively complexes the iron, if the pH is sufficiently low, effectively preventing the iron from forming a deposit on system surfaces, and eliminating the final acidizing step in the stimulation of producing wells.

This was demonstrated experimentally. In one experiment, two identical 2-L ClO_2 solutions were prepared, with the exception that HCl was utilized in the first and citric acid was used in the second. Both had ClO_2 concentrations of ~ 2200 mg/L. Both had pH of ~ 3.2. The corrosion rate was monitored by corrater, which provided an electronic measurement of instantaneous corrosion rate. At the end of 1 hour, the carbon steel corrosion probes were photographed and are shown in Figure 17-1. The HCl-generated ClO_2 solution has resulted in some buildup of iron corrosion products. The citric acid-generated ClO_2 solution shows no such buildup. In addition, the measured corrosion rate of the citric acid-generated ClO_2 solution was substantially lower than the HCl-generated ClO_2 solution.

Figure 17-1. Corrosion of Carbon Steel Tips

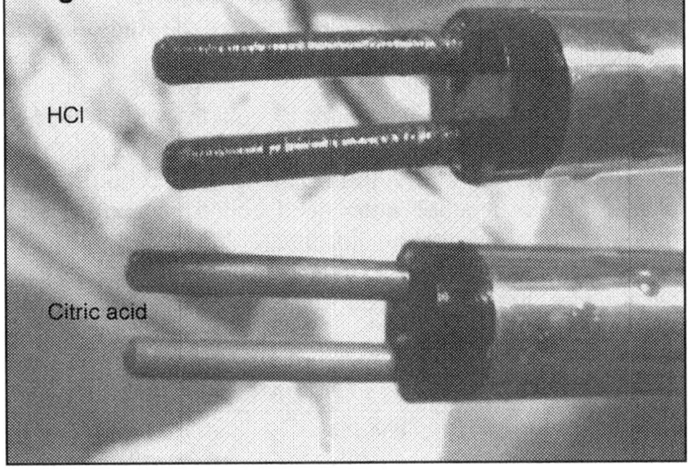

Despite the known corrosion problems, ClO_2 has seen significant use in oilfield applications. Various approaches have been patented [Smeck 1978; Key et al. 1982; Schnepf 1986; Sacco 1990] and used in the field [Sacco 1984; Knickrehm et al 1987; McCafferty et al. 1990; Pitochelli and Mason 1994; Romaine et al. 1996].

In addition to ClO_2, chlorite ion has also seen some use for the control of sulfide ion.

C. Literature Review – Control of H_2S and Sulfide with Chlorite

The use of sodium chlorite applied at pH 4 was described in 1933 as a means of sweetening (reducing sulfur levels in) petroleum distillates [Taylor 1933]. The acidified sodium chlorite oxidizes the sulfur compounds to water-soluble compounds.

One report describes the effective use of "stabilized" chlorine dioxide [Oakes 1984] for sulfide control in drilling fluid. From a chemical standpoint, however, stabilized chlorine dioxide appears to be nothing more than a buffered chlorite solution [Junli et al. 2001].

Another report describes the successful use of sodium chlorite for sulfide control in high pressure injection wells [Sacco 1990].

The effective use of chlorite with an amphoteric ammonium betaine corrosion inhibitor to mitigate the corrosivity of the chlorite, applied at acidic to neutral pH was described in another report [Howson 1992].

There are undoubtedly other reports on this topic, but they are very likely proprietary due to the highly competitive nature of the oilfield environment.

Two areas of continuing interest in both the dosage required and the corrosion which results led to the following two experiments.

D. Experiment 1 – Dosage Requirements of Sodium Chlorite for Sulfide

In 2001, a sample of a petroleum refinery waste water containing spent caustic (pH > ~ 10) with about 2700 mg/L sulfide was tested with a variety of concentrations of aqueous sodium chlorite and chlorine dioxide [Simpson 2001]. Testing of the sample for soluble sulfide ion was done using the methylene blue method. Testing of H_2S in the vapor phase was done using a Draeger tube.

Treatment with Aqueous Sodium Chlorite: Surprisingly, when successive solutions were treated with increasing concentrations of aqueous sodium chlorite, in each case the solution remained clear but turned yellow within a few seconds, suggesting the formation of ClO_2. (However, this is not likely the case, given the very rapid reaction of ClO_2 with sulfide. The data show that a substantial amount of sulfide remained after initial treatment, and a yellow color was produced.) H_2S was measured in the vapor above the solution. Once sufficient aqueous sodium chlorite had been added, the no sulfide could be measured, either in the liquid phase or the vapor phase. When a small amount of acid was added to the final solution, a large amount of finely divided white precipitate formed. It is likely that this precipitate was elemental sulfur or polysulfide, although this material was not analyzed. This solution was then filtered, and the filtrate was again measured for soluble sulfide using the methylene blue method. These results are shown in Table 17-1.

Table 17-1. 25% Sodium Chlorite for Sulfide Reduction

Sample	25% $NaClO_2$ mLs	25% $NaClO_2$ mg/L	pH final	Sulfide mg/L	Gaseous H_2S-ppm
1	0	0	10.3	~ 2700	over range
2	2	8000	6.2	700	over range
3	4	16000	6.2	250	over range
4	7	28000	5.8	13	over range
5	9	36000	5.9	0	0

A plot of this data is shown in Figure 17-2. This graph shows a direct reduction in sulfide, as measured in solution, with aqueous sodium chlorite. In addition, the aqueous sodium chlorite addition results in a pH depression substantially less than if ClO_2 had been added, and thus the impact on corrosion observed due to pH effects will be substantially less.

Figure 17-2. Sulfide Reduction with 25% Sodium Chlorite

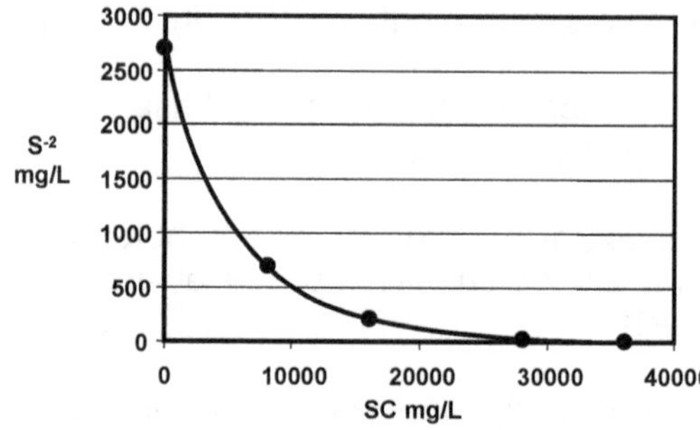

The data show that the higher the sulfide concentration, the greater effect the aqueous sodium chlorite had in sulfide reduction.

For example, 8000 mg/L aqueous sodium chlorite (as 25% $NaClO_2$) was required to achieve the first 2000 mg/L reduction in [S^{-2}], from 2700 mg/L to about 700 mg/L. As the [S^{-2}] decreases, it takes proportionately more aqueous sodium chlorite to achieve a similar reduction.

To reduce [S^{-2}] to near zero by this method would require about 28,000 mg/L aqueous sodium chlorite (= 7,000 mg/L $NaClO_2$), or about 2.8%. Therefore, there is a decreasing rate of effectiveness.

The weight ratio of treatment with aqueous sodium chlorite to achieve a zero [S^{-2}] is approximately 9.3:1 on a weight to weight basis of 25% $NaClO_2$:S^{-2} (2.4:1 $NaClO_2$:S^{-2}).

E. Experiment 2 – Relationship Between pH, [Chlorite], and Corrosion

It was desirable to investigate the corrosivity of the solution in which aqueous sodium chlorite was used to consume H_2S. A design of experiments study was done whereby the concentration of sulfide was held constant at 25 mg/L. The aqueous sodium chlorite concentration and pH were used as independent variables. The corrosion rate of the solution was measured by linear polarization probes. The data was analyzed by DOE Pro of Air Academy Associates, Colorado Springs, CO.

Figure 17-3. Y-hat Surface Plot of Corrosion Rate vs. pH and [ClO2-]

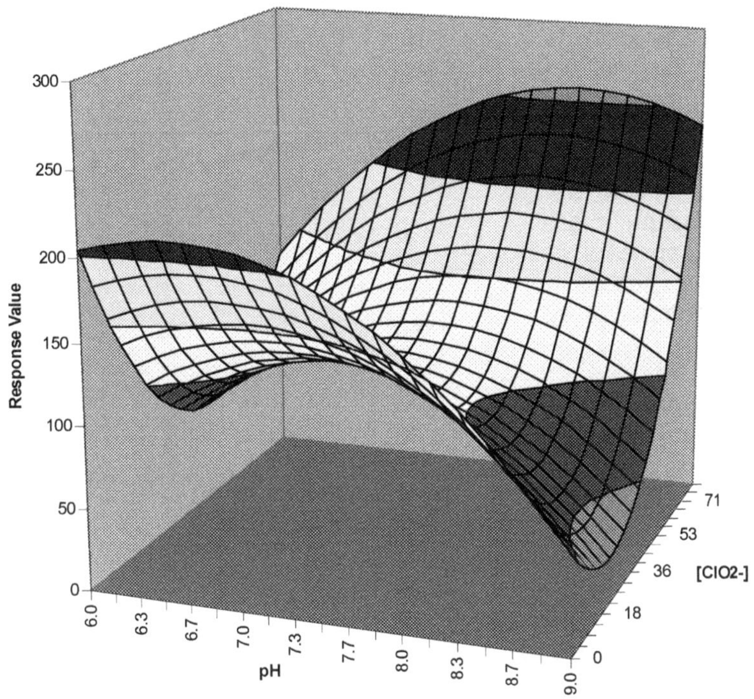

The response surface of the interaction between aqueous sodium chlorite and sulfide with pH is shown in the Figure 17-3.

This data shows that if the pH can be kept above 9—9.5, the corrosion can be minimized without supplemental corrosion inhibitors, the use of which could reduce the corrosion even further. In addition, Experiment 1 suggests that, under the right conditions, the use of chlorite to oxidize sulfide can be done in a way that does not result in the formation of elemental sulfur, contrary to published work.

F. Experiment 3 – Dosage Requirements of Sulfide for Sodium Chlorite

In this experiment, the possibility of using sulfide to consume trace quantities of chlorite which might be present in an aqueous stream was investigated. Specifically, the goal was to determine whether ORP could be used to monitor such a process.

A solution of 1.6 grams of $Na_2S \cdot 9H_2O$ in 100 mL water was made (2130 mg/L sulfide [as S^{-2}]).

Then, a solution of 1.34 grams of 25% $NaClO_2$ in 170 mL of water was made. The pH of this solution was buffered to ~ 7.0 with 2.0 mL of pH 7.0 buffer concentrate.

The chlorite solution was placed on a magnetic stirrer, and the sulfide solution was added dropwise. After every few drops, the solution pH was measured, along with ORP. A plot of this data is shown in Figure 17-3. The final pH was 8.0.

Figure 17-4. ORP vs. Sulfide Addition

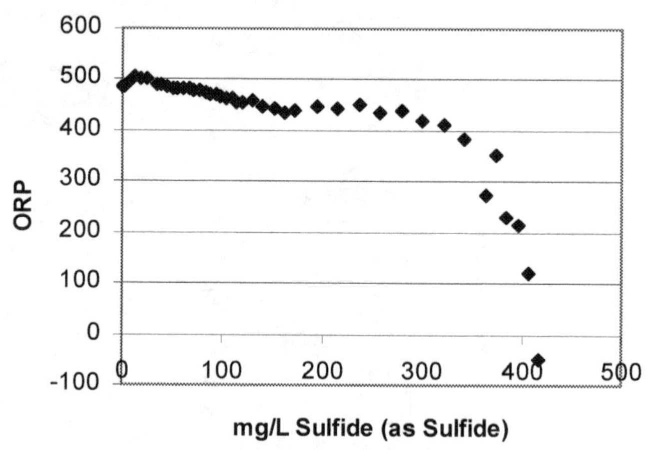

It takes about 417 mg/L sulfide to completely kill 1340 mg/L chlorite ion, or about 0.31 mg/L sulfide per mg/L chlorite ion. This process appears, at least at the concentrations studied in this experiment, to be controllable by ORP.

This data suggests that there is a pH dependency on the rate of reaction of chlorite ion with sulfide, with the lower pH resulting in a more rapid consumption.

G. Conclusions

The published literature regarding use of chlorite to control sulfide is rather sparse, and the information that exists is somewhat contradictory. It is likely that the conditions under which the experiments were carried out was responsible for the inconsistent results.

The data presented here suggests that chlorite, if applied correctly, can be used to control sulfides at a minimum corrosion rate [Simpson 2005b]. Clearly with the increased pH, there will be an increased driving force for the deposition of retrograde salts such as calcium carbonate, and any product formulated to apply this technology will likely require good scale control additives.

H. References

[Allen 1949] Allen, L., "The Effect of Nitro Compounds and Some Other Substances on Production of Hydrogen Sulfide by Sulfate-Reducing Bacteria in Sewage," Proceedings of the Society of Applied Bacteriology, 2, 26(1949).

[Carpenter 1932] Carpenter, W., "Sodium Nitrate Used to Control Nuisance," *Water Works and Sewage*, 79, 175(1932).

[Davidova et al. 2001] Davidova, I., Hicks, M., Fedorak, P., and Sufita, J., "The Influence of Nitrate on Microbial Proceses in Oil Industry Production Waters," *Journal of Industrial Microbiology and Biotechnology*, 27, 80(2001).

[Dillon 1990] Dillon, E., "Composition and Method for Sweetening Hydrocarbons," US Patent 4,978,512, December 18, 1990.

[Edmondson and Holder 1991] Edmonson, J. and Holder, E., "Inhibitor Treatment Program for Chlorine Dioxide Corrosion," US Patent 4,994,195, February 19, 1991.

[Fales 1929] Fales, A., "Treatment of Industrial Wastes from Paper Mills and Tannery on Neponsit River," *Journal of Industrial and Engineering Chemistry*, 21, 216(1929).

[Goeres 1995] Goeres, D., "Control of Souring in an Oil Field Surface Facility: Biotic and Abiotic Effects of Nitrite," MS Thesis, Montana State University, Bozeman, MT, December 1995.

[Heukelekian 1943] Heukelekian, H., "Effect of the Addition of Sodium Nitrate to Sewage on Hydrogen Sulfide Production and BOD Reduction," *Sewage Works Journal,* 15(2), 225(1943).

[Holden and Simpson 2005] Holden, G. and Simpson, G., "Synergistic Composition for Odor Control, patent pending 2005.

[Howson 1992] Howson, M., "Removal of Sulfides Using Chlorite and an Amphoteric Ammonium Betaine," US Patent 5,082,576, January 21, 1992.

[Hunniford and Davis 1990] Hunniford, D. and Davis, H., "Process for Removal of Dissolved Hydrogen Sulfide and Reduction of Sewage BOD in Sewer or Other Waste Systems," US Patent 4,911,843, March 27, 1990.

[Jenneman et al. 1986] Jenneman, , G., McInerney, M., and Knapp, R., "Effect of Nitrate on Biogenic Sulfide Production, *Applied Environmental Microbiology*, 51(6), 1205(June 1986).

[Junli et al. 2001] Junli, H., Lihua, C., and Zhenye, Z., "The Pattern of ClO_2 Stabilized by Na_2CO_3/H_2O_2," *Water Research*, 35(10), 2570(2001).

[Key et al. 1982] Key M. and Crawford, W., "Inhibited Oil Field Drilling Fluid," US Patent 4,310,425, January 12, 1982.

[Knickrehm et al. 1987] Knickrehm, M., Caballero, E., Romualdo, P., and Sandidge, J., "Use of Chlorine Dioxide in a Secondary Recovery Process to Inhibit Bacterial Fouling and Corrosion," Corrosion 87, paper No. 383, San Francisco, CA, March 9-13, 1987.

[McCafferty et al. 1990] McCafferty, J., Tate, E. W., and Williams, D. A., "Field Performance in the Practical Application of Chlorine Dioxide as a Stimulation Enhancement Fluid," SPE, 65th Technical Conference, New Orleans, Sept 23-26, 1990.

[McCoy 1974] McCoy, W., The Chemical Treatment of Cooling Water, Chemical Publishing Company, New York, 1974.

[Oakes 1984] Oakes, D., "Method for Reducing Hydrogen Sulfide Concentrations in Well Fluids," US Patent 4,473,115, September 25, 1984].

[Pitochelli and Mason 1994] Pitochelli, T. and Mason, J., "Large-Scale Field Applications of Chlorine Dioxide," Oxidative Treatment of Pollutants in Wastewater, Intertech Conferences, Philadelphia, PA, March 21-24, 1994.

[Prues et al. 1985] Prues, W., Lee, E., and Kossel, C., "Chemical Mitigation of Corrosion by Chlorine Dioxide in Oilfield Waterfloods," *Materials Performance*, 24(5), 45-50(1985).

[Romaine et al. 1996] Romaine, J., Vickers, B., and Smith, P., "Long Term Use of Oxidizing Agents in Utah Salt Water Disposal Results in Increased Injection Rates and Reduced Operating Costs," 1996 SPE/DOE Tenth Symposium on Improved Oil Recovery, Tulsa OK, April 21-24, 1996.

[Sacco 1984] Sacco, F., "The Use of Chlorine Dioxide in a Late Life Waterflood," Symposium on Advances in Oilfield Chemicals and Chemistry, Division of Petroleum Chemistry, Inc., American Chemical Society St. Louis Meeting, April 8-13, 1984.

[Sacco 1990] Sacco, F., "Process for Producing or Cleaning High Pressure Water Injection Wells," US Patent 4,945,992, August 7, 1990.

[Schnepf 1986] Schnepf, S.., “Method of Treating Oil Field Produced Fluids With Chlorine Dioxide,” Canadian Patent 1,207,269, July 8, 1986.
[Simpson 2001] Simpson, G., “Treatability Study,” unpublished report, 2001.
[Simpson 2005a] Simpson, G., “Control of Fe and FeS in Water Systems,” patent pending, 2005.
[Simpson 2005b] Simpson, G., “Method and Composition for Control of Off-Gassing and Corrosivity of Hydrogen Sulfide,” patent pending, 2005.
[Smeck 1978] Smeck, I., “Treatment of Sulfide-Bearing Waters With Chlorine Dioxide,” US Patent 4,077,879, March 7, 1978.
[Taylor 1933] Taylor, M., “Sweetening Petroleum Distillates,” US Patent 1,908,273, May 9, 1933.
[Teeters 1992] Teeters, S., “Corrosion Inhibitor,” US Patent 5,084,210, January 28, 1992.

Control of Mercaptan Odors with ClO_2

A. Mercaptans – What They Are and the Problems They Cause

"Mercaptans" is the common name for a class of organic compounds more correctly called thiols. That is, they are analogous to alcohols, with the exception that the oxygen in the alcohol group is replaced by a sulfur atom. Thus CH_3CH_2OH is ethyl alcohol, and CH_3CH_2SH is ethyl thiol, or ethyl mercaptan.

Mercaptans are commonly implicated in odor problems. The exceptionally low odor threshold, shown in Table 18-1, the noxious odor, and the high volatility of these compounds makes them exceptionally prevalent as an odor causing species. In fact, a number of mercaptans are commonly added to natural gas, which is odorless in its natural state. This provides natural gas in homes the odor peculiar to that gas for safety reasons, to make sure that leaks of the gas are readily apparent.

Table 18-1. Odor Threshold of Common Mercaptans

Compound	Structure	Odor Threshold (ppm)
Methyl Mercaptan	CH_3SH	0.0011
Ethyl Mercaptan	CH_3CH_2SH	0.00019
Propyl Mercaptan	$CH_3CH_2CH_2SH$	0.000075
Crotyl Mercaptan	$CH_3CHCHCH_2SH$	0.000029
Allyl Mercaptan	CH_2CHCH_2SH	0.00005
Benzyl Mercaptan	$C_6H_5CH_2SH$	0.00019
Amyl Mercaptan	$CH_2(CH_2)_3CH_2SH$	0.0003
Thiophenol	C_6H_5SH	0.000062

Mercaptans are generated in the chemical and/or biological processes of a number of industries. These include municipal sewage plants, petroleum refineries, chemical plants, pulp and paper, steel, rendering, and food processing.

B. Literature Review – ClO_2 and Mercaptans

Although very little work has been reported on the reactivity of ClO_2 with mercaptans, a few reports have surfaced.

In early work the reaction of chlorine dioxide and sodium chlorite with amino acids containing sulfur at pH 3.5—4 was described [Masschelein 1979]. At this pH, chlorite reacts as either chlorous acid or it may form chlorous acid and disproportionate to chlorine dioxide, which then reacts with the organic sulfur compounds. The final reported products included both sulfoxides and sulfones, (R-S=O, R-SO_2), which were the same products that formed from the reaction of ClO_2 with mercaptans and organic sulfides. No mention of any reactions at neutral or alkaline pH was made.

In another report, results of experiments in which ethyl mercaptan was oxidized by varying amounts of ClO_2 were presented [Page 1993]. At a pH of 3.0, the oxidation of ethyl mercaptan was accomplished with a 1:1 ratio of ClO_2 to ethyl mercaptan. Ethyl disulfide was formed and the reaction was completed within 1 minute. A higher ratio of ClO_2 to ethyl mercaptan produced ethyl sulfonic acid, sulfite ion, and sulfate ion.

At a pH of 9.0, ClO_2 also completely oxidized ethyl mercaptan in 1 minute, but ethyl disulfide was produced along with smaller amounts of oxidized sulfur species.

If a stoichiometric excess of ethyl mercaptan was used (2:1 ethyl mercaptan:ClO_2), even at a pH of 3.0, 100% of the ethyl mercaptan was achieved, but only small amounts of the final breakdown products was achieved.

If a solution of ClO_2 were added to a strong solution of ethyl mercaptan, a detonation occurred inside the vessel. Although there was no conclusion as to whether the detonation occurred as a result of aqueous or vapor phase reaction, subsequent work showed that the detonation was likely a vapor-phase reaction, given the high volatility of both the ClO_2 and the mercaptan [Pitochelli 2001].

In a series of experiments, various ratios of ClO_2 to ethyl mercaptan were mixed, and the reaction products measured. Not until the fifth experiment, however, was the reaction mixture measured using ion chromatography (IC), with the identification of the ethyl sulfonic acid. Results are shown in Table 18-2.

Table 18-2. Reaction Products of Varying ClO_2/Ethyl Mercaptan Ratios

Expt No.	ClO_2:RSH ratio	pH	% RSH destroyed	%SO_3^{-2} produced	%SO_4^{-2} produced	%RSO_3H produced	% total recovery
1	1:1	3	100	28	11	-	39
2	1:1	9	100	18	7	-	25
3	1:1	2	100	29	11	-	40
4	1:2	3	100	11	8	-	19
5	1:1	3	100		14	79	93

ClO_2 was reported to oxidize ethyl mercaptan at various pHs to produce sulfites, sulfates, and sulfonic acids, with complete oxidation of mercaptans.

In summary, the following points were made:

- The reaction of mercaptan with ClO_2 is very energetic. The reaction will cause detonation if strong mercaptan solutions are used.
- Optimum reaction conditions are acidic. The natural pH of an aqueous solution of the mercaptan used (ethyl mercaptan) is about 3.
- The reaction was complete in about 1 minute.
- Experiment 5 involved use of IC, which showed the presence of sulfonic acid. Thus, the recovered sulfur was nearly 100% in that experiment.
- The report suggested that chlorite will not react with mercaptans, except at low pH. At low pH, the chlorite ion is known to form chlorous acid, $HClO_2$, which is much more reactive than the chlorite ion [Simpson 2005].

The oxidation of mercaptan was proposed to occur via a 5-electron transfer, with very little of the ClO_2 undergoing a 1-electron transfer.

Another report describes work where NMR and GCMS were used to determine reaction products of the reaction between chlorite and mercaptan [Ramadas and Srivivasan 1995]. The purpose of their research was not to destroy mercaptan, but instead to investigate the use of chlorite for synthesizing disulfides. A number of different aliphatic and aromatic thiols were tested. The reaction of mercaptan with chlorite was described as "mild", and yields of 73—97% with high purity were reported.

They noted that even in the presence of other oxidizable functional groups, the disulfide formed exclusively.

In later work, both ClO_2 and aqueous chlorite ion were reacted with an aqueous solution of sodium ethyl mercaptide or sodium ethanethiolate, the sodium salt of ethyl mercaptan, $NaSCH_2CH_3$ [Pitochelli 2001].

With ClO_2, three ratios were studied, including 0.5:1, 1:1, and 3:1, on a weight: weight basis, at natural pH (~ 3.0). Results are shown in Table 18-3.

Table 18-3. ClO_2 (mg/L) + RSH (mg/L) - Observations

Expt	ClO_2	RSH	Observations
1	0.5	1	Evolution of white vapor; solution turned milky white; slight mercaptan odor
2	1	1	Evolution of white vapor; solution turned milky white, but after ~ 15 seconds turned clear; mercaptan odor much less
3	3	1	Evolution of white vapor; solution clear but noticeably yellow; mercaptan odor gone

The white color of the solution was thought to be due to very small droplets of relatively non-polar organic disulfide, which would not be expected to be very water soluble. With a stoichiometric excess of mercaptan, there was insufficient ClO_2 to fully oxidize the sulfur. When a stoichiometric excess of ClO_2 was added, a reaction product of sulfone was proposed, which was thought to be water soluble.

In the same report, the reaction of the sodium ethyl mercaptide with aqueous sodium chlorite was described. In a manner similar to previous experiments with ClO_2, chlorite was reacted with mercaptan at 1:1, 2:1 and 3:1 on a weight: weight basis, except the pH of the solution was maintained at 12. Results are shown in Figure 18-4.

Table 18-4. ClO_2^- (mg/L) + RSH (mg/L) at pH 12 - Observations

Expt	ClO_2^-	RSH	Observations
1	1	1	No immediate change; solution began to get cloudy after 20—30 seconds; in 5 minutes, solution was cloudy and most of the mercaptan odor was gone.
2	2	1	Solution turned slightly milky after 1.5 minutes; after several more minutes, oil droplets floated to the surface; solution was milky after 4 hours.
3	3	1	Solution began to cloud within 1 minute; oil droplets separated and floated to the surface; solution was milky after 4 hours.

In a subsequent experiment, a 1:1 mix was prepared and the pH adjusted to 8.2. A milky color developed immediately upon mixing in the chlorite. After sitting for several minutes, a water-insoluble oil separated. Extraction with methylene chloride proved unsuccessful. The oil layer that separated between the water and organic layers was dissolved in methanol and analyzed by GCMS. The oil was a mixture of diethyldisulfide and diethylsulfide.

C. Conclusions

Volatile mercaptans and their associated odors can be controlled with both chlorine dioxide and chlorite. While the reaction between ClO_2 and mercaptan is more rapid than between chlorite ion and mercaptan. If the ClO_2:mercaptan ratio is low, disulfide oil is formed and will separate from aqueous solution. If the ClO_2:mercaptan ratio is higher, sulfonate and possibly sulfones, along with inorganic sulfite and sulfate will form.

D. References

[Page 1993] Page, J., "Oxidation of Mercaptans with Chlorine Dioxide," International Technical Center Technical Report, May, 1993.

[Pitochelli 2001] Pitochelli, T., "Reaction of Sodium Chlorite with Mercaptans," unpublished report, 2001.

[Ramandas and Srivivasan 1995] Ramandas, K. and Srivivasan, N., "Sodium Chlorite–Yet Another Oxidant for Thiols to Disulfides," Synthetic Communications, 25(2), 227(1995).

[Simpson 2005] Simpson, G., "Reaction Chemistry of Chlorine(III)," Chapter 4, Practical Chlorine Dioxide: Volume I – Foundations, Greg D. Simpson & Associates, publishers, 2005.

Controlling VFA Odors in Closed Paper Mills with ClO_2

A. Volatile Fatty Acids – What Are They and What Causes Them?

Volatile Fatty Acids (VFA) are small, volatile organic acids with a chain length that varies from C_3—C_5. These include propionic acid (C_3), isobutyric acid (C_4), butyric acid (C_4), isovaleric acid (C_5), and valeric acids (C_5). These VFAs have been identified in several paper mills as contributing to objectionable, sour odors [O'Connor et al. 1998].

Volatile fatty acids result from fermentation-type reactions of anaerobic bacteria. VFAs have been shown to be produced from the anaerobic degradation of protein and fiber [Davis and Smith 2001]. Nowhere is this more prevalent than in paper mills, especially in closed mills, i.e., those having zero discharge. VFA odors are a concern because an increasing number of papers mills are closing their water systems as a result of environmental concerns, discharge limitations, and the large capital cost of adding a wastewater treatment facility.

Turning a mill water system into a closed system results in a number of problems. Two major obstacles to overcome include achieving a water balance and maintaining microbiological control.

- **Water balance.** Achieving a water flow balance is a major obstacle that must be overcome when a mill is closed. Most mills address this problem by adding or using an existing tank or chest as a buffer which is used to keep the water flow balanced. In such mills, this loop recycles until an upset, during which water flows in or out of the tank as needed to maintain the proper water balance.

- **Microbiological Control.** A number of factors conspire to make microbiological control difficult. Some of these are addressed as follows:

    - Stagnant areas. In stagnant areas of a whitewater system, only a few hours is required for significant production of VFAs or odorous sulfur compounds. It is imperative that the water flow be kept moving so that no stagnant areas are created. Any chronically stagnant areas should be identified and targeted for treatment. Housekeeping becomes a very important factor for control of VFA odors.

    - Wastewater (not to be confused with sanitary sewer, which is on a separate circuit) recycle. In a closed mill, anything spilled must be reused. The wastewater loop must either be eliminated, or it must be flushed continuously to avoid becoming stagnant.

    - Chest design. Many chests are not designed to promote continuous flow. Square chests have stagnant areas in the corners. Round chests with tapered bottoms promote flow much better. Chests with stagnant areas should be monitored to determine whether there is a tendency for corners of the chest to become stagnant, and if so, the chest should be targeted for treatment.

The closed whitewater system is warm and wet and has an abundance of nutrients. These factors promote the growth of a wide variety of microbiological organisms. The odor problems in one such mill and the steps taken to correct the problem are described in the following case history.

B. Case History

Background of Mill Odor Problem: One northeastern (NE) closed recycle fiberboard mill was closed in 1972. The mill was located on a river, between a power plant and a municipal sewage treatment plant. The prevailing wind typically blew in the direction of a community which lay across the river. Because of the odors which emanated from the mill-side of the river, this upscale community filed a class-action lawsuit, and since the community had no way of knowing the specific source of the offensive odor (nor could they identify the odor, despite numerous odor panels), all three plants were named as defendants. As offensive odors were generated in all three facilities, each facility began to work toward curtailing any odors produced at their facility.

It is an important to distinguish those odors that were apparent within the mill from those odors which were apparent in the nearby community. Odors apparent in the community have not been directly attributed to VFAs or other compounds which have been volatilized over the dryer section of the paper machine, and they could have emanated from either the power plant or municipal sewage treatment plant.

A significant source of odors in the mill resulted from fermenting stock which fell under a pump or lay stagnant in a sump for some period of time. Apparently, only a few hours were sufficient for the production of sour VFA odors. Therefore, housekeeping in such mills became critical for control of VFA odors.

In this mill, a sour odor was evident all of the time. Contact of the stock with human skin would impart a strong sour odor that was difficult to wash off. While the odor was quite offensive, it was not oppressive. So plant personnel learned to live with the problem.

After numerous complaints from the community, a great deal of work was done to identify the compounds which caused the odor problem in this mill. Odor panels were assembled by experts and ultimately it was the short-chain VFAs which were identified as being the source of odors, at least within the mill. Evidence for the production of organic acids was supported by the high caustic feed required to maintain the stock in a neutral pH range.

Use of Hydrogen Peroxide: As VFAs have been shown to be produced by anaerobic bacterial action, aeration of the stock seemed to be a logical approach to solving the problem. In the presence of oxygen, the growth of anaerobic bacteria is inhibited. As aeration of the stock was not practical, an alternate approach was to add hydrogen peroxide. Hydrogen peroxide was selected initially because it is a common oxygen source used in municipal sewage treatment plants on those occasions where the activated sludge system becomes anaerobic. It is not uncommon for large tanker trucks of peroxide to be unloaded directly into an activated sludge system that has become septic.

One initial injection point was selected for treatment. Feed of peroxide was implemented, but cost considerations resulted in a low feedrate. This treatment appeared to help the odor problem. Then, an accidental spill of a significant amount of the peroxide resulted in the immediate and noticeable reduction of odors. As a result of this observation, four additional injection points were selected and the feed of peroxide was increased. Peroxide usage for the month of April, 2000, was over 16,000 gallons. Odors clearly improved.

After a few weeks slime began to grow on the paper machine. ClO_2 was considered as a supplemental biocide, due to its known benefits at controlling slime on paper machines, particularly in alkaline whitewater applications [Baker 1980; Baker 1981a; Baker 1981b; Nelson 1982; Nelson 1982].

Application of ClO_2 at the Mill: A field trial of ClO_2 was implemented as a means of controlling slime on the paper machine.

Prior to implementation of feed of ClO_2, bacterial baseline testing was done. Figure 19-1 shows the results of bacterial testing of the process water system prior to injection of ClO_2. Aerobic bacteria (AER), sulfate-reducing bacteria (SRB), and acid-producing bacteria (APB) were measured using serial dilution vials.

Figure 19-1. Process Water Bacteriology Before ClO_2

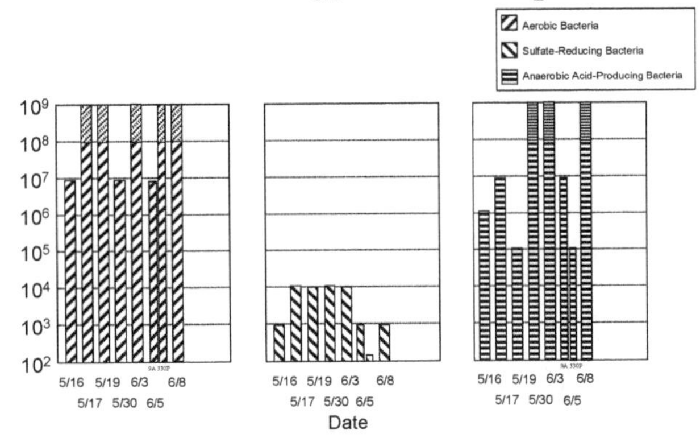

AERs were 10^7 to $\geq 10^8$, SRBs were 10^3 to 10^4, and APBs varied from 10^5 to $\geq 10^8$. Based on positive results with peroxide in terms of odors, the bacterial target for slime control was set to 10^5 to 10^7 AERs and APBs, which tracked each other pretty well. SRBs were considered to be acceptable if they could be controlled at 10^3 to 10^4.

Even at the high volume of hydrogen peroxide used, the control of VFAs appeared to be on the ragged edge, because getting water on ones hand and allowing the water to dry still resulted in a noticeable but weaker sour odor. It was much better than before use of peroxide, but further improvement was still desired.

Figure 19-2. ClO_2 Injection Point

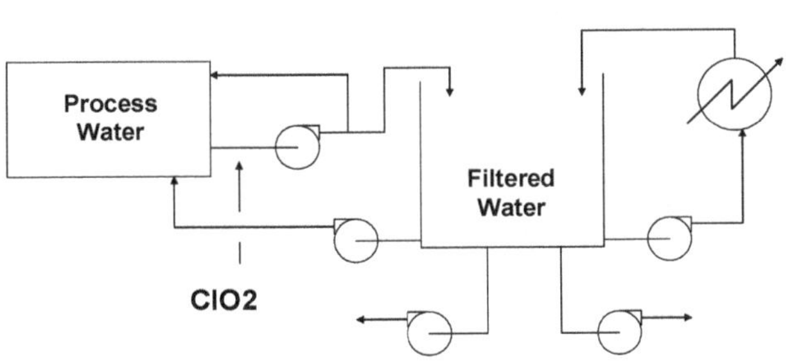

A single injection point selected for ClO_2 was to the suction side of the Process Water Pumps (Figure 19-2). The Process Water Tank (PWT) served primarily as a wide spot in the road, a tank that helped to maintain the water balance required in the mill. Most of the time the water from the process water tank recirculated from the PWT to the PWT pump and back to the PWT. It was only when the water level in the Beater Room Filtered Water Tank fell below a certain level that a control valve opened and water flowed to or from this tank to the system.

The PWT was targeted for treatment because under normal circumstances the water/solids in the PWT recycled for extended periods of time and was not treated.

Feed of ClO_2 was implemented on 6/10, a Saturday morning, at 8:00 AM. The treatment dosage was approximately 8 mg/L, selected because of prior [ClO_2] vs. bacterial reduction beaker studies showed this level of treatment (~ 10% of demand) would accomplish the goals outlined previously. The dosage remained constant throughout the test period.

The bacterial results after implementation of ClO_2 feed are shown in Figure 19-3. The dotted vertical line shows when feed of ClO_2 was implemented. Within 2 days, the bacteriology of the water changed, and bacterial reductions of as much as 10^4 were observed.

Results for both the aerobic and anaerobic acid producers showed similar reductions, and SRBs were reduced to 10^2.

Figure 19-3. Process Water Bacteriology With ClO_2

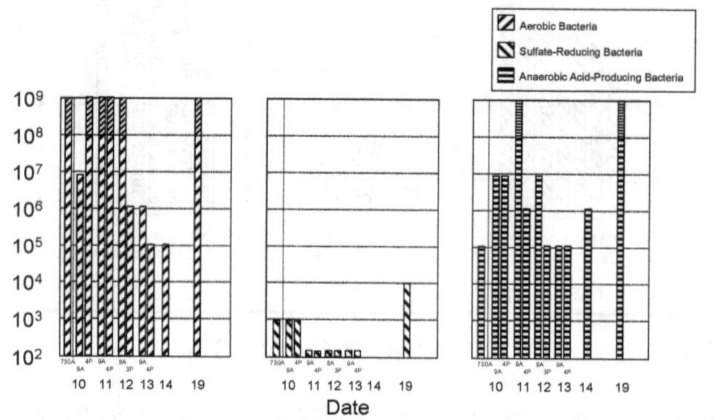

Although the data is not shown here, the bacterial levels were measured at a number of points around the mill. Even though ClO_2 was injected at only one point, the same kind of bacterial reduction was observed all over the mill. As ClO_2 was being fed well below the demand of the water, the widespread reduction in bacterial activity suggested that the chlorite ion was participating in some way to help control VFA-producing bacteria. This phenomenon is discussed in a later section.

Over the weekend, chemical feed was lost, and the system remained untreated for an estimated 18 hours. Note the rapid bacterial rebound of all bacteria. Once such an event transpired, it apparently took several days to recover.

C. Results and Discussion

There were two areas were of potential concern. The first was the possible corrosive effects of ClO_2, given the fairly high feedrate relative to that of potable water. The second was that, given the closed nature of the mill, the degradation of ClO_2 to Cl^- would ultimately result in an increasing level of chloride, which is known to be aggressive towards stainless, particularly if temperatures were elevated and the metal was stressed. Both corrosion rate and chloride concentration were monitored extensively during the trial.

Corrosion: A Rohrback Cosasco Model 9030 continuously recording corrater was installed prior to use of ClO_2 on a stream of water from the Machine Room Filtered Water Chest, which was reported to be the hottest water available (~ 110 °F). Thus the corrosion rate was projected to be higher here than anywhere else in the plant, assuming an equilibrium distribution of corrosion-causing species. A data-acquisition computer was set up to receive signals from the Rohrback instrument. In this way data could be stored and plotted over long periods of time.

The corrater and data acquisition system was moved just downstream of the injection point for ClO_2 prior to implementation of ClO_2. In this way, the corrosion would be monitored in the place most likely to have high corrosion.

General corrosion rates were very low, < 0.05 mpy (mils per year, i.e., a way of expressing the corrosion rate in thousandths of an inch penetration per year), and the pitting index was near zero for the entire time. The data was so consistently low that concerns developed regarding the integrity of the Rohrback instrument. To test this, the instrument was unhooked from the probes and hooked to a test probe, designed to provide a 5.0 ± 0.2 mpy general corrosion rate and a pitting index of 0 ± 0.2. Testing showed the instrument to be in good working order. However, to be sure, the probe was replaced by a new probe. Within a short period of time, the corrosion rate dropped to the steady-state low levels seen with the previous probe. The conclusion was that the corrosion caused by ClO_2 in this system was very low on 304 stainless.

Cycling Up of Chlorides: Chloride ion is known to cause corrosion in stainless steels. The corrosivity of water containing chloride ion appears to be a fairly complex relationship between several factors, the most important including pH, the concentration of Cl^-, temperature, and stress of the metal.

Chlorides were measured by the Hach photometric method, and occasionally samples were sent to an outside laboratory to insure that there was no interference by other species in the water and thus provide confirmation of the Hach results. The samples were very turbid and had to be diluted to get the concentration of chloride into a range appropriate for testing. In addition, the high solids in the sample required allowing the sample to settle for some period of time to provide water of sufficient clarity for testing.

Surprisingly, although mass-balance calculations showed that the chloride concentration should increase steadily with time, the concentration of chloride ion did not change significantly during the course of the trial. One explanation for this phenomenon, given the strong ionic character and high solubility of chloride salts, and the fact that the system is closed, is that the finished paper acted as an ion exchange medium. It is known that paper acts as an ion exchange medium toward cations. It is apparent that a similar phenomenon occurs with anions.

Bacteriological Observations: In any given system, there are many types of bacteria that compete for predominance by producing toxins (antibiotics) that inhibit the growth of competitive organisms. The bacterial species which predominate at a given time is a function of the type and amount of nutrient, the amount of oxygen present, and whether there are other, more highly competitive organisms that utilize these same nutrients. These bacteria include AER, APB (many of which may be facultative, i.e., they can thrive both in the presence and in the absence of oxygen) and SRB. In a system such as this one, which has historically been treated with non-oxidizing biocides and no oxidizing biocide, the anaerobic bacteria, i.e., APB and SRB, tended to predominate, as illustrated in Figure 19-4.

Figure 19-4. Relative Population with Non-Oxidizing Biocides

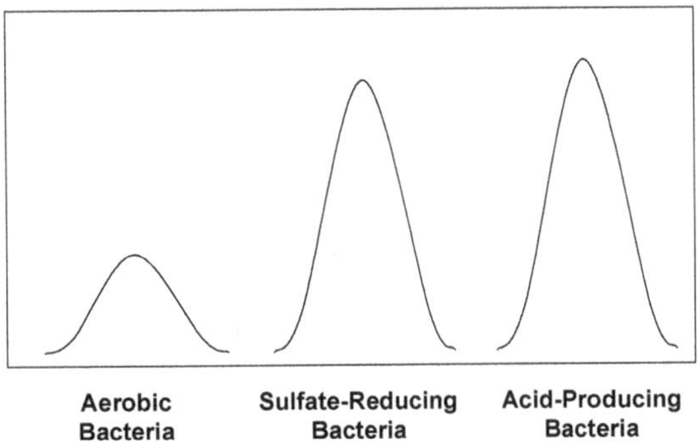

At this mill slime had never before been a problem. The reason appeared to be that notorious aerobic slime producers such as Pseudomonas or other slime-formers had been kept in check by the lack of oxygen and competitive anaerobic bacteria. Instead, at this mill the anaerobic acid producers had predominated.

The peroxide spill resulted in the immediate and noticeable reduction of odors. Figure 19-5 shows, in a simplistic way, what the addition of peroxide did to the relative bacterial populations. Sufficient nutrient existed in the system to feed each type of bacteria. While the system was essentially anaerobic prior to the use of peroxide, aerobic bacteria did not have sufficient oxygen to predominate. With peroxide being a source of oxygen, the relative population of bacteria shifted, creating conditions where aerobic bacteria could predominate. At that point, anaerobic bacteria were brought under control, their growth being inhibited by oxygen and antibiotics of competitive organisms.

Table 19-5. Relative Population with Peroxide and Non-Oxidizing Biocides

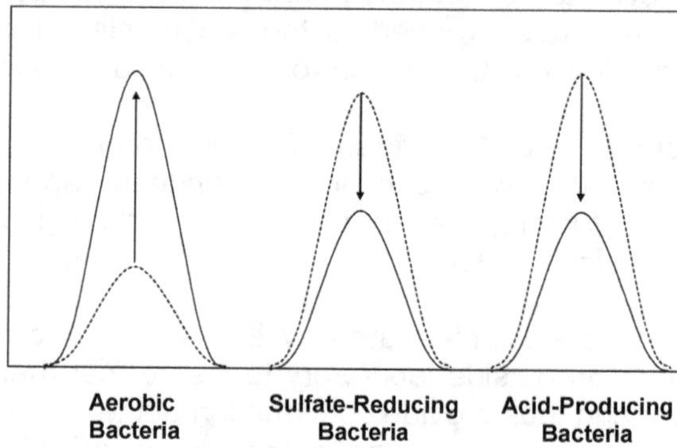

One consequence of addition of peroxide was that, with the increase in aerobic bacteria, slime formers began to grow and bacterial slime began to form on the paper machine and become troublesome. The problems associated with slime formation on paper machines are well documented [Cravens 1966; Sanford 1986; Latshaw 1994; Cowley 1995].

Figure 19-6. Relative Population with Peroxide and Non-Oxidizing Biocides

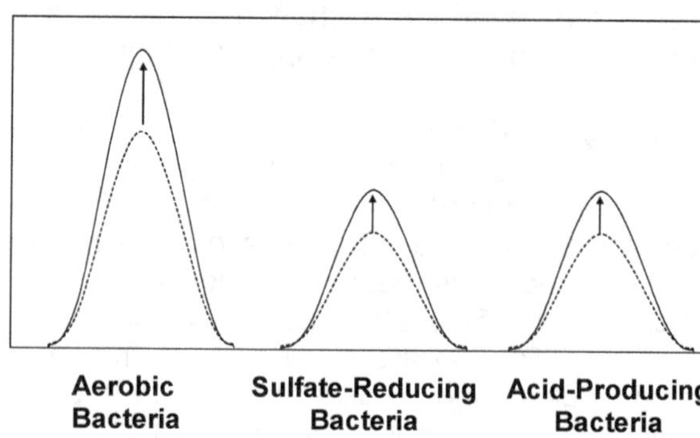

Another observation was that the slime created places where the anaerobic bacteria could grow in a more protected environment, i.e., underneath and within the aerobic slime mass. As a result, it appeared that in addition to a shift in bacterial population, overall numbers of bacteria increased, even with the feed of a substantial amount of non-oxidizing biocides, as shown in Figure 19-6.

D. Confirming Observations at another Mill

This theory appeared to be supported by observations at a southwestern (SW) closed paper mill. The phenomenon occurring at the SW mill appeared to be the reverse of what happened in this case history involving the NE closed mill. Historically, the SW mill used ClO_2 coupled with non-oxidizing biocides. However, slime was present in part because the major chemical supplier was using a non-oxidizing biocide that was sulfur based (i.e., it consumed ClO_2). In addition, ClO_2 was not being applied to areas of the paper machine where the slime formed.

In an attempt to solve the slime problem, a totally organic biocide program in whitewater system was used and ClO_2 was eliminated.

It is interesting to note that their slime problem was brought under control, but shortly after the change, the mill began to develop a sour odor similar to that of the first mill before peroxide was used. This anecdotal information appears to support the observations regarding the bacteriology of closed mills.

E. Conclusions

The goal of treatment of a closed mill with peroxide and ClO_2, such as those described above is shown conceptually in Figure 19-7. In such systems, the goal is not to disinfect, which would be exorbitantly expensive but to reduce bacterial numbers sufficiently low to prevent VFA production and help with slime prevention while at the same time minimizing the use of peroxide.

Figure 19–7. Relative Population with ClO_2 and Non-Oxidizing Biocides

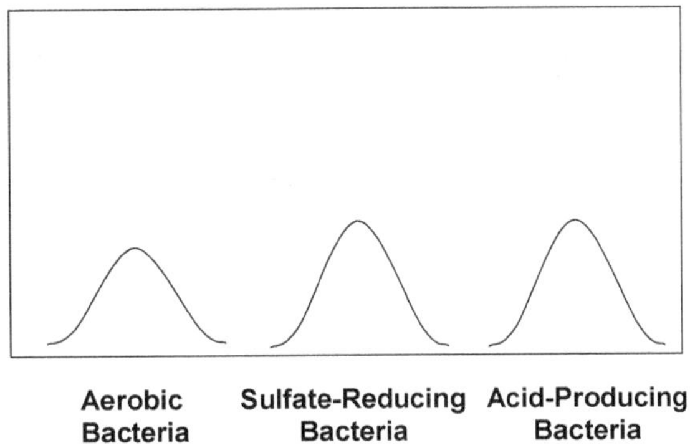

The combination of peroxide and ClO_2 should be fed at a rate sufficient to control bacteria to a level that will prevent formation of slime, H_2S, or VFAs, and at a rate low enough so that chloride does not become high enough to cause significant corrosion to the stainless. The relative feedrate of each will likely be system dependent.

Figure 19-8. Process Water Bacteriology Before ClO_2

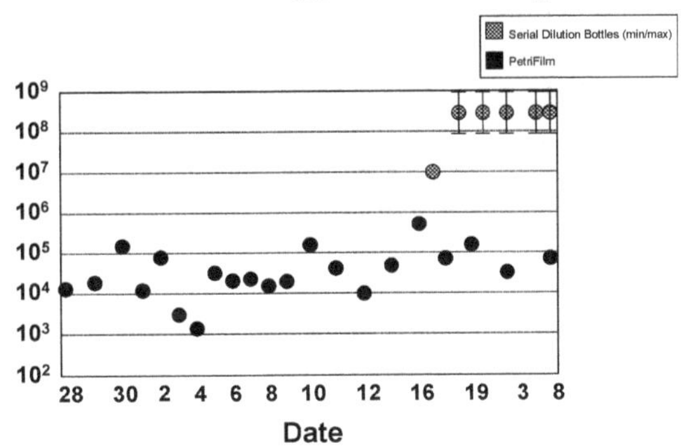

It is likely that ClO_2, possibly coupled with an organic biocide and a source of oxygen such as hydrogen peroxide or liquid oxygen, can successfully control both VFA odors and slime on the paper machine, as long as the correct feedpoints have been identified.

Comparison of Bacteriological Testing Methods: Another goal of this work was to compare various microbiological testing methods to determine how each mirrored mill water bacteriology. The following figures show testing results for PetriFilm™ vs. Aerobic serial dilution vials before (Figure 19-8) and during treatment with ClO_2 (Figure 19-9). Scrutiny of these figures shows that prior to treatment with ClO_2, the serial dilution method showed bacterial numbers 100—10,000 cfu/ml higher than that measured by the PetriFilm™ method. After implementation of ClO_2, as VFA producing bacteria were brought under

Figure 19-9. Process Water Bacteriology With ClO_2

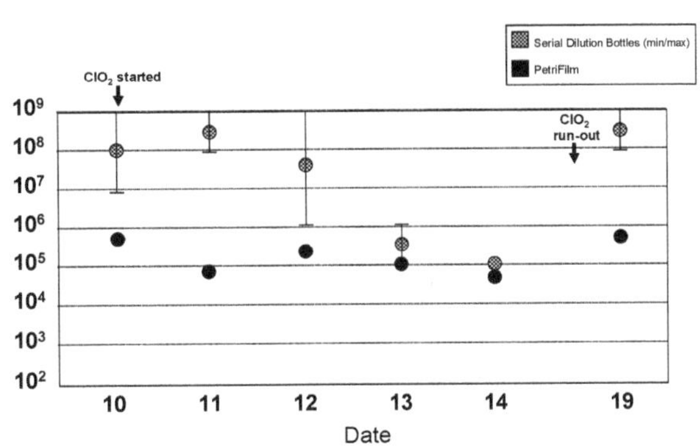

control, results of the serial dilution method (ranges are shown here because multiple samples were tested on a given day, and the spread is shown) began to drop to levels near the results obtained from the PetriFilm™. To understand this phenomenon a bit better, it is instructive to look at how each test works.

PetriFilm™. 3-M Corporation markets this quick and easy method of bacterial testing. A sheet of paper is coated with a microbiological nutrient that is very similar to that used in Standard Plate Counts, the industry standard. 1.0 ml of water to be tested is added to a 9 ml sterile water sample (10^1). Then, 1.0 ml of this is added to another 9 ml sterile water sample (10^2). This is done typically 4 or 5 times, so that the last dilution is 10^4 or 10^5 dilutions are made. Then, 1.0 ml of a given dilution is added to the PetriFilm™ media, and the water is spread in a circle with a plastic spreader provided with the PetriFilm. After 2 days of incubation, the red colonies can be counted. That count times the number of the bottle, indicates bacterial activity. For example, if the third sterile water solution was plated, and 32 red dots were counted, the bacterial activity would be $32X10^3$ CFU/mL.

Phenol-Red Aerobic Serial Dilution Bottles. In this method, 1.0 ml of water to be tested is drawn into a sterile syringe. This solution is injected into a serial dilution bottle containing 9 mLs of media appropriate to test the bacteria of interest. The syringe is filled several times with the solution in the serial vial to mix the solution and to insure that the syringe has the solution rinsed out. Then, 1.0 ml of this solution is withdrawn and injected into the next serial dilution vial and the process is repeated.

The medium present in the phenol-red bottles promotes the growth of ***aerobic acid producers***. The color change is related to a change in pH caused by bacterial metabolism of medium nutrients, and not by the utilization of the indicator by the bacteria in its life process.

The conclusion is that the growth of slightly different types of bacteria, ones which are apparently related to the problem of VFA production, is promoted by the phenol red nutrient. The phenol red dilution bottles, while probably no more accurate than PetriFilm™, seem to track changes in the bacteria that cause odor problems in whitewater of closed recycle mills much better than does the PetriFilm™.

Note that the pH of the whitewater system runs 6.8—7.0 in this mill. Caustic was used to keep the pH elevated, to neutralize the acids produced by the bacteria. With implementation of ClO_2, the amount of caustic has been reduced, and all but one of the peroxide addition points have been eliminated (there were 5 at one time).

In addition, use of the sterile water used in the PetriFilm™ test was evaluated to see if the sterile water could be used to provide an inexpensive way to do dilutions to measure bacterial activity with serial dilution vials instead of having to use a number of relatively expensive serial dilution bottles. There was a 2—4 order of magnitude difference between the results of using serial dilution bottles only vs. using the 4th sterile bottle (corresponding to a 10^4 dilution) as a starting point for the serial dilution bottles. The sterile water manufacturer revealed that these sterile water samples use a buffer to keep the pH of a sample within close guidelines. The presence of this buffer interferes with the pH depression which results from the bacteria in the serial dilution vial method and which is required to cause the color change which indicates the end-point. This information indicates that while unbufferred sterile water (which the company also markets) could be used as a starting point for the serial dilution vials, the ones used in the test could not.

Impact of the Chlorite Ion: The fact that control was achieved over the entire whitewater system even though ClO_2 was injected into only one location initially has some pretty interesting implications. It should be clear that with the high solids loading of the water, no ClO_2, even as high as 8 mg/L, will survive for long. Therefore, it has to be the chlorite ion that is doing the work (the literature suggests that oxidation by ClO_2 results in a one electron change where by the ClO_2 is converted to chlorite ion, 50—80% of the time, generally).

ClO_2 reacts to form chlorite ion, which is relatively non-reactive and non-biocidal. As the chlorite ion contacts an acidic sludge, deposit, or slime layer where bacteria are producing acids (VFAs) as a metabolic byproduct, the chlorite ion is activated to chlorous acid, which also has exceptional disinfection properties. The chlorous acid formed inactivates the bacteria and disrupts the acid production. The use of chlorite alone for slime control in has already been been identified as being effective for acid whitewater systems [Buonanno and Shore 1962].

Will this phenomenon occur if the bulk water pH is in the slightly alkaline range (7.5—8.0)? It is likely. Very localized low pHs produced by these bacteria will still exist, and the chlorite can do its work. Therefore, it is likely that even as the pH increases above 7 chlorite can still reduce VFA production.

F. References

[Baker 1980] Baker, E., "Chlorine Dioxide Proves an Effective Biocide in Alkaline Whitewater Systems," *Pulp and Paper*, 54(8), 140(1980).

[Baker 1981a] Baker, E., "Using Chlorine Dioxide for Slime Control in Alkaline Paper Machine Systems," *TAPPI Journal*, 64(8), 91(1981).

[Baker 1981b] Baker, E., "Slime Control in Alkaline Paper Machine Systems - Chlorine Dioxide," Proceedings of the Papermakers Conference, 49 (1981).

[Buonanno and Shore 1962] Buonanno, J. and Shore, J., "Slime Control Agents and Disinfectants in Paper Mill White Water," U.S. Patent No. 3,046,185 (July 24, 1962).

[Cowley 1995] Cowley, G., "Chlorine Dioxide as a Biocide in Paper Applications," Third International Chlorine Dioxide Symposium, New Orleans, LA., September 15, 1995.

[Cravens 1966] Cravens, B., "Stabilized Chlorine Dioxide for Microorganism Control," *TAPPI Journal*, 49(8), 53A(August 1966).

[Davis and Smith 2001] Davis, C. and Smith, K., "Controlling Odor from Volatile Fatty Acids in Closed, Recycled Paper Mill Systems," *Pulp and Paper*, December 2001.

[Latshaw 1994] Latshaw, C., "Chlorine Dioxide: Effective, Broad-Spectrum Biocide for White-Water Systems," *TAPPI Journal*, 78(4), 163(April 1994).

[Nelson 1982] Nelson, T., "Appleton Papers finds Chlorine Dioxide to be an Alternative to Conventional Biocides in Alkaline Systems," *TAPPI Journal*, 65(6), 69(1982).

[Nelson 1982] Nelson, T., "Chlorine Dioxide: An Alternative to Conventional Biocides in Alkaline Systems," TAPPI Annual Meeting, 261(1982).

[O'Connor et al. 1998] O'Connor, B., Buchanan, B., and Kovacs, T., "Compounds Contributing to Odours from Pulp and Paper Mill Biosolids," Proceedings, TAPPI International Environmental Conference & Exhibit, 1998.

[Sanford 1986] Sanford, T., "Chlorine Dioxide Controls Deposits in Alkaline System," *Southern Pulp and Paper*, 20(July 1986).

Control of RO Membrane Biofouling with ClO_2

A. The Problem of RO Membrane Biofouling

Declining water supplies around the country have forced cities to look at recycled municipal wastewater as a major source of water for certain commercial - institutional and light industrial applications. The City of San Antonio Texas implemented a program of using well-disinfected tertiary treated municipal wastewater as makeup to the cooling towers of buildings, hospitals, universities, and small industrial facilities [Wilcut et al. 2003]. While work remains to be done, approximately one third of the possible end users have opted to use this recycled wastewater. Others are more reluctant for a variety of reasons, including concerns over receiving water that may be of lesser quality than potable water. As corrosion and deposit formation potential is generally lower at lower cycles of concentration (COC), some equipment owners are also reluctant to risk increasing the potential for corrosion and deposit formation in their system which would be higher with higher COC. Despite these concerns, some cities are promoting the use of wastewater for industrial use.

Heavy industry has also been impacted. The California Department of Health Service, in Administrative Code Title 22, provides that industrial facilities that fall within a specified distance of a municipal wastewater plant to use the treated wastewater as plant process water [Breshnahan 1996].

The reuse of municipal wastewater will only increase. An item of absolutely critical importance is that the wastewater be treated in such a way as to assure that the quality of the water is sufficient for the desired goal. Various technologies will be employed to achieve these goals. Reverse osmosis (RO) technology will be an integral part of many of these applications.

RO technology is used extensively in a number of applications. In addition to water reuse, another major use of RO is for pretreatment of water used to produce the high purity feedwater of power plant steam generation systems. Generally RO is used to take the load off of demineralizers which polish the water for further use in these high pressure steam generators. The importance of RO technology to the power industry is highlighted by the projected 9.7% annual growth rate of RO technology to reach a market size of ~$2.5 billion by 2005 [Anon 2001].

The Achilles Heel of all RO units is biofouling [Flemming et al. 1997]. Bacteria that survive any disinfection process can find a home in the membrane. Bacterial survivors that reach the membrane may attach and begin to produce a polymer coating that generally protects the bacteria from external threats such as bacterial toxins or oxidants. This polymer is composed primarily of polysaccharides and effectively acts as "glue," providing a sticky place where other bacteria and any dirt, debris, or scale that gets by the prefilter can lodge. The biofilm can also promote the formation of scale or deposits.

As biological material builds up on the RO membrane, the differential pressure increases and RO effluent water quality generally suffers. Run lengths between required cleanings decrease, and this results in an increase in the cost of operation. Some have suggested that the costs of membrane cleaning lie somewhere between 5—20% of the operating costs [Fane 1997].

Common cleaning chemistries include various acids, bases, surfactants, detergents, chelants, oxidants, enzymes and various blends of these chemicals [Madaeni et al. 2001]. The optimum cleaning practice is likely membrane and site specific, as the type of foulant is a function of the composition of the influent water.

Although these cleaning methodologies are somewhat effective, none are completely effective at removing the biofilm and none are 100% effective at inactivating the bacteria that produce the biofilm [Berg et al. 1981]. Concerns over the potential for degradation of membranes by oxidants have limited the use of oxidizing disinfectants for membrane cleaning. Cellulose acetate membranes appear to be the exception, as these membranes are reportedly quite resistant to oxidation in the neutral pH range [Vos et al. 1968; Glater et al. 1981a; Glater et al. 1981b; Glater et al. 1983a; Glater et al. 1983b].

The search for a chlorine-resistant membrane has been a major undertaking which continues today [Glater et al. 1994]. Alternatively, an effective biocide that is less aggressive towards the membrane might be found. Such a biocide should be able to penetrate and remove the biofilm at a relatively low dose. The biocide should be able to inactivate bacteria in the bulk water, be easy to apply, and be cost-effective. Chlorine dioxide stands alone when these criteria are selected.

Most common ClO_2-generating chemistries produce some free chlorine. The generator can be tweaked, to provide a slight excess of chlorite, which, for a well designed generator, should eliminate free chlorine in the effluent. However, there is still the potential for production of some free chlorine, which is known to be very aggressive to certain types of RO membranes. Only a few types of generators are available commercially that produce no free chlorine by design.

Chlorine dioxide offers some unique performance characteristics that make it suitable for further examination as a membrane cleaner or disinfectant for RO membrane biofilm control. Three of these crucial performance characteristics which merit further investigation of the use of ClO_2 on an RO membrane are:

1. **Selectivity** – ClO_2 is a much weaker oxidant than other commonly used oxidants. The oxidation potential of ClO_2 in aqueous solution is 0.95V, compared to 1.33, 1.49, and 2.07 for bromine, chlorine and ozone, respectively. Thus ClO_2 is less reactive, ignoring kinetics, which we can do for the vast majority of reactions, than any one of the three oxidants mentioned. The Center of Biofilm Engineering has proposed a theory that the weaker oxidant is better for biofilm control, as the weaker oxidant can effectively penetrate the biofilm without reacting with biofilm constituents [Costerton 1997].

2. **Biofilm-Penetrating Ability** – Unlike most other oxidants, ClO_2 can readily penetrate a biofilm. As an illustration, the mainly polysaccharide-composed biofilm can be considered as a layer of organic on the surface. This phenomenon can be easily demonstrated by taking a solution of aqueous ClO_2, adding an equal amount of an immiscible organic (e.g., n-hexane) slowly so that the solution does not mix, and allowing the solution to stand unmixed for a few minutes. The ClO_2 will become distributed throughout both phases within a few minutes, as observed by both phases becoming yellow. This is further illustrated by the method of making ClO_2 in a non-aqueous solvent [Pitochelli 1995; Pitochelli 1997; Pitochelli 1998].

3. **Continued-Action** – When ClO_2 reacts, most of the time (50—80%) it accepts one electron and reverts to chlorite ion, ClO_2^-, from which it was generated. This chlorite ion, when exposed to acids, can re-form ClO_2. Given the fact that the polysaccharides found in biofilm can be acidic [Kanaani et al. 1992; Sutherland 1990], the chlorite ion can still be effective at controlling and removing existing biofilm [Clark and Langley 1990]. An in-depth discussion of this phenomenon is given elsewhere [Simpson 1995].

B. Literature Review of ClO_2 and Its Impact on RO Membranes

There have been a few reports concerning the effect ClO_2 has on RO membranes in the literature.

One group reported on the effect of chlorine on cellulose acetate (henceforth CA) membranes [Vos 1966; Vos et al. 1968]. Above 50 mg/L Cl_2 was shown to attack CA membranes after 1 week of continuous exposure. At 10 mg/L for 15 days, no detectable change in performance was noted. ClO_2 was shown to be unreactive towards CA membranes [Vos et al. 1968].

Most of the early work reported in the literature was done at the California Water Resources Center [Glater et al. 1981a; Glater et al. 1981b; Glater et al. 1983a; Glater et al. 1983b; McCray et al. 1981; McCutchan and Glater 1980; McCutchan et al. 1982; McCutchan et al. 1983; Zachariah and Glater 1985]. While this work was important, several important factors should be considered. First, in the early 1980s, generating chlorine-free ClO_2 was difficult, as the method for speciating the oxychlorine species had not yet been fully developed [Aieta et al. 1984]. The second factor was that the method used to produce ClO_2 was to acidify sodium chlorite (HCl + $NaClO_2$), which, at that time, was thought to produce chlorine-free ClO_2. The authors noted that the ClO_2 was prepared and used with the assumption that no free chlorine was present but the solution was not tested [McCutchan and Glater 1980]. It has subsequently been found that some free chlorine can be produced by this method [Pitochelli 1995].

The third factor is that acidifying sodium chlorite produces chlorous acid ($HClO_2$), which disproportionates to form ClO_2. The disproportionation reaction can take a relatively long time (i.e., days) [Pitochelli 2005] which means that it may not have been ClO_2 alone ($E° = 0.95$ V), but a mixture of ClO_2 and chlorous acid ($HClO_2$; $E° = 1.56$ V) that was investigated [Simpson 2005a]. Given the much stronger oxidizing power of chlorous acid over ClO_2, it is not unreasonable to assume that chlorous acid would be much more aggressive towards an RO membrane than would ClO_2.

The last factor was the relative concentrations studied. In several reports the following oxidants and dosages were compared [McCutcheon and Glater 1980; Glater et al. 1981; McCray et al. 1981; McCutcheon et al. 1982; McCutcheon et al. 1983]:

Table 20-1. Oxidant and Dosage Studied

Oxidant	Dosage mg/L
Chlorine	3.0
Bromine	7.0
Iodine	11.0
Chlorine Dioxide	3.0

Apparently the dosages were selected based on relative molecular weight to the dosage of chlorine. That is, 3.0 mg/L Cl_2 was considered to be equivalent to 3.0 mg/L ClO_2, whereas in the real world, the highest concentration of ClO_2 that would be equivalent to 3.0 mg/L Cl_2 would be about 20—25 % of that amount, or 0.6—0.7 mg/L. (In systems with a high chlorine demand, this spread would widen.) Similarly, bromine equivalency would also be less than that shown.

In these reports, several membranes were tested at concentrations of oxidant and at arbitrary pHs of 8.6, 5.6, and 3.0. Membranes tested included those shown below:

Table 20-2. Key to Membrane Manufacturer and Polymer Type		
Code	Manufacturer	Polymer type
A1	UOP	Aromatic Polyamide (composite)
A2	DuPont	Homogeneous Polyamide
C1	UCLA	E-400-25 CA
C2	Environgenics	CA-CTA 50/50 (homogeneous)
C3	Hydranautics	CA
C4	Hydranautics	CA
C5	Environgenics	CA-CTA (composition unknown)
U1	UOP	Poly(ether/urea) (composite)
V1	Hydranautics	Homogeneous CA (coated with vinyl acetate)
X1	FilmTec	Composition unknown (composite)
X2	FilmTec	Polysulfone (TFC)

The CA membranes were resistant to all of the oxidants tested.

The U-1 membrane appeared to be sensitive to all oxidants tested, being very sensitive to bromine and to ClO_2.

The other non-CA membranes tested, A-2 and X-2, showed good oxidant stability with ClO_2, except at pH 8.6. The reasons for this are not clear, except that the disinfection ability of ClO_2 appears to improve as the pH increases. One possible explanation may be that as the pH increases, ClO_2 tends to adsorb more strongly to surfaces than at lower pHs [Simpson 1997].

A later researcher made similar observations and performed experiments on various RO membranes under more real-world conditions [Adams 1990]. Membranes he tested included DuPont B-15, FilmTec FT-30, Desalination System 2B, and UOP MP and L membranes.

In general, Adams stated that the common practice of using the mg/L oxidant X hrs of operation (analogous to the Ct factor used in potable water to measure disinfection efficiency) to look at membrane life when exposed to oxidants is misleading. He also stated that high concentrations of ClO_2 are aggressive to membranes. However, he also stated that at concentrations < 1.0 mg/L, membranes were found to have a greater tolerance for ClO_2 than for Cl_2. This is important because ClO_2 is known to retain its excellent biocidal characteristics at that concentration. In addition, because ClO_2 is a gas, it can permeate through the membrane and can provide biocidal protection downstream of the RO unit.

Other researchers investigated the use of acidified chlorite to clean Polysulfone membranes [Bohner and Bradley 1990]. They found that ClO_2 can be used to disinfect RO membranes, although ClO_2 was not found to be effective at removing inorganic deposits.

These researchers later reported on the use of ClO_2 as a sanitizer in ultrafiltration systems, commonly found in dairies [Bohner and Bradley 1991]. In particular they were interested in determining the corrosivity various formulations containing ClO_2 would be if used intermittently. They found acidified chlorite (acidified with HCl) was very aggressive to stainless (low pH and high chlorides), but that ClO_2 made from acid/bleach/chlorite was not corrosive. Their work concluded that ClO_2 can be used safely for sanitation of UF systems at a concentration of 15 mg/L and near neutral pH.

In a later report, one author said,

> "One of the most promising alternatives to chlorine is ClO_2. Studies show the great potential of chlorine dioxide due to its biocidal effectiveness, lack of harmful by-products, and its relatively mild effect on polymeric membrane structures. Removal of ClO_2 may not even be needed ahead of the membranes, providing a continuous disinfection throughout the entire membrane system at low dosages (1—2 mg/L)." [Saad 1992]

In a 2004 study, results of a six-month long test with up to 100 mg/L ClO_2 on thin film polyamide membrane samples were reported [Wise et al. 2004]. The ClO_2 was prepared by acidification of chlorite, but the acid used was sulfuric and not hydrochloric, thus insuring that the potential for formation of chlorine was reduced. A very low change in the salt passage resulted.

In a 2005 study to investigate the tendency of ClO_2 to degrade membranes during long-term exposure, laboratory experiments using 0.05—0.9 mg/L chlorine-free ClO_2 were conducted over about 1000 hours [Pitochelli et al. 2005]. Dow FilmTec FT-30/BW-30 membranes were exposed to continuous chlorine dioxide concentrations between 0.05 and 0.1 mg/L for up to 600 hours with a slight decline in permeate flowrate (although the permeate flowrate was still equal to or greater than untreated reference membranes) but with no increase in salt passage. The ClO_2 concentration was increased to 0.5 mg/L for about 150 hours and then to 0.9 mg/L for the last 250 hours. During the entire period, there was no increase in salt passage, and the flux, which appeared to decline during the first 600 hours, began to increase over the last 400 hours.

The work suggests that ClO_2, applied continuously at levels at or below ~ 1 mg/L, maintains flux with stable salt passage, and that with such use, not only are the TFC RO membranes undamaged, but the performance of these membranes may actually be enhanced with use of ClO_2 [Pitochelli et al. 2005].

C. Case History 1 – Cleaning Badly Fouled RO Membranes with ClO_2

Arizona Public Services (APS) supplied power to the southwestern power grid from their Four Corners Power Plant located 21 miles WSW of Farmington, N.M. A coal-fired system used water from the San Juan River which are pumped into a zero discharge lake (Morgan Lake) on the Navajo Reservation. Water & Power Technologies (WPT) of Salt Lake City, Utah, was contracted to pre-treat water through an RO system fed out of this lake. Water for the high pressure boilers was produced from a demineralizer that polished the RO water.

Background: Two RO systems have been installed that process an average 400 gpm through FilmTec 5010 BW30-365 polyamide membranes and TriSep X201 TSA polyurea membranes in the first pass ROs and TriSep ACM2 membranes in the second pass ROs. A water analysis of the lake water in mid February of 2003 reported a pH of 8.4 and conductance of 1100 microSiemens (mS). This lake water is used for cooling and remains moderately warm in winter and very warm in summer. The lake is concentrating with use and provides abundant nutrients for microbiological growth.

Biox, Inc., an industrial and oilfield water quality consulting firm in Farmington, N.M., employs a variety of chemistries including chlorine dioxide to control bacteria in oilfield and industrial waters. Biox was recommended to employ chlorine dioxide for bacterial control in the facility. Measurement of bacterial levels was done by use of serial dilution vials containing API RP-38 media to measure sulfate-reducing bacteria (SRB) and phenol red media to measure aerobic bacteria (AER).

Multiple inoculations of raw water showed 10,000 colony-forming units per mL (CFU/mL) of AER and 100 CFU/mL of SRB in oxygenated water. A chlorine dioxide demand test on raw water was measured to be ~ 1 mg/L. Another lake in the area with no industrial heating had a chlorine dioxide demand of 0.4—0.5 mg/L.

Early in the project, an attempt was made to control bacteria with sodium hypochlorite but the limited kill achieved was attributed to the elevated pH of the raw water. Ferric sulfate was being used as a flocculant ahead of twelve multimedia vessels using anthracite and garnet/sand as filter medium.

A continuous feed of chlorine dioxide to the raw water was recommended. Spot disinfection was being done throughout the facility to remove biofilm which was measured to be ~ 1 mm thick in some piping runs. Bacterial populations measured on back flush water of the multi-media filters showed consistently > 10^5 CFU/mL aerobic and 10^3—10^4 CFU/mL SRB. Biofilm was present visually throughout the facility when pipe was cut, when sensing instrumentation was changed out, and on RO elements during autopsy.

WPT chose to study ClO_2 before committing to generator installation. For purposes of demonstrating chlorine dioxide's effectiveness in bacterial control, three areas in the WPT facility were treated as isolated entities (Figure 20-1):

1. Prior to multimedia filters
2. Prior to the Suction Surge Tank
3. Prior to the membrane E-100, 300 1st array RO (TriSep X201 TSA membranes) in first and second pass array of E-100. (eight vessels, 1st array; four vessels, 2nd array)

Figure 20-1. RO System Schematic

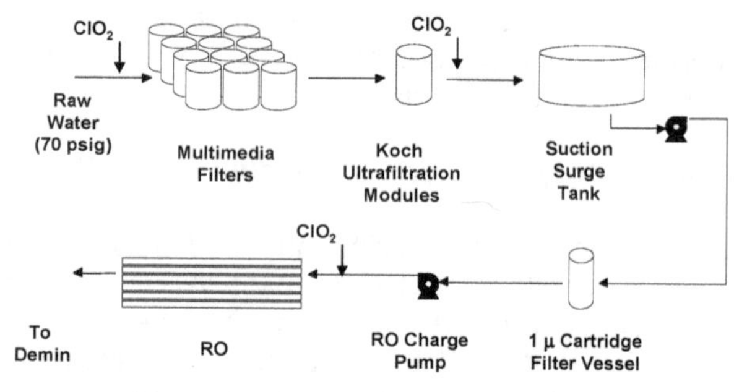

ClO_2 Generation: Concerns over incomplete conversion of chlorite to ClO_2 and the possibility of traces of free chlorine existing in the produced ClO_2 solution led to two important operational decisions.

First, the generator was deliberately "detuned". That is, rather than operate on an exact balance of acid, bleach, and chlorite to maximize production of ClO_2, the generator was detuned so that a few percent excess chlorite was fed to the generator over what would normally be required to optimize the generation process. The presence of excess chlorite minimized the possibility of free chlorine existing in the generated ClO_2 solution.

Secondly, a third party analyst was contracted to monitor halogen species of the generated ClO_2 solution.

Multimedia filters: The twelve multimedia filters were shut in and backflushed before treating with chlorine dioxide. About two minutes into the backflush cycle, bacterial samples were caught and inoculated for aerobic and anaerobic growth. A ClO_2 solution of 200—300 mg/L was introduced to the bottom of each isolated vessel until the vessel was about ¾ full. After one hour of contact time, each vessel was backflushed to drain until water sampling showed clean water off the top of the media. Bacterial sampling was done about 5—6 minutes into the cycle after testing confirmed the absence of a measurable residual of ClO_2. The post treatment sampling showed no bacterial growth.

The procedure was repeated two weeks later. Although the multimedia filters showed complete kill with ClO_2 treatment, bacterial levels rebounded quickly with undisinfected raw water.

Ultrafiltration Piping Feed to RO Pumps: This section of piping was selected for ClO_2 cleaning because several cut-ins for repair had visually shown a large amount of biofilm. A bypass loop was set-up to allow ClO_2 to be introduced and shut-in for kill without contacting the RO membranes. This loop was treated, shut-in, and flushed four times over eight days. After treating this loop, the ROs were run with extra filtration in the line to prevent possible release of large particles of biofilm into the staged RO pumps. Bacterial monitoring after this treatment showed no growth out of the piping. Some areas of this piping could be manually inspected for wall "slickness" and revealed no observable biofilm..

Membrane Treatment of 1st & 2nd Pass Array of RO First Array Pass: A previous bad experience with chlorine resulted in plant personnel having a strong aversion to having any chlorine contact the membranes. Therefore, the following criteria were placed on the ClO_2 treatment regimen:

- A very low to no free chlorine in the generated chlorine dioxide solution.
- The solution pH should be kept at 9.0.
- ClO_2 should be applied at a concentration $\leq$ 1.0 mg/L. ClO_2 was produced at ~ 500 mg/L. The ClO_2 solution was fed into a tank, and mixed with permeate water to achieve a ClO_2 residual of < 1.0 mg/L in the tank.
- Cleaning should be reverse normal flow.

The clean in place (CIP) skid was used with 1100 gals of pH 9 permeate water. All testing was done by either DPD-glycine [Hach], starch-iodide titration [Simpson 2005b] or Lissamine Green B [Palintest]. Chlorine dioxide was added by cup, quart, or gallon containers from a 500 mg/L solution. Samples into and out of the RO array were caught and tested. Cleaning was done an average of six hours per day for five days. During and after chlorine dioxide cleaning, the wash water tank was recirculated and dumped ten or more times.

Flush water was described variously as "stinky," "washing machine water gray," "mud," and "dirty lemonade with pulp." One flush with 300 gals of permeate water was so turbid that white tank bottom could not be seen through water 4" deep.

After treatment, the conductivity of permeate from this stage of the treated RO settled to about 76 mS conductivity (up from 12—13 mS). Then, as time passed, the cross pressure began to increase and the conductivity decreased.

The data appear to support the conclusions of others, that biofilm formation on RO membranes performs like a secondary membrane which acts as a patch to cover microscopic holes in the membrane [McDonough et al. 1994]. As biofilm begins to regrow, the biofilm provides a kind of patch that results in improved membrane integrity and thus provides an explanation for the high conductivity observed just after the cleaning process. Possible removal of biofilm/bacteria which plugs membrane holes can account for this initial rise in conductivity. This observation seems to be born out by increased cross pressure and lowered conductivities with time in service after cleaning.

It may be this phenomenon that has been observed when ClO_2 has been applied at low doses to an already-fouled membrane. The ClO_2 stripped the biofilm, uncovered the pre-existing holes. Salt leakage increased, and the conclusion reached was that ClO_2 damaged the membrane. When ClO_2 feed stopped, however, the biofilm repaired itself and leakage stopped.

Although this possibility was considered initially and steps taken to minimize the possibility of its occurrence, another possible cause for the initial rise in conductivity could be the result of incomplete conversion of the hypochlorite to ClO_2. With this incomplete conversion the trace free Cl_2 could build up in the cleaning tank. While the ClO_2 was consumed by the biogrowth, the presence of trace free Cl_2 could have caused additional damage to the membrane. Testing showed this not to be the case.

The net effect of this cleaning process was that cross pressures after cleaning were lower by a factor of three over the best prior historical cleaning. Photo micrographs of previous element autopsies gave the impression of "invasive tree roots breaking concrete."

Summary: Three conclusions can be made from this work.

1. **An ounce of prevention is worth a pound of cure.** It is much easier, less time consuming, and much less costly to prevent or control bacterial biofilm buildup in an RO system than it is to try to clean up a severely fouled RO system. A good analogy for water treatment professionals is that of scale formation in boilers. It is much easier and much more cost effective to complex hardness (requires stoichiometric treatment) going to an industrial boiler than it is to try to clean the scale up once it forms (treatment is not stoichiometric). Once the scale forms on boiler tubes, one can never completely remove the scale.

2. **ClO_2 can be used for cleaning MODERATELY fouled membranes.** However, using ClO_2 in this way will require extensive manpower and downtime (eight-twelve days in this case) and thus should be used to remediate existing facilities during low demand and/or turnaround. (Severely fouled membranes will generally have too much damage to be cleaned effectively.)

3. **Use of ClO_2 can reduce cost and improve effectiveness of RO operation *when used correctly.***

D. Case History 2 –Using ClO_2 to Prevent RO Membrane Biofouling

Background: A large seawater desalination facility (DESAL) was built to produce about 25 million gallons of drinking water per day from approximately 44 million gallons of seawater. Effluent water from the nearby power plant provided influent water to DESAL. At certain times of the year, primarily April through October, the power plant effluent water temperature was too high for the RO membranes at DESAL. During this period of time, power plant influent water was pumped and mixed with the power plant effluent water to insure the temperature remained in a range suitable for the RO membranes. The 19 million gallons of wastewater from DESAL was blended at a minimum of 20:1 to as high as 70:1 back into the power plane effluent water. A schematic of the DESAL facility is shown in Figure 20-2.

Figure 20-2. Desalinization Facility Overview

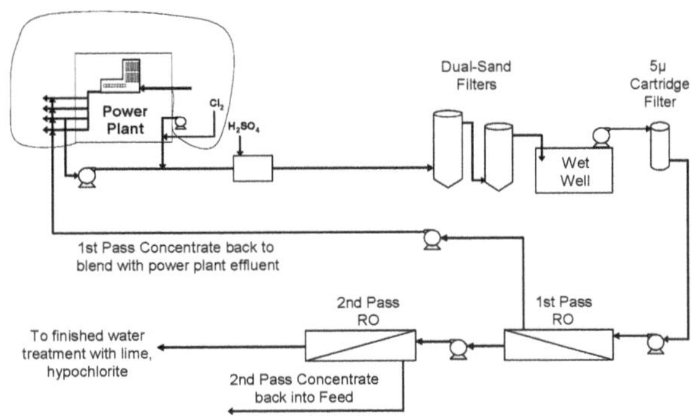

A number of interrelated problems have plagued the facility. The relative importance of each problem varied depending on a number of factors. As a result, the relative priority of each factor changed with circumstances. These areas of concern are described below:

1. **Trihalomethane production:** Naturally occurring organics in seawater such as humic and fulvic acids can react with chlorine to produce trihalomethanes, which include chloroform, chloro-dibromo methane, dichloro-bromo methane, and bromoform [Aieta and Berg 1986]. The current national limit for total trihalomethanes (TTHMs) in potable water is 80 μg/L (ppb). The brominated forms are thought to have a greater health risk than their chlorinated counterparts [Cooper et al. 1985; Robeck 1981]. If bromide ion is present, chlorine will oxidize the bromine to hypobromite ion, with the result that by far the greater part of the TTHMs will be the more highly brominated forms. Given the need for oxidant to control micro and macro biological growth in the influent line, TTHMs has been unacceptably high.

2. **Macrofouling of First Stage Filters:** A variety of organisms, and most specifically the Asian green mussel, grew in the influent line. This non-native invasive mollusk has caused substantial plugging problems in the first stage filters and other water lines. The problems were exacerbated by the lack of filtration of influent water to DESAL, particularly during the warmer months, when cooler seawater from the power plant influent reservoir was blended with the warmer power plant effluent to provide a water temperature suitable for RO membranes. The pump used to provide the cooler water lay outside any strainers that power plant uses. As a result, unfiltered seawater laden with veligers of these mussels can entered the DESAL influent water pipe, attach to walls of piping, and grow to a size where, once they detach from pipe walls, they may be of sufficient size and density that they cannot be removed from the first stage filters by the normal self-cleaning operation of the filters. Cleaning of these filters must be accomplished manually, resulting in unplanned-for increases in maintenance costs.

3. **Biofouling of Dynasand® Filters:** Dynasand®, a patented, two-stage, continuously self-cleaning sand filter lay at the heart of the desalination plant [Suozzo and Suozzo 1998]. When this filter operation was optimized, the plant ran with few problems. Therefore, the key part of any program was to keep these filters clean. This meant keeping them clean of mollusk shells too large for the filters to remove as part of their self cleaning process. It also meant keeping the sand filters free of microbiological biofilm so that the continuous cleaning process could work effectively.

4. **Biofouling of 5 μ Cartridge Filters:** The 5 μ cartridge filters were the last line of protection before the RO membranes. Replacement of these filters had been a major challenge. Any

material that did not get caught on the Dynasand® filters would foul the 5 µ cartridge filters. The differential pressure would then increase, requiring frequent replacement of these filters. This had been a major expense historically in terms of actual cartridge filter costs as well as time and effort required to replace these filters.

Chlorine had been used as the primary micro- and macro- fouling control chemical. The use of chlorine had been somewhat successful, but its use may have caused more problems than it solved. Specifically, use of chlorine had resulted in extremely high TTHMs, and its use had not prevented macrofouling of the first stage filters. Dechlorination had also been required to insure that no free chlorine contacted the RO membranes. The addition of another chemical to remove chlorine added to the complexity and cost of the overall operation.

The unique chemistry of chlorine dioxide was proposed as a means of addressing some of these problems. However, the use of ClO_2 introduced additional concerns, specifically, the introduction of the chlorite ion, which might prove toxic to aquatic life, and the potential adverse impact ClO_2 might have on the RO membrane. Therefore, a trial was proposed to investigate the use of ClO_2, and to compare the expected benefits with the possible drawbacks. The ClO_2 would be applied on the pilot plant unit, so that a short term, intensive investigation could be done and these things could be investigated.

ClO_2 Field Trial: Due to the small amount of ClO_2 (< ~ 2—3 lb/day) required for testing on the pilot unit, an electrolytic generator which could be operated to produce chlorine-free ClO_2 was selected [PureLine Treatment Systems]. The protocol was to apply ClO_2 at a dosage which would clean up any biological growth in the sand filters, so that a measurable residual could be achieved on the effluent of the sand filter. Then, the generator output would be adjusted so that only a trace of ClO_2 was found in the sand filter effluent, to minimize the ClO_2 residual which contacted downstream equipment. During the field trial, a number of measurements were made both manually and continuously. Parameters tested manually included chlorine dioxide and its primary disinfection by-product, chlorite ion (ClO_2^-) at every point. Parameters measured continuously included turbidities on the effluent of the Dynasand® filters, differential pressures on the cartridge filter and RO unit, and RO permeate conductivity.

Figure 20-3. Pilot Plant

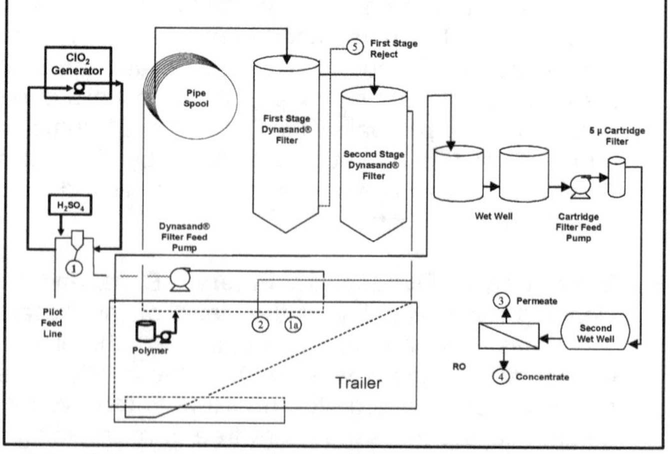

The unit was installed and run initially to insure it was operating properly. Then the unit was shut down until the Dynasand® filters could be placed into service. Once the Dynasand® filters were placed into service the generator was set to provide a fairly high level of ClO_2 to clean up the filters prior to the trial. As the filters cleaned up, the residual out of the filters increased. At that point, the generator production rate was then reduced to provide a low residual out of the Dynasand® filters.

Once the RO was put into service, a complete data set was taken twice per day during the test period, although some parameters such as ClO_2 residuals were taken more frequently.

In general, the pilot plant was arranged to simulate as nearly as possible, the residence times, relative flows, and feedrates that would be required by the plant (Figure 20-3). Sample points and descriptions are given in Table 20-3.

Table 20-3. Sample Points and Descriptions

Sample Point No	Description
1	Sample point for chemical analyses
1a	Sample point for ClO_2
2	Post Sand Filter – Pre RO
3	RO Permeate
4	RO Concentrate
5	First Stage Reject

There were several goals in this trial. The goals, in no particular order of importance, were as follows:

1. To reduce TTHMs
2. To minimize biofouling of the sand filters
3. To extend the life of the cartridge filters
4. To provide a non-toxic effluent, as measured by third party toxicity testing
5. To provide water that has no adverse impact on the RO membranes
6. To minimize or eliminate macrofouling of the primary stage filters

Results: In pre-trial work, a chlorine dioxide demand study was done to identify the dosage required. In this test, an increasing dosage of an aqueous solution of ClO_2 of known concentration was added to a number of samples until a slight trace of ClO_2 was measured in the sample after a few minutes. The demand was observed to be erratic, varying between about 0.4 and about 2 mg/L. This variation is very unusual and is not at all typical.

At the same time, a sample of the influent water, before any chemical addition, was obtained and treated with the level of chlorine that was actually being applied in the plant and a separate sample was treated with the projected dosage of ClO_2, based on demand studies. Both samples were sent for THM analysis. Results showed that the chlorine-treated sample had 845 μg/L TTHMs, while the ClO_2-treated sample had 1 μg/L TTHMs. This kind of spread is somewhat unusual, although spreads of this magnitude have been observed by others [Lykins and Griese 1986].

Figure 20-4. ClO_2 Data
Prior to RO Startup

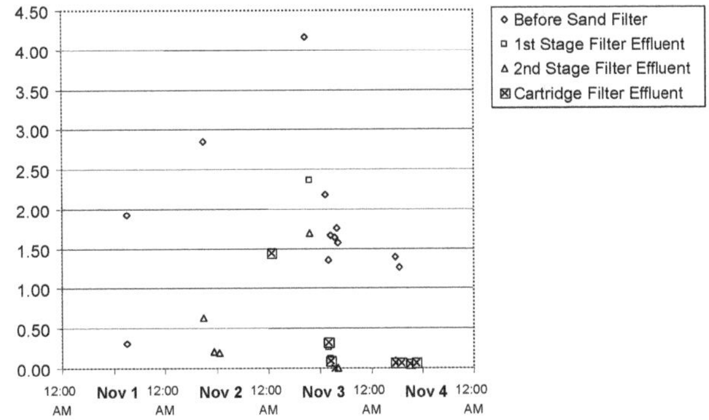

Two items noted early on during this test were 1) the relatively high ClO_2 demand of this water when compared to other seawater demands measured previously and 2) the apparently variable nature of this demand. The measured residual varied considerably during this test, unlike what has been observed in other seawater applications. Although the reasons for this are not known specifically, several possibilities exist. First it could be a function of the seawater. To investigate this, a sample of seawater behind a nearby motel was collected and the demand was measured. A demand consistent with values measured periodically in the inlet water to the pilot unit was found. The second and more likely possibility deals with the erratic performance of the centrifugal pump, which was designed to pump water and not water containing a fair amount of air / ClO_2 gas. Problems with this pump were noted on several occasions.

ClO_2 results prior to the RO being put into service are shown in Figure 20-4. The generator was adjusted to provide a residual of about 4 mg/L prior to the first stage filter. The residual out of the first stage filter increased over a period of hours, and once it stabilized, the dosage was reduced to provide a low level of ClO_2 (0.0—0.3 mg/L) in the effluent of the second stage filter. Generally, a 1.5—2.0 mg/L differential was observed between ClO_2 in and ClO_2 out of the sand filters, and this spread did not change much during the field trial.

Figure 20-5 shows the ClO_2 and chlorite residuals before the sand filter. The residual was reasonably constant as was the generation rate.

Figure 20-6 shows the ClO_2 and chlorite residual after the 5 μ cartridge filters, but before the RO. Except for a few data points toward the end of the trial, the residuals are very low, generally < ~ 0.3 mg/L. In real world applications, the use of oxidation-reduction potential instrumentation

Figure 20-5. ClO_2 & ClO_2^- Before Sand Filter

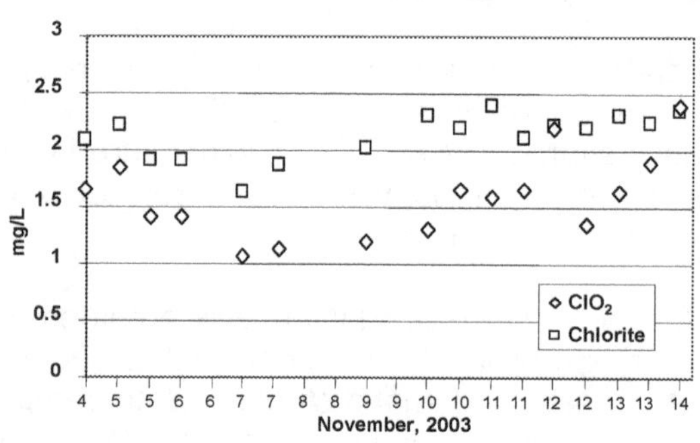

Figure 20-6. Post Cartridge – Pre RO

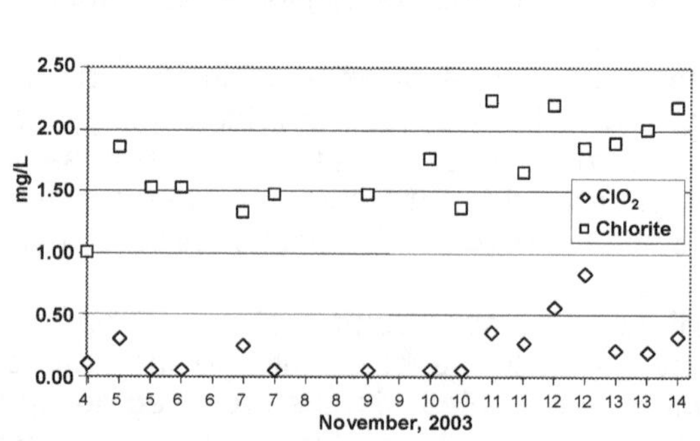

(ORP) for such applications would be appropriate. The ORP vs. [ClO_2] curve is very steep for [ClO_2] < about 0.5 mg/L, so the use of ORP could be used to control the feedrate of such an oxidant very well indeed.

Figure 20-7. RO Permeate

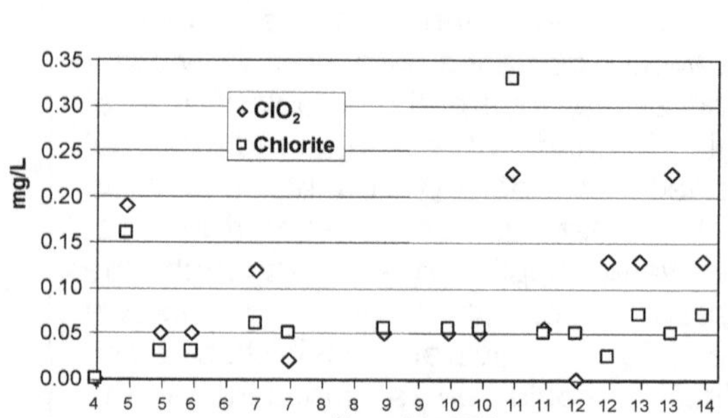

Figure 20-7 shows the ClO_2 and chlorite residuals in the RO permeate. Some chlorite is shown here. With the amperometric titration method used, one can distinguish between ClO_2 and chlorite ion. First one titrates to measure ClO_2. During this titration step, the ClO_2 is converted to chlorite ion. Subsequent steps in the procedure measures chlorite. Thus, any chlorite that is found in a sample is a combination of that which originated from the ClO_2 present initially, and other chlorite, which might include an inefficient generation system (which can't occur in this situation because of the way ClO_2 is being made) or from ClO_2 which had reacted previously. Thus, any chlorite present in excess of that which came from the ClO_2 in the sample would be due to chlorite present in the water going to the RO unit and which had passed through the RO membrane. Since the chlorite ion is at or below that of the original ClO_2, we can conclude that no chlorite ion is getting through the membrane. One data point exception was observed on Nov 11th. As none of the other points in the system show anything like this in terms of a high residual, this data point is somewhat questionable.

The RO concentrate showed a chlorite residual of about 3.5—5 mg/L. Even though this much chlorite was observed, dilution at 1:20 for toxicity testing was done and submitted for third party testing. The RO concentrate flowed to the power plant effluent basin where it was diluted from 20:1 to 70:1 prior to discharge. Therefore a 20:1 dilution was allowed for toxicity testing. The diluted effluent was not found to be toxic to the organisms tested (Table 20-4).

Table 20-4. Toxicity Test Results

Organism	Toxicity
Mysidopsis bahia	pass
Menidia beryllina	pass

Figure 20-8 shows the ClO_2 and chlorite measured in the first stage reject water. In general, the same sort of trend is followed as that of the post cartridge-pre RO data (Figure 20-5). Periodic inspection of the first stage reject water by the pilot plant operator revealed some interesting anecdotal information. The highly experienced pilot plant operator noted that the first pass reject water out of the pilot unit was dirtier than he had ever seen it before. The black gumbo type substance, normally found at the bottom of the first stage filters by the divers as they manually clean the filter vessels, which is normally removed during the self cleaning process of the filters was enhanced by the ClO_2.

Figure 20-8. First Stage Reject

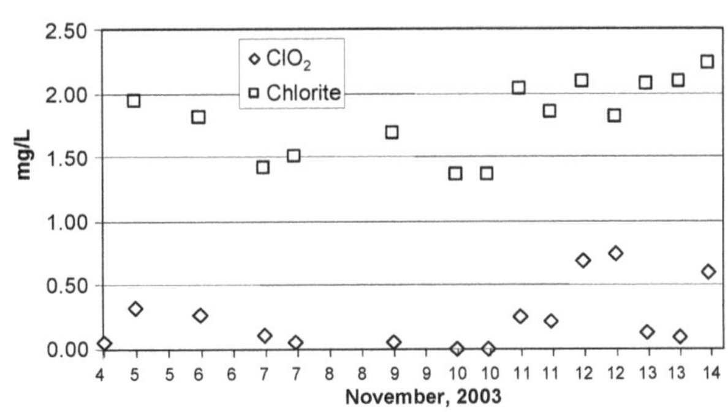

A plot of the turbidity along with the ClO_2 residual out of the sand filters did not appear to show any correlation.

Figure 20-9 shows the differential pressure of the 5 μ cartridge filters along with the ClO_2 residual out of the cartridge filter. As expected the differential pressure shows a slow steady trend upwards, indicating that the filter is slowly fouling. However, it should be pointed out that the differential pressure of the cartridge filter never reached a point which required replacement of the filter

Figure 20-9. Cartridge Filter Differential Pressure & ClO_2 Residual (x 10)

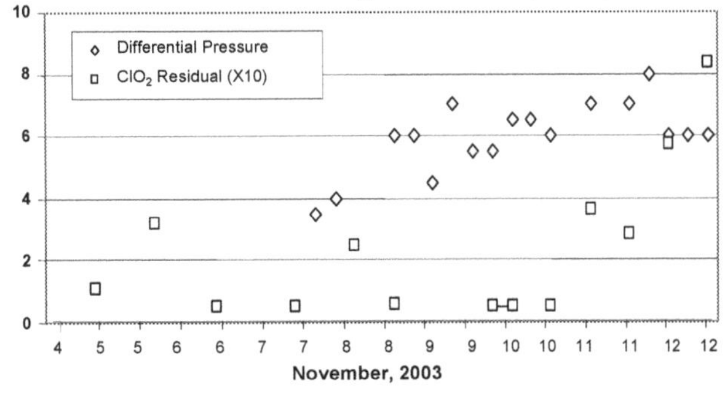

over the ~ 2 week trial period. At the same time, filter replacements in the facility were being required every 3—5 days. This indicates that use of ClO_2 can significantly extend the life of the cartridge filter, which in this case was a major unplanned-for expense (approaching $1MM annually).

Figure 20-10. RO Differential Pressure & ClO_2 Residual (x 10)

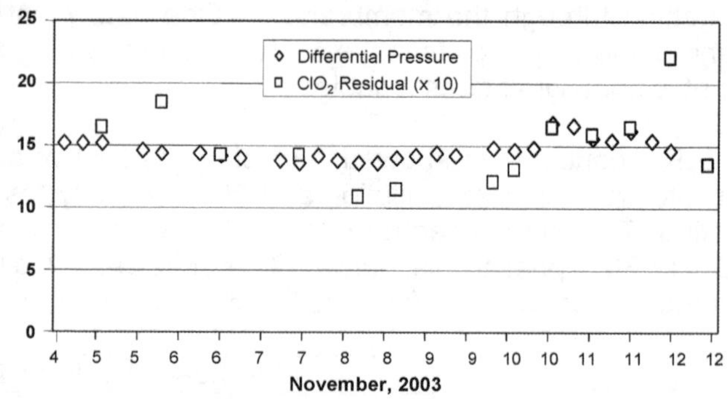

Figure 20-10 shows the differential pressure of the RO compared with ClO_2 residual. The differential pressure did not change significantly over the trial period. In addition, the ClO_2 residual which was exposed to the RO membrane varied from about 0.1—0.22 mg/L. Only an extended run of a number of months would show with any degree of confidence whether chlorine-free ClO_2 had any long term impact on the membrane.

Figure 20-11. RO Permeate Conductivity & ClO_2 Residual (x 100)

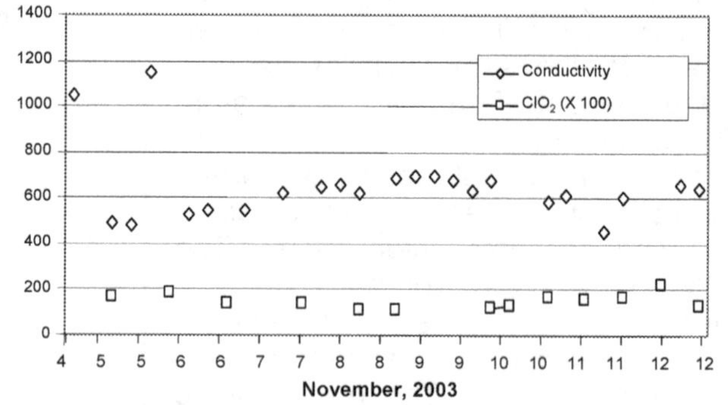

Figure 20-11 shows the RO permeate conductivity compared with ClO_2 residual. There seems to be a slight upward trend in the early days of the trial, although the conductivity appears to plateau during the last half of the trial.

Neither plant THM data nor RO membrane autopsy data were made available to us, because of pending litigation between several of the groups involved in the design, construction and operation of the plant.

Summary: From the perspectives of the goals of this trial, the use of ClO_2 generally improved, sometimes significantly, the operation of the pilot plant, as described below:

TTHM Reduction – Although the data obtained by the plant during the first week of the trial was not available, there is a considerable body of evidence that supports the contention that use of ClO_2 can effectively address this problem [Lykins et al. 1989; McVeigh et al. 1985; Miltner 1976; Stevens 1982]. The pre-trial study that was done is in agreement with the vast body of work that has been done in potable water plants that supports use of ClO_2 for TTHM reduction.

Control of Biofilm in Dynasand® Filters – ClO_2 is known to have well established biofilm preventing/controlling/removing properties [Simpson and Miller 2000; Simpson 2001]. This, coupled with the observations of the pilot plant operator, indicate significant cleaning was on-going.

Extend the Life of the 5 μ Cartridge Filters – The slower than projected increase in differential pressure strongly suggests that ClO_2 was effective in preventing biofouling of these filters, as replacement of the cartridge filter used in the pilot plant was not required over the duration of the field trial.

Provide a Non-Toxic Effluent – Toxicity testing of the RO reject water, when diluted by 20:1 was not found to be toxic to the species used in the toxicity test by third party testing.

Minimize Macrofouling of the Primary Stage Filters – While the field trial was not designed to demonstrate the efficacy of ClO_2 towards Asian green mussels, the efficacy of ClO_2 towards Asian clams [Cameron et al. 1987; Cameron et al. 1989a; Cameron et al. 1989b; Goss et al. 1977], a benthic organism, the zebra mussel [Pitochelli and Mason 1994; Rybarik et al. 1995], and other mollusks [Kemp and Okimoto 1982] is well established.

A periodic, i.e., once per week or so treatment of the influent line for 2 hours at ~ 1.0 mg/L above demand will encourage the relocation of any mollusks which may have attached. Periodic application will prevent any organisms from achieving a size which will prevent them from being backwashed during the routine operation of the Dynasand® filters. The approach of using low levels of ClO_2 to prevent settlement of mollusks has been utilized in a number of applications since the mid 1990s, and is described elsewhere [Pickrell et al. 1999].

No Adverse Affect on RO Membranes – Autopsy information was not provided, however, none of the data suggested that any damage had occurred. A longer term trial would be required to be able to say with any degree of certainty that chlorine-free ClO_2, applied at very low dosages, would have any negative impact on the RO membrane integrity. However, there is an increasing body of evidence from both research and field work that suggests such is the case.

For example, in the mid to late 1990s, intermittent use of ClO_2 was successfully applied 0.1—2.0 to the mill water of a recycle paper mill, some of the water being treated by RO to produce boiler feedwater. In this 5-year application, no damage to the RO membrane was observed [Trefry 2003].

In summary, early work showed there was some damage to RO membranes by ClO_2, the purity of which has subsequently been questioned. Later work by others showed little to no apparent damage to membranes if the applied concentration of ClO_2 is controlled to < about 1 mg/L. Actual data from field use of ClO_2 has shown very positive benefits, with little apparent damage to RO membranes.

E. Conclusions

Two case histories were presented, one using ClO_2 for cleaning of fouled RO membranes. The second was for low level application to prevent biofouling in RO membranes.

Cleaning Fouled RO Membranes with ClO_2: The principle thing learned from this case history is that badly fouled RO membranes already have tears or holes created by bacteria. The biofilm is known to act as a secondary membrane, and in such systems, when the biofilm is removed, leakage occurs. It is better to prevent biofilm formation on the membrane than to try and cleanup a badly fouled membrane, although this can be done if the membrane is not too badly fouled.

Control of RO Membrane Biofouling with ClO_2

Biofilm Prevention in RO Membranes with ClO_2: Although ClO_2 can be used to clean up moderately fouled RO membranes, it can be used in a much better, less costly, and more effective preventative tool as a routine part of the biofouling prevention program. To reduce costs associated with overall operation and in particular to minimize biofouling it is important to:

1. Provide water that is as well disinfected as possible – this will involve continuous use of either ozone or ClO_2, as these oxidants provide a much more complete kill than other oxidants, and will thus minimize any viable bacteria that get to the RO membrane.

2. Periodically provide ClO_2 disinfection and soak program to multimedia filters during the backwash procedure. This will minimize the potential for the multimedia being a source of bacterial growth.

3. Provide a continuous feed of chlorine-free ClO_2, at a rate of $\leq$ 1.0 mg/L (lower levels may be effective, but this will likely be system dependent). For example, application of very low ClO_2 residuals has shown to be effective at not only preventing, but cleaning up established biofilm in cooling tower fill, even when gray water is used as tower makeup [Nickels et al. 2000; Simpson and Miller 2000]. Low levels of ClO2 have also been effective in extended tests at preventing biofilm formation on RO membranes [Mainz and Simpson 2004; Pitochelli et al. 2005].

4. With some routine, the frequency to be determined and dictated by the needs of the specific application, a periodic cleaning procedure using ClO_2 sequentially with a good alkaline RO cleaner followed by a reverse-flow ClO_2 cleaning step, should be employed.

5. Chlorine dioxide chemistry is fairly complex. An in-depth knowledge of ClO_2 chemistry, generation, and analytical idiosyncrasies is essential for the production, application, and measurement of chlorine-free ClO_2. Therefore, any group or person that is consulted with for assistance in this area should have an established track record of experience with references.

F. References

[Adams 1990] Adams, W., "The Effects of Chlorine Dioxide on Reverse-Osmosis Membranes," *Desalination*, 78(3), 439 (1990).

[Aieta et al. 1984] Aieta, E.M., Roberts, P.V., and Hernandez, M., "Determination of Chlorine Dioxide, Chlorine, Chlorite, and Chlorate in Water," *Journal of the American Water Works Association,* 76, 64(January 1984).

[Aieta and Berg 1986] Aieta, E. and Berg, J., "A Review of Chlorine Dioxide in Drinking Water Treatment," *Journal of the American Water Works Association*, 78(6), 62(1986).

[Anon 2001] Anon, "Power Industry Expected to Pace Growth of Reverse Osmosis Water Treatment," *Power Engineering*, 105(9), 27(2001).

[Averett 2003] Averett, B., personal communication, December 3, 2003

[Averett et al. 2003] Averett, W., Simpson, G., and Miller, J., "Cleaning RO Membranes with Chlorine Dioxide," Southwest Chemistry Conference, sponsored by TXU Energy, Dallas, Texas, July 28 – August 1, 2003.

[Berg et al. 1981] Berg, J., Matin, A., and Roberts, P., "Growth of Disinfection Resistant Bacteria and Simulation of Natural Aquatic Environments in the Chemostat," *Water Chlorination, Environmental Impact and Health Effects*, Jolley, et al., eds., 4(81), 1137(1981).

[Bohner and Bradley 1990] Bohner, H. and Bradley, R., "Effective Control of Microbial Populations in Polysulfone Ultrafiltration Membrane Systems," *Journal of Dairy Science*, 73, 2309(1990).

[Bohner and Bradley 1991] Bohner, H., and Bradley, R., "Corrosivity of Chlorine Dioxide Used as a Sanitizer in Ultrafiltration Systems," *Journal of Dairy Science*, 74, 3348(1991).

[Breshnahan 1996] Bresnahan, W., "Reuse of Treated Municipal Wastewater in Oil Refinery Cooling Towers," *Industrial Water Treatment*, 22(May/June, 1996).

[Cameron et al. 1987] Cameron, G., Symons, J., Spencer, S., and Ma, J., "Comparison of Various Disinfectants for Control of the Asiatic Clam (*Corbicula fluminea*)," Proceedings, American Water Works Association Annual Conference, 1987.

[Cameron et al. 1989a] Cameron, G., Symons, J., Spencer, S., Ma, J., "Minimizing THM Formation During Control of the Asiatic Clam: A Comparison of Biocides," *Journal of the American Water Works Association*, 81(10), 53(1989).

[Cameron et al. 1989b] Cameron, G., Symons, J., Bushek, D., and Kilkarni, R., "Effect of Temperature and pH on the Toxicity of Monochloramine to the Asiatic Clam," *Journal of the American Water Works Association*, 63(October 1989).

[Clark and Langley 1990] Clark, J. and Langley, D., "Biofilm Control," United States Patent 4,929,365. May 29, [1990].

[Cooper et al. 1985] Cooper, W., Zika, R., and Steinhauer, M., "Bromide-Oxidant Interactions and THM Formation: A Literature Review," *Journal of the American Water Works Association*, Management and Operations, 116(April 1985).

[Costerton 1995] Costerton, J. "Overview of Microbiological Biofilms," *Journal of Industrial Microbiology*, 15, 137(1995).

[Costerton 1997] Costerton, W., private communication, September 1997.

[Fane 1997] Fane, A., Proceedings, Symposium on Characterization of Polymers with Surface, Lappeenranta, Finland, 51(1997).

[Flemming et al. 1993] Flemming, H., Schaule, G., and McDonogh, R., "How do Performance Parameters Respond to Initial Biofilm Formation on Separation Membranes?" *Vom Wasser*, 80, 177(1993).

[Flemming et al. 1997] Flemming, H., Schaule, G., Grieve, T., Schmitt, J., and Tamachkiarowa, A., "Biofouling – the Achilles Heel of Membrane Processes," *Desalination*, 113, 215(1997).

[Glater et al. 1981a] Glater, J., McCutchan, J., McCray, S., and Zachariah, M., "Halogen Interactions with Typical Reverse Osmosis Membranes, " AWWA Water Reuse Symposium II, Washington, DC, 2, 1339(August 1981).

[Glater et al. 1981b] Glater, J., McCutchan, J., McCray, S., and Zachariah, M., "The Effect of Halogens on the Performance and Durability of Reverse Osmosis Membranes," ACS Symposium Series, 153(1}, 173 (1981).

[Glater et al. 1983a] Glater, J., Zachariah, M., McCray, S., McCutchan, J., "Reverse Osmosis Membrane Sensitivity to Ozone and Halogen Disinfectants," *Desalination*, 48, 1(1983).

[Glater et al. 1983b] Glater, J., McCutchan, J., McCray, S., and Zachariah, M. "The Effect of Water Pretreatment Chemicals on the Performance and Durability of Reverse Osmosis Membranes." UCLA Report No. 831 (1983)

[Glater et al. 1994] Glater, J., Hong, S., and Elimelech, M., "The Search for a Chlorine-Resistant Reverse Osmosis Membrane," *Desalination* 95, 325(1994).

[Goss et al. 1977] Goss, L., Jackson, J., Flora, H., Isom, B., Gooch, C., Murray, S. Burton, C. and Bain, W., "Control Studies on Corbicula for Steam-Electric Generating Plants," Proceedings of the First International Corbicula Symposium, 139(1977).

[Hach] Hach Company, Loveland, CO.

[Jurado 1993] Jurado, J., "A Method of Evaluating Protein Removal by Disinfectants Used in Brewery CIP," *Journal of the American Society of Brewing Chemists*, 51(4), 149(1993).

[Kanaani et al. 1992] Kanaani, Y., Adin, A., and Rav-Acha, Ch., "Biofilm Interactions in Water

Reuse Systems: Adsorption of Polysaccharide to Kaolin," *Water Science and Technology*, 26(3-4),673(1992).

[Kemp and Okimoto 1982] Kemp, P., and Okimoto, N., "The Use of Subtoxic Levels of Chlorine Dioxide to Prevent Surface Condenser Fouling in Once-Through Seawater Cooling Exchangers," Proceedings 43rd International Water Conference, IWC 82-8, October 25 - 27, 1982.

[Lykins and Griese 1986] Lykins, B. and Griese, M., "Using Chlorine Dioxide for Trihalomethane Control," *Journal of the American Water Works Association*, 78(6), 88(June, 1986).

[Lykins et al. 1989] Lykins, B., Goodrich, J., Hoff, J., and Kothari, N., "Chlorine Dioxide for Drinking Water Disinfection," 1989 Montana Section of AWWA/ Montana WPC Association Joint Conference, Great Falls, Montana, March 29-31, 1989.

[Madaeni et al. 2001] Madaeni, S., Mohamamdi, R., and Moghadam, M., "Chemical Cleaning of Reverse Osmosis Membranes," *Desalination*, 134, 77(2001).

[Mainz and Simpson 2004] Mainz, E. and Simpson, G., "Method for Treating Reverse Osmosis Membranes with Chlorine Dioxide," patent pending, 2004.

[Martens 2004] Martens, E., private communication, July 2004.

[Mason 2003] Mason, J., private communication, February, 2003.

[McCray et al. 1981] McCray, S., Glater, J., and McCutchan, J., "The Effect of pH and Halogens on the Stability of Reverse Osmosis Membranes," UCLA-ENG-8115, WRC 73, July, 1981.

[McCutchan and Glater 1980] McCutchan, J., and Glater, J., "The Effect of Feedwateer Pretreatment Chemicals on the Performance and Durability of Reverse Osmosis Membranes," UCLA-ENG-8004, Water Resource Center 71, January 1980.

[McCutchan et al. 1982] McCutchan, J., Glater, J., McCray, S., Nobe, K., and Zachariah, M., "Saline Water Demineralization by Means of a Semipermeable Membrane," Water Resources Center, Desalination Report No 75, July 1982.

[McCutchan et al. 1983] McCutchan, J., Glater, J., McCray, S., Zacharian, M., and Nobe, K., "The Effect of Water Pretreatment Chemicals on the Performance and Durability of Reverse Osmosis Membranes," Office of Water Research and Technology Report, April, 1983.

[McDonough et al. 1994] McDonough, R., Schaule, G., and Flemming, H., "The Permeability of Biofouling Layers on Membranes," *Journal of Membrane Science*, 87, 199(1994).

[McVeigh et al. 1985] McVeigh, J., Lee, R., and Kelleher, D., "Chlorine Dioxide Use in Pretreatment for Control of Trihalomethane Generation," Proceedings of the AWWA Annual Conference, Washington, D.C., 1985.

[Miltner 1976] Miltner, R., "The Effect of Chlorine Dioxide on TrihaloMethanes in Drinking Water," MS Thesis, University of Cincinnati, 1976.

[Nickels et al. 2000] Nickels, A., Harriman, C., Nickels, A., Freedman, A., and Harris, A., "A Novel Method of Applying Chlorine Dioxide in a Recirculating Cooling Water System," AWT Annual Convention and Exposition, Water Technologies 2000, Honolulu, HI, October 31 - November 4, 2000.

[Palintest] Palintest USA, Erlanger, KY.

[Pickrell et al. 1999] Pickrell, W., Germer, M., Miller, J., and Simpson, G., "Plants Increase Production Through Use of Chlorine Dioxide," Practical Industrial Water Treatment Technology for the New Millennium, Houston, TX, January 27 - 29, 1999.

[Pitochelli 1995] Pitochelli, A., "Chlorine Dioxide Generation Chemistry," 3rd International Symposium: Chlorine Dioxide: Drinking Water, Process Water, and Wastewater Issues, New Orleans, LA., September 14 - 15, 1995.

[Pitochelli 1995] Pitochelli, A., "Generation of Chlorine Dioxide in a Non-Aqueous Medium," US Patent 5,405,549, April 11, 1995.

[Pitochelli 1997] Pitochelli, A., "Generation of Chlorine Dioxide in a Non-Aqueous Medium," US Patent 5,693,252, December 2, 1997.

[Pitochelli 1998] Pitochelli, A., "Generation and Storage of Chlorine Dioxide in a Non-Aqueous Medium," US Patent 5,707,546, January 13, 1998.
[Pitochelli 2005] Pitochelli, A., Private Communication, October 2005.
[Pitochelli and Mason 1994] Pitochelli, T. and Mason, J., "Large-Scale Field Applications of Chlorine Dioxide," Intertech Oxidative Treatment of Pollutants in Wastewater Conferences, Philadelphia, Pa., March 21 - 24, 1994.
[Pitochelli et al. 2005] Pitochelli, A., Mainz, E., and Griffith, D., "Continuous Chlorine Dioxide Use to Prevent Biofilm Formation on RO Membranes," *Ultrapure Water*, 40(May/June 2005).
[Pureline Treatment Systems] Pureline Treatment Systems, Lake Forest, CA.
[Robeck 1981] Robeck, G., "Chlorine, Is There a Better Alternative?," *Science of the Total Environment*, 18, 235(1981).
[Rybarik et al. 1995] Rybarik, D., Byron, J., and Germer, M., "Chlorine Dioxide Treatment for Zebra Mussel Control," Proceedings Fifth International Zebra Mussel and Other Aquatic Nuisance Organisms Conference, Toronto, Canada, February 1995.
[Saad 1992] Saad, M., "Biofouling Prevention in RO Polymeric Membrane Systems," *Desalination*, 2(88), 85(1992).
[Simpson 1995] Simpson, G., "Biofilm: Removal and Prevention with Chlorine Dioxide," Third International Symposium: Chlorine Dioxide: Drinking Water, Process Water, and Wastewater Issues, New Orleans, LA., September 14 - 15, 1995.
[Simpson 1997] Simpson, G., unpublished results, 1997.
[Simpson 2001] Simpson, G., "Control of Biofilm in Cooling Towers with ClO_2," Fourth International Symposium on Chlorine Dioxide, Las Vegas, Nevada, February 15 - 16, 2001.
[Simpson 2005a] Simpson, G., "Analytical," Chapter 8, Practical Chlorine Dioxide: Volume I – Foundations," Greg D. Simpson & Associates, Publishers, 2005.
[Simpson 2005b] Simpson, G., "The Reaction Chemistry of Chlorine (III)," Chapter 4, Practical Chlorine Dioxide: Volume I – Foundations," Greg D. Simpson & Associates, Publishers, 2005.
[Simpson and Miller 2000] Simpson, G. and Miller, J., "Control of Biofilm with Chlorine Dioxide," AWT Annual Convention and Exposition, Water Technologies 2000, Honolulu, HI, October 31 - November 4, 2000.
[Simpson et al. 1993] Simpson, G. D., Miller, R. F., Laxton, G. D., and Clements, W. R., "A Focus on Chlorine Dioxide: The "Ideal" Biocide," paper no. 472, Corrosion 93, New Orleans, La., March 8 - 12, 1993.
[Stevens 1982] Stevens, A., "Reaction Products of Chlorine Dioxide," *Environmental Health Perspectives*, 46, 101(1982).
[Suozzo and Suozzo 1998] Suozzo, J. and Suozzo, K., "Wastewater Management System," U.S. Patent 5,843,308, December 1, 1998.
[Sutherland 1990] Sutherland, I. W., *Biotechnology of Microbial Exopolysaccharides*, Cambridge University Press, [1990].
[Trefry 2003] Trefry, M., Private communication, December 2003.
[Vos et al. 1966] Vos, K., Burris, F., and Riley, R., "Kinetic Study of the Hydrolysis of Cellulose Acetate in the pH Range of 2-10," Journal of Applied Polymer Science, 10, (May 1966).
[Vos et al. 1968] Vos, K., Nusbaum, I., Hatcher, A., and Burris, F., "Storage, Disinfection, and Life of Cellulose Acetate Reverse Osmosis Membranes," *Desalination*, 5, 157(1968).
[Weaver-Meyers et al. 2000] Weaver-Meyers, P., Stolt, W., and Kowaleski, B., "Controlling Mold on Library Materials with Chlorine Dioxide: An Eight-Year Case Study," *Abbey Newsletter*, 24(4), (Dec 2000).
[Wilcut et al. 2003] Wilcut, E., Sims, J., and Tvedt, T., "San Antonio Water System's Cooling Tower Audit Program Results in Significant Water Savings," Cooling Technology Institute Annual Conference, Technical Paper, TP03-14, San Antonio, Texas, February 10 - 13, 2003.
[Wise et al. 2004] Wise, B., Marker, L., and Muellere, P., "Effectiveness of Chlorine Dioxide in Sanitizing Thin-Film Membrane Systems," *Ultrapure Water*, 13 (September 2004).

[Zachariah and Glater 1985] Zachariah, M. and Glater, J., "A Mechanistic Study of Halogen Interactions with Polyamide Reverse Osmosis Membranes," *Reverse Osmosis and Ultrafiltration*, S. Sourirajan, ed., 345(1985).

Chapter 21

Control of Biofilm with ClO_2

Chapter 21

A. Biofilm – What It Is

Anyone who has tried to walk across a mountain stream by stepping from rock to rock that has water flowing over it likely has first hand experience with biofilm. I did this as a young boy and ended up in the stream, soaked from head to toe, due to the slippery nature of the stepping stones. This slick material, commonly referred to as "slime," or more correctly referred to as biofilm, is produced by many bacteria. These bacteria attach to surfaces and begin to produce a polymeric byproduct that helps to protect them from a potentially hostile environment. This polymeric material is comprised primarily of polysaccharides.

The study of biofilm is a relatively new field, with most of the work being done since about 1980. Although there are several premier places where biofilm research is being done, one of the foremost is the Center of Biofilm Engineering (CBE), a multidisciplinary research facility located at Montana State University, Bozeman, MT, where a number of scientific disciplines are brought together to work jointly to help understand the phenomenon of biofilm growth, development, and control. CBE has an excellent web site, www.erc.montana.edu, where numerous technical papers written by CBE scientists can be found.

Biofilm growth occurs in three stages: induction, growth and plateau phases (Figure 21-1). During the induction phase, bacteria attach to surfaces and begin to produce biofilm. The induction period is a function of the roughness and composition of the surface to which bacteria attach [Characklis 1973a]. A major factor influencing the induction period include the type and amount of available bacterial nutrient [Characklis 1979].

Figure 21-1. The Three Phases of Biofilm Development

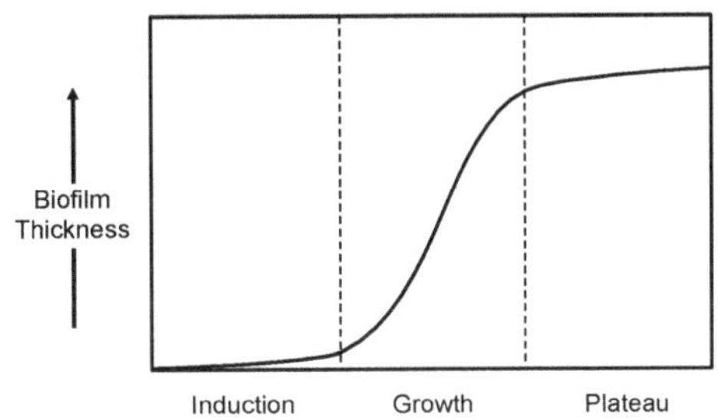

During the growth phase, the biofilm grows at a logarithmic rate which depends on factors such as the type and amount of nutrient available to bacteria, wall surface temperature, bulk water temperature, and wall shear stress. In general, the rate of biofilm development increases with the available bacterial nutrient.

The plateau phase begins when the biofilm thickness reaches a steady state, with the rate of bacterial production being roughly the same as the rate of bacterial loss through sloughing, dislodging, etc. The maximum thickness achieved by the biofilm is strongly dependent on shear stress and less dependent on nutrient loading rate in low shear stress experiments [Characklis and Marshall 1990].

Rather than being continuous, biofilm is patchy, the channels existing throughout the biofilm serving to provide a means of mass transport of materials through the biofilm (Figure 21-2).

Figure 21-2. Structure of Biofilm

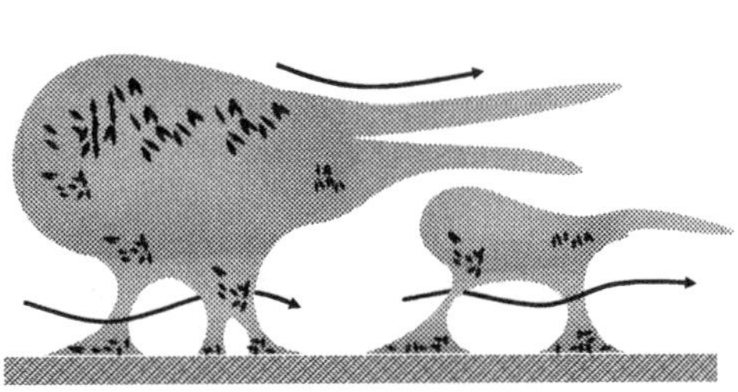

B. Biofilm - The Problems It Causes

Potable Water Systems. Biofilm causes several problems in potable water systems. These include water flow restrictions, nitrification, corrosion and deposition. Biofilm in potable water systems also provides a harbor for pathogenic bacteria.

Biofilm Restricts Water Flow: One problem that can result from biofilm growth is a restriction of water flow. Once the thickness of the biofilm exceeds that of the viscous sublayer [Bryers and Characklis 1979], the visco-elastic nature of the biofilm can cause significant restrictions in water flow - reductions far greater than can be explained by a simple reduction in effective pipe diameter [Characklis et al. 1980; Characklis et al. 1991]. In some cases, these frictional losses are very significant [Picologlou et al. 1980]. The growth caused by iron bacteria can also cause similar phenomenon, and are referred to in severe cases as "water calamities" [Starkey 1945]. One author noted that "it has been reported that these microorganisms have reduced the effective area of a 6" pipe to a 2" pipe in a matter of weeks" [Shair 1975].

Biofilm Encourages Nitrification: In potable water systems which use chloramine to provide the measurable disinfectant residual, the chloramine can decay in the distribution system to produce trace amounts of ammonia. The biofilm present on walls of distribution system can provide a place where bacteria which are able to metabolize the ammonia can proliferate. The metabolism by bacteria of the trace amounts of ammonia produce nitrite and nitrate, both of which are of concern in potable water.

Cooling Water Systems. Industrial systems generally have greater nutrient levels than potable water systems. This makes the biofilm grow more rapidly and magnifies the problems. These problems include restrictions in heat transfer, a variety of corrosion problems, deposit/scale formation and health concerns resulting from pathogenic bacteria that the biofilm may harbor. Problems may also include the flow restrictions discussed above.

Biofilm Interferes with Heat Transfer: Bacterial biofilm cause numerous problems in industrial plants. These problems include heat transfer restrictions [Schlesinger 1977; Characklis et al. 1991]. Even biofilm too small to be seen with the naked eye can result in substantial losses in heat transfer [Carter and McGraw 1979].

Biofilm Can Cause Corrosion: Another problem caused either directly or indirectly by bacterial biofilm is corrosion. The impact of corrosion caused by bacteria has been grossly underestimated, as work has shown that corrosion caused by microbes can be at least as significant as that caused by galvanic action [Costerton and Geesey 1986]. Corrosion can be caused by direct bacterial action [Costello 1969; Tatnall 1981; Costerton and Geesey 1986]. Corrosion by bacterial enzyme action [Boivin et al. 1990] has also been documented, and can occur even after cell death [Characklis et al. 1991]. It has been found that in order to control microbially induced corrosion (MIC), it may be necessary to remove the biofilm completely, even though no viable bacteria remain within [Characklis et al. 1991; Roe et al. 1996].

Corrosion through differential aeration cells [Updegraff 1955] can also occur as a result of the patchy nature of biofilm [Characklis and Marshall 1990]. In addition, the anaerobic conditions created under a healthy biofilm promote the growth and development of the highly corrosive sulfate-reducing bacteria, *Desulfovibrio desulfuricans* (SRB) [McKoy 1980; Hamilton and Maxwell 1986]. In fact, a biofilm layer only 10—12 organisms thick can result in anaerobic conditions sufficient to promote the growth of SRB [Sanders and Hamilton 1986; Watkins and

Costerton 1987; Pacheco et al. 1987].

Biofilm Can Cause Deposit Formation: Indirectly biofilm causes other problems. A biofilm provides a tacky surface where microscopic particulates of marginally soluble mineral scale, corrosion products, or silt can accumulate [Geesey et al. 1988; Reeves 1989]. Under high heat transfer rates, as the biofilm denatures, the deposit becomes more inorganic in nature because water constitutes a large portion of the biofilm. It is not uncommon for this type of problem to be erroneously attributed to scale or deposit formation.

Biofilm Provides a Harbor for Pathogenic Bacteria: Bacterial biofilm provides a place to harbor pathogenic organisms, and as such pose a serious health threat [Potera 1999]. Biofilm allows organisms such as Legionella to grow in a protected environment, and some believe that this is a major mechanism for growth of Legionella [Marrao et al. 1993; Bentham et al. 1993; Bentham and Broadbent 1995]

Paper Systems: Of all the industrial systems, pulp and paper systems may have the greatest level of nutrients of all the systems. Biological growth in these systems result in a high potential for corrosion, odor production, and slime formation, particularly on the paper machine.

Biofilm Causes Problems on the Paper Machine: The high level of nutrients creates an environment that can promote the growth of aerobic bacteria. When aerobic bacteria are favored, aerobic slime formers can grow rapidly on surfaces near the paper machine. When pieces of the slime sloughs off, holes and tears in the paper can result.

Biofilm Causes Odor Problems in Paper Mills: If conditions favor the growth of anaerobic bacteria, anaerobic reaction of nutrients by bacteria can produce a sour odor (see, for example, Chapter 19 in this text).

C. The Limitations of Chlorine for Biofilm Control

Gaseous chlorine has been used historically for disinfection. It is relatively inexpensive and it works almost all of the time. However, there are some circumstances that are system specific, under which chlorine does not perform adequately. For example, while acceptable as a disinfectant at neutral pHs, chlorine is known to lose disinfection efficiency with increasing pH. Specifically the effectiveness of hypochlorite ion (OCl^-), the predominant chlorine species formed as the pH increases above about 7.5, has been measured to be 1/20th [Robeck 1981], 1/80th [Johnson and Overby 1971], and 1/300th [White 1986] as effective as hypochlorous acid (HOCl), depending on the organism tested. Other conditions, and the performance deficiencies of chlorine in these systems, have been described in the literature and are mentioned below.

One early paper was done to investigate the impact hypochlorite had on biofilm [Characklis 1971]. This work showed that the use of hypochlorite resulted in a slow process of dissolution or breaking off of small pieces of biofilm rather than a massive sloughing phenomenon. This was supported by the measured total organic carbon (TOC) being significantly greater than that of soluble organic carbon (SOC). In addition, the data suggested that some of the polymeric material being released from the biofilm was acting as a "drag reducer." That is, some of the polymeric material being released immediately after treatment with hypochlorite was thought to be working to minimize frictional resistance to flow. Strong oxidants remove small pieces of the biofilm, and the deposits or biofilm which remain accelerate regrowth.

A number of researchers have demonstrated in laboratory studies that chlorine does not penetrate a biofilm well [Marshall and Walker 1990; de Beer et al. 1994; Chen and Stewart 1996; Xu et al. 1996]. Instead, chlorine has been found to be more reactive with system components, as illustrated by the fact that chlorine reacted with the testing apparatus, unlike monochloramine [Chen et al. 1993]. In addition, hypochlorite ion has been found to be so reactive with the biofilm constituents that it is effectively removed by reaction before it can penetrate the biofilm [Boivin et al. 1992].

Corresponding field results showed that chlorine does not control biofilm. One group noted that simply maintaining a chlorine residual does not eliminate all bacteria in a potable water distribution system [LeChevallier et al. 1990]. Another group found that maintaining a free chlorine residual of 1—2 mg/L was not sufficient to prevent the growth of biofilm, and this was in a potable water distribution system with minimal organic contamination as compared to industrial cooling systems [Nagy et al. 1982]. In other work, research showed that 4.3 mg/L free chlorine was not sufficient to eliminate coliform occurrence in a New Jersey municipality [LeChevallier et al. 1987]. In another municipality, free chlorine levels as high as 10—12 mg/L chlorine were ineffective at controlling coliform bacteria; 15 mg/L free chlorine was required [Earnhardt 1980]. Others showed in an industrial account that 2 mg/L free chlorine at pH 8.5 was not effective at preventing the growth of sessile bacteria [Franklin et al. 1991]. Treatment with 16 mg/L chlorine was found to be required for biofilm removal, with continued treatment at 2 mg/L being required to prevent its recurrence.

Others have found that the nature of the surface substrate is an important factor in determining the relative difficulty of biofilm removal. For example, one group recovered coliforms from a potable water supply after cleaning biofilm from a redwood water storage tank with a 200 mg/L chlorine solution [Seidler et al. 1977]. Another researcher found that the relative effectiveness and difficulty of removal of biofilm was different for biofilm grown on granular activated carbon (GAC) [LeChevallier 1990b]. Still others found that 15—20 mg/L free chlorine was necessary to control biofilm on reverse osmosis membranes [Ridgway et al. 1984].

These reports suggest that maintaining a free chlorine residual is not sufficient to inactivate all bacteria in a cooling water system. Some of the bacteria which survive the disinfection process can attach and produce biofilm. The use of free chlorine has not been found to be sufficient to either prevent remove biofilm. Thus, when using chlorine, it is not possible to clean up a biofouled system simply by resuming a discontinued microbiological control program. A fully developed biofilm can be removed only through intensified on-line or off-line treatment [Kozelski 1983].

Reconciling Research Results with Historical Evidence: Given the preponderance of evidence that says chlorine doesn't penetrate biofilm well, how does one reconcile this with the fact that chlorine has been used historically with generally good results? That is, why does chlorine work well under most conditions, but in others it does not work well? In order to understand this, one must look at what happens under normal conditions and then look at what happens under conditions where chlorine does not perform at an acceptable level.

In systems with low levels of bacterial nutrient, the rate of bacterial growth is relatively slow. The rate of removal of biofilm with chlorine is relatively constant, being limited by the ability of chlorine to get to the surface of the biofilm. Under such conditions, chlorine has the ability to remove the biofilm at a rate faster than the rate of biofilm production (Figure 21-3).

As the level of metabolizable bacterial nutrient increases, so does the rate of biofilm production. Thus, in systems with high concentrations of nutrient, the rate of biofilm production may be greater than the rate of biofilm removal with chlorine (Figure 21-4).

Figure 21- 3. Biofilm Removal with Cl_2 – Low Level of Bacterial Nutrient

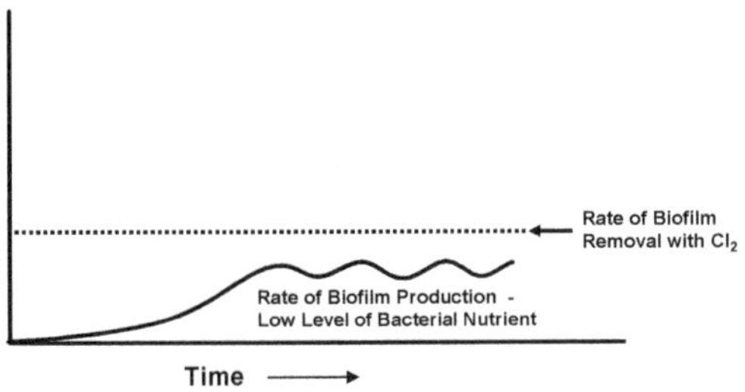

Figure 21- 4. Biofilm Removal with Cl_2 – High Level of Bacterial Nutrient

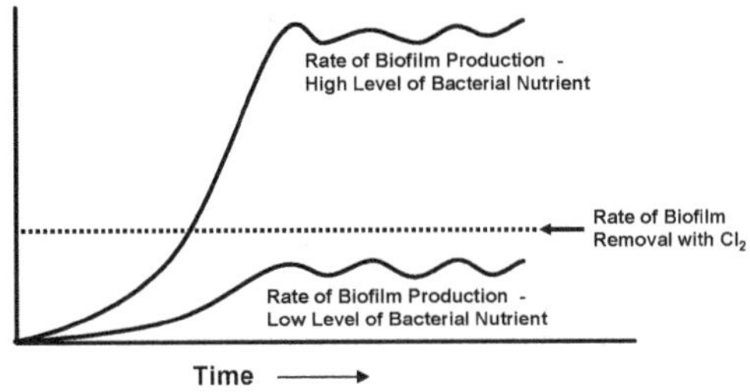

D. Chloramine for Biofilm Control

Researchers at the CBE have proposed that weaker oxidants may provide for better control of biofilm, because the weaker the oxidant, the less reactive the oxidant with biofilm constituents [Costerton 1997]. Thus the weaker oxidant penetrates the biofilm and is able to get to the bacteria within. This is confirmed by work that has shown that sessile bacteria, i.e., those attached to surfaces, are 150—3000 times more resistant to hypochlorous acid than bacteria in the bulk water, while they are only 10—100 times more resistant to chloramine than bacteria in the bulk water [LeChevallier et al. 1988].

Chloramine is a weaker oxidant than chlorine. Work has shown that chloramine was able to penetrate a biofilm better than chlorine and inactivate bacteria within [Chen et al. 1993; Yu and McFeters 1994]. However, monochloramine did not remove the biofilm [Chen et al. 1993; Yu et al. 1993; Huang et al. 1995]

E. Literature Review – ClO_2 for Biofilm Control

The disinfection requirements of an open recirculating industrial cooling system are markedly different from those of a potable water treatment facility. The goal of potable water treatment is the inactivation of specific water borne pathogens. The goal of disinfection for industrial cooling systems, in addition to pathogen (Legionella) control, is the removal, minimization, or prevention of biofilm, which retards heat transfer, causes biofouling, provides a place of agglomeration for marginally soluble or insoluble salts, and provides a place which nurtures and promotes the growth of highly corrosive anaerobic bacteria [Characklis and Marshall 1990].

In the pulp and paper industry, slime control in alkaline whitewater systems is a major problem. There are numerous reported case histories of ClO_2 being used for this purpose [Baker 1981a; Baker 1981b; Baker 1980; Nelson 1982a; Nelson 1982b]. A major strength of ClO_2 in the pulp and paper industry is its ability to disperse and remove existing biofilm and to prevent its recurrence, even with intermittent treatment. In most paper mills, non-oxidizing biocides are used. Non-oxidizing biocides are typically bacterial poisons and are less effective than ClO_2 at controlling biofilm.

There have also been numerous examples of ClO_2 being used to control biofilm in cooling systems. In virtually all of these examples, ClO_2 was used because chlorine was ineffective. For example, ClO_2 has been used for biofilm control with great success in biofouled power plant utility cooling systems [Kemp and Okimoto 1982; Sussman and Ward 1977], in petrochemical plant cooling systems [Sussman and Ward 1977; Simpson et al. 1993a; Simpson et al. 1993b], in petroleum refinery cooling systems [McGuire and Dishinger 1984; Sussman and Ward 1977; Simpson et al. 1993a; Simpson et al. 1993b], in ethylene glycol contaminated systems [Pacheco et al. 1989], in systems contaminated with grain alcohol [Freund et al. 1984], in cooling systems contaminated with ammonia [Sussman and Ward 1977; Simpson et al. 1993a; Simpson et al. 1993b; Ward et al. 1978], and for slime control in food plant processing cooling systems [Costillow et al. 1984; Sussman and Ward 1977; Synan 1979]. In addition, ClO_2 is used to control biofilm in the oil production industry [Sacco 1984; Knickrehm et al. 1987; McCafferty et al. 1990; Pitochelli and Mason 1994]. Several laboratory studies of the use of ClO_2 for biofilm control have been undertaken [LeChevallier 1990; Mayack et al. 1984].

In addition, ClO_2 has been shown to be very effective at biofilm removal, even in cooling systems where ClO_2 has been applied at very low residuals on an intermittent basis [Simpson 1995; DeBenedetto et al. 1999; Pickrell et al. 1999; Nickels et al. 2000; Simpson and Miller 2000; Simpson 2001].

In addition to controlling bacterial biofilm better than chlorine, ClO_2 does not suffer from the other drawbacks associated with use of chlorine. For example, several researchers have suggested that ClO_2 is a better choice in highly contaminated systems, i.e., those with high chlorine demand [Pacheco et al. 1987; Marshall and Walker 1990].

Others have reported on their experience in controlling microbially induced corrosion in industrial cooling systems [Pacheco et al. 1987; Simpson et al. 1993b]. In particular, use of chlorine resulted in a high level of corrosion without the benefit of biofilm removal or control. Replacement of chlorine with ClO_2 resulted in both biofilm control and reduced corrosion.

F. Why ClO_2 is so Effective at Biofilm Control - The Role of the Chlorite Ion

Chlorine dioxide, although it has a greater oxidation capacity than chlorine, usually accepts only one electron in an oxidation-reduction reaction, forming chlorite ion. This raises the question as to why chlorine dioxide is so effective at disinfection and biofilm removal. For disinfection, some have suggested that because it is a relatively non-polar, undissociated molecule, ClO_2 can permeate bacterial cell walls much easier than highly polar molecules such as chlorine or hypochlorite ion [Ingols and Ridenour 1948]. The affinity of ClO_2 for reduced sulfur compounds, including the sulfhydryl group, has also been suggested as one mode of bacterial disinfection [Green and Stumpf 1946]. Whatever the mode of action, it appears that chlorine dioxide disrupts some vital cell process, causing rapid cell death.

For biofilm removal, other factors appear to be involved. For an open recirculating cooling system in a petroleum refinery, petrochemical plant or gas plant, the recirculation rate is in the range of 10—15% of the system volume. This means that the entire volume of the cooling system will be turned over 4—6 times in an hour. If a material were to be fed to the system on the basis of recirculation rate at 1 mg/L, the concentration of this material would be 4—6 mg/L after one hour, neglecting blowdown or other losses. For HVAC systems, this ratio is considerably higher, i.e., it might take only 2—3 minutes for the entire cooling water volume to be treated once.

For chlorine dioxide, 50—80% of the chlorine dioxide converts to chlorite in the disinfection process [Miltner 1976; Aieta et al. 1984; Werdehoff and Singer 1987]. Thus, at the end of the intermittent treatment, around 2—4 mg/L of chlorite ion might remain in the system for a significant period of time. This chlorite, a relatively poor disinfectant and oxidant, was previously thought to remain in the system and be lost only through blowdown. However, during numerous field trials testing of the cooling water for chlorite showed a decay in the chlorite level that was much more rapid than could be explained solely by blowdown considerations.

In addition, during these field trials bacterial recovery or rebound was slower than expected. This had been attributed to a complete removal of bacterial biofilm, making it more difficult for the bacterial population to reestablish, as partial removal of biofilm during treatment was known to accelerate the bacterial rebound rate [Characklis 1979]. Subsequent inspections revealed that while much of the biofilm had been removed, some biofilm was still intact. Questions remained about the slow bacterial recovery.

Several researchers noted that the presence of chlorite had a significant effect on bacterial recovery and biofilm growth. One researcher noted that while chlorite is not very effective as a disinfectant, it has bacteriostatic properties, and retards the rate of bacterial rebound in potable water systems [Masschelein 1991].

One possible reason for the relatively slow re-growth of bacteria after sterilization by chlorine dioxide lies in its superior ability to penetrate and disperse a biomass. By effectively killing and stripping off the vast majority of biofilm, bacteria are much slower to re-establish than when the biofilm is left intact [Characklis and Marshall 1990].

Others demonstrated biofilm removal by adding "stabilized" chlorine dioxide (in reality, buffered chlorite) to a system and then adding a bacterial nutrient [Clark and Langley 1990]. This nutrient resulted in an increase in bacterial activity which in turn resulted in a greater amount of acid production in the biofilm. Biofilm composition has been shown to contain acidic polysaccharides [Kanaani et al. 1992; Sutherland 1990]. With the addition of buffered chlorite, the acidic biofilm was contacted and chlorous acid that resulted either acted as a disinfectant alone, or it disproportionated to ClO_2. Biofilm removal resulted.

In one potable water system, chlorine dioxide was consumed in the distribution system, the residual disappearing. Yet the residual reappeared at the end of the distribution system [Hoehn et al. 1990]. Reasons were not known, although one explanation was that traces of residual chlorine reacted with the chlorite to produce a measurable residual of chlorine dioxide [Hoehn 1995]. Based on the above observations, it may have been the result of biofilm action which regenerated the chlorine dioxide.

G. Summary

ClO_2 is much more effective at controlling biofilm than chlorine. Chlorine reacts with biofilm constituents and is consumed before it can penetrate the biofilm. In systems that are nutrient limited, this does not matter a great deal. In systems that are nutrient rich, it may become very important.

ClO_2 reacts to form the chlorite ion, which is generally unreactive. However, the chlorite ion provides a reservoir for on-going action to control biofilm, by remaining in solution until it contacts acidic components of the biofilm, where it is activated to chlorous acid, a reasonably good disinfectant. Disinfection and some biofilm removal results.

H. References

[Aieta et al. 1984] Aieta, E., Roberts, P., and Hernandez, M., "Determination of Chlorine Dioxide, Chlorine, Chlorite, and Chlorate in Water," *Journal of the American Water Works Association*, 76(1), 64(January 1984).

[Baker 1980] Baker, E., "Chlorine Dioxide Proves an Effective Biocide in Alkaline Whitewater Systems," *Pulp and Paper*, 54(8), 140(1980).

[Baker 1981a] Baker, E., "Slime Control in Alkaline Paper Machine Systems - Chlorine Dioxide," Proceedings of the Papermakers Conference, 49(1981).

[Baker 1981b] Baker, E., "Using Chlorine Dioxide for Slime Control in Alkaline Paper Machine Systems," *TAPPI Journal*, 64(8), 91(1981).

[Bentham and Boradbent 1995] Bentham, R. and Broadbent, C., "Field Trial of Biocides for Control of Legionella in Cooling Towers," *Current Microbiology*, 30, 167(1995).

[Bentham et al. 1993] Bentham, R., Broadbent, C., and Marwood, L., "The Influence of the Sessile Population in the Legionella Colonization of Cooling Towers," Legionella: Current Status and Emerging Perspectives, Barbaree, Breiman and Dufour, eds., American Society for Microbiology, 1993.

[Boivin et al. 1990] Boivin, J., Laishley, E., Bryant, R., and Costerton, J., "The Influence of Enzyme Systems on MIC," Paper No. 128, Corrosion 90, Las Vegas, NV, April 23-27, 1990.

[Bryers and Characklis 1979] Bryers, J. and Characklis, W., "Measurement of Primary Biofilm Formation," EPRI Condenser Biofouling Control Symposium, Atlanta, GA, March 25-29, 1979.

[Carter and McGraw 1979] Carter, D. and McGraw, P., "The Monitoring and Control of Microbicides," *Water Technology*, September/October 1979.

[Characklis 1971] Characklis, W., "Effect of Hypochlorite on Microbial Slimes," Proceedings of the 26th Industrial Waste Conference, Purdue University, Lafayette, IN, 171, May 4-6, 1971.

[Characklis 1979] Characklis, W., "Biofilm Development and Destruction in Turbulent Flow," CTI Paper No. 204A, Cooling Tower Institute, January 22-24, 1979.

[Characklis and Marshall 1990] Characklis, W. and Marshall, K., Biofilms, Wiley Interscience Publications, John Wiley & Sons, Inc., New York, 1990.

[Characklis et al. 1980] Characklis, W., Zelver, N., and Picologlou, B., "Biofilm Growth and Hydraulic Performance," *Journal of Hydraulics Div of ASCE*, 106, 733(May 1980).

[Characklis et al. 1991] Characklis, W., Little, B., Stoodley, P., and McCaughey, M., "Microbial Fouling and Corrosion in Nuclear Power Plant Service Water Systems," Paper No. 281, Corrosion 91, Cincinnati, OH, March 11-15, 1991.

[Chen and Stewart 1996] Chen, X. and Stewart, P., "Chlorine Penetration into Artificial Biofilm is Limited by a Reaction-Diffusion Interaction," *Environmental Science and Technology,* 30(6), 2078(1996).

[Chen et al. 1993] Chen, C., Griebe, T., and Characklis, W., "Biocide Action of Monochloramine on Biofilm Systems of *Pseudomonas aeruginosa*," *Biofouling*, 7, 1(1993).

[Clark and Langley 1990] Clark, J. and Langley, D., "Biofilm Control," U.S. Patent 4,929,365. May 29, 1990.

[Costello 1969] Costello, J., "The Corrosion of Metals by Micro-organisms, A Literature Survey," *International Biodeterioration Bulletin*, 5(3), 101(1969).

[Costerton 1997] Costerton, W., private communication, September 1997.

[Costerton and Geesey 1986] Costerton, J. and Geesey, G., "The Microbial Ecology of Surface Colonization and of Consequent Corrosion," Biologically Induced Corrosion, NACE, Houston, 223(1986).

[Costilow et al. 1984] Costilow, R., Uebersax, M., and Ward, P., "Use of Chlorine Dioxide for Controlling MicroOrganisms During the Handling and Storage of Fresh Cucumbers," *Journal of Food Science*, 49, 396(1984).

[de Beer et al. 1994] de Beer, D., Srinivasan, R., and Stewart, P., "Direct Measurement of Chlorine Penetration into Biofilms during Disinfection," *Applied and Environmental Microbiology*, 60(12), 4339(1994).

[DeBenedetto et al. 1999] DeBenedetto, R., Freedman, A., Boschetti, D. and Hays, G., "Cooling Towers and Secondary Wastewater Makeup: Four Years' Operating Experience," International Water Conference, Paper No IWC-99-42, 1999.

[Earnhardt 1980] Earnhardt, K., "Chlorine Resistant Coliform – The Muncie, Indiana, Experience," Proceedings AWWA Water Quality Technology Conference, Miami Beach, FL, December 7-10, 1980.

[Franklin et al. 1991b] Franklin, M., Nivens, D., Vass, A., Mittelman, M., Jack, R., Dowling, N., and White, D., "Effect of Chlorine and Chlorine/Bromine Biocide Treatments on the Number and Activity of Biofilm Bacteria and on Carbon Steel Corrosion," *Corrosion*, 47(2), 128(1991).

[Freund et al. 1984] Freund, J., Booker, J., and Ward, W., "Biocontrol - Key to Heat Exchanger Efficiency in Grain Alcohol Plant," Proceedings, International Water Conference, 1984.

[Geesey et al. 1988] Geesey, G., Jang, L., Jolley, J., Hankins, M., Iwaoka, T., and Griffiths, P., "Binding of Metal Ions by Extracellular Polymers of Biofilm Bacteria," *Water Science and Technology*, 20(11/12), 161(1988).

[Green and Stumpf 1946] Green, D. and Stumpf, P., "The Mode of Action of Chlorine," *Journal of the American Water Works Association*, 38, 1301(November, 1946).

[Hamilton and Maxwell 1986] Hamilton, W. and Maxwell, S., "Biological and Corrosion Activities of Sulfate Reducing Bacteria within Natural Biofilms," Biologically Induced Corrosion, NACE, Houston, 131(1986).

[Hoehn 1995] Hoehn, R., "Chlorine Dioxide Treatment of Drinking Water: Pros and Cons," Association of Environmental Engineering Professors Lecture, American Water Works Association Annual Conference, Anaheim, California, June 19, 1995.

[Hoehn et al. 1990] Hoehn, R., Dietrich, A., Farmer, W., Orr, M., Lee, R., Aieta, E., Wood, D., and Gordon, G., "Household Odors Associated with the use of Chlorine Dioxide," *Journal of the American Water Works Association*, 166(April 1990).

[Huang et al. 1995] Huang, C., Yu, F., McFeters, G., and Stewart, P., "Nonuniform Spatial Patterns of Respiratory Activity within Biofilms during Disinfection," *Applied and Environmental Microbiology*, 61(6), 2252(June 1995).

[Ingols and Ridenour 1948] Ingols, R. and Ridenour, G., "Chemical Properties of Chlorine Dioxide," *Journal of the American Water Works Association*, 40, 1207(1948).

[Johnson and Overby 1971] Johnson, J. and Overby, R., "Bromine and Bromamine Disinfection Chemistry," *Journal of the Sanitary Engineering Division, American Society of Civil Engineers*, 617(October 1971).

[Kanaani et al. 1992] Kanaani, Y., Adin, A., and Rav-Acha, Ch., "Biofilm Interactions in Water Reuse Systems: Adsorption of Polysaccharide to Kaolin," *Water Science and Technology*, 26(3-4), 673(1992).

[Kemp and Okimoto 1982] Kemp, P. and Okimoto, N., "The Use of Subtoxic Levels of Chlorine Dioxide to Prevent Surface Condenser Fouling in Once-Through Seawater Cooling Exchangers," International Water Conference, 1982.

[Knickrehm et al. 1987] Knickrehm, M., Caballero, E., Romualdo, P., and Sandidge, J., "Use of Chlorine Dioxide in a Secondary Recovery Process to Inhibit Bacterial Fouling and Corrosion," Paper No. 383, Corrosion 87, San Francisco, Ca., March 9-13, 1987.

[Kozelski 1983] Kozelski, K., "Field Experience with a Simple Cooling Water Biofilm Monitoring Device," Paper Number IWC-83-46, International Water Conference, Pittsburgh, PA, October 24-26, 1983.

[LeChevallier 1990] LeChevallier, M., "Disinfection of Bacterial Biofilms," *Water Chlorination*, 6, 905(1990).

[LeChevallier et al. 1987] LeChevallier, M., Lee, R., and Moser, R., "Examination and Characterization of Distribution System Biofilms," Research Report, American Water Works Service Company, Inc., 1987.

[LeChevallier et al. 1988] LeChevallier, M., Cawthon, C., and Lee, R., "Inactivation of Biofilm Bacteria," *Applied and Environmental Microbiology*, 54(10), 2492(October 1988).

[LeChevallier et al. 1990] LeChevallier, M., Lowry, C., and Lee, R., "Disinfecting Biofilms in a Model Distribution System", *Journal of the American Water Works Association*, 87(July 1990).

[Marrao et al. 1993] Marrao, G., Verissimo, A., Bowker, R., and da Costa, M., "Biofilms as Major Sources of *Legionella* sp. in Hydrothermal Areas and their Dispersion into Stream Water," *FEMS Microbiology Ecology* 12, 25(1993).

[Marshall and Walker 1990] Marshall, A., and Walker, G., "Increased Awareness of Biological Control and Its Implications," Paper No 354, Corrosion 90, Las Vegas NV, April 23-27, 1990.

[Masschelein 1991] Masschelein, W. J., "Chlorine Dioxide," Chemical Oxidation, Technology for the Nineties, First International Symposium, Vanderbilt University, Nashville, TN, February 20-22, 1991.

[Mayack et al. 1984] Mayack, L., Soracco, R., Wilde, E. and Pope, D., "Comparative Effectiveness of Chlorine and Chlorine Dioxide Regimes for Biofouling Control," *Water Research*, 18(5), 593(1984).

[McCafferty et al. 1990] McCafferty, J., Tate, E. W., and Williams, D. A., "Field Performance in the Practical Application of Chlorine Dioxide as a Stimulation Enhancement Fluid," SPE, 65th Technical Conf, New Orleans, Sept 23-26, 1990.

[McGuire and Dishinger 1984] McGuire, L. and Dishinger, T., "Chlorine Dioxide Solves Biofouling Problems in a Refinery Cooling Tower Used for Phenol Destruction," CTI, Houston, 1984.

[McKoy 1980] McKoy, J., Microbiology of Cooling Water, Chemical Publishing Company, New York, New York, 1980.

[Miltner 1976] Miltner, R., "Measurement of Chlorine Dioxide and Related Products," Proceedings 1976 American Water Works Association Water Quality Technology Conference, San Diego, CA, 1976.

[Nagy et al. 1982] Nagy, L., Kelly, A., Thun, M., and Olson, B., "Biofilm Composition, Formation and Control in the Los Angeles Aqueduct System," Proceedings of the 1982 American Water Works Association Water Quality Technology Conference, Nashville, TN, December 5-8, 1982.

[Nelson 1982a] Nelson, T., "Appleton Papers finds Chlorine Dioxide to be an Alternative to Conventional Biocides in Alkaline Systems," *TAPPI Journal*, 65(6), 69(1982).

[Nelson 1982b] Nelson, T., "Chlorine Dioxide: An Alternative to Conventional Biocides in Alkaline Systems," Annual Meeting, TAPPI, 261(1982).

[Nickels et al. 2000] Nickels, A., Harriman, C., Nickels, A., Freedman, A., and Harris, A., "A Novel Method of Applying Chlorine Dioxide," AWT Annual Conference, November 4, 2000.

[Pacheco et al. 1987] Pacheco, A., Dishinger, T., and Tomlin, J., "Experiences in Controlling Microbiologically-Induced Corrosion," Paper No. 376, Corrosion 87, San Francisco, CA, March 9-13, 1987.

[Pacheco et al. 1989] Pacheco, A., Durham, H., Dhillon, R., and Edward, C., "The Use of Chlorine Dioxide to Control Microbiological Growth in an Ethylene Glycol Contaminated Cooling Tower...A Case History," TP89-14, Cooling Tower Institute, 1989.

[Pickrell et al. 1999] Pickrell, W., Germer, M., Miller, J., and Simpson, G., "Plants Increase Production through Use of Chlorine Dioxide," First International Conference: Practical Industrial Water Treatment Technology for the New Millennium, Houston, Texas, January 27-29, 1999.

[Picologlou et al. 1980] Picologlou, B., Zelver, N., and Characklis, W., "Biofilm Growth and Hydraulic Performance," *Journal of Hydraulics Division of ASCE*, 738 (May, 1980).

[Pitochelli and Mason 1994] Pitochelli, T. and Mason, J. Y., "Large-Scale Field Applications of Chlorine Dioxide," Oxidative Treatment of Pollutants in Wastewater, Intertech Conferences, Philadelphia, Pa., March 21-24, 1994.

[Potera 1999] Potera, C., "Forging a Link Between Biofilms and Disease," *Science*, 283, 183(March 19, 1999).

[Rauh 1976] Rauh, J., "Chlorine Dioxide. A Cooling Water Microbiocide," Proceedings of the Annual Industrial Pollution Conference, Water Wastewater Equipment Manufacturers Association, 4th, LIII, 20, McLean, VA, 1976.

[Reeves 1989] Reeves, N. K., "A Study of Bacteria-Influenced Scale Deposition," SPE Joint Rocky Mountain Region/Low Permeability Reservoirs Symposium and Exhibition, Denver, CO, March 6-8, 1989.

[Robeck 1981] Robeck, G., "Chlorine, Is There a Better Alternative?" *Science of the Total Environment*, 18 235(1981).

[Ridgway et al. 1984] Ridgway, H., Justice, C., Whittaker, C., Argo, D., and Olson B., "Biofilm Fouling of RO Membranes – Its Nature and Effect on Treatment of Water for Reuse," *Journal of the American Water Works Association*, 76(6), 94(June 1984).

[Roe et al. 1996] Roe, F., Lewandowski, Z., and Runk, T., "Simulating Microbiologically Influenced Corrosion by Depositing Extracellular Biopolymers on Mild Steel Surfaces," *Corrosion*, 52(10), 744(October 1996).

[Sacco 1984] Sacco, F., "The Use of Chlorine Dioxide in a Late Life Waterflood," Symposium on Advances in Oilfield Chemicals and Chemistry, Division of Petroleum Chemistry, Inc., American Chemical Society, St. Louis, April 8-13, 1984.

[Sanders and Hamilton 1986] Sanders, P. and Hamilton, W., "Biological and Corrosion Activities of Sulfate Reducing Bacteria in Industrial Process Plant," <u>Biologically Induced Corrosion</u>, NACE, Houston, 47(1986).

[Schlesinger 1977] Schlesinger, H. A., "Economics of Alternatives for OTEC Biofouling Protection," Ocean Thermal Energy Conversion Biofouling and Corrosion Symposium, Seattle, WA, Oct. 10-12, 1977.

[Seidler et al. 1977] Seidler, R., Morrow, J., and Bagley, S., "Klebsielleae in Drinking Water Emanating From Redwood Tanks," *Applied and Environmental Microbiology*, 33, 893(1977).

[Shair 1975] Shair, S., "Iron Bacteria and Red Water," *Industrial Water Engineering*, March/April, 1975.

[Simpson 1995] Simpson, G., "Biofilm: Removal and Prevention with Chlorine Dioxide," The Third International Symposium on Chlorine Dioxide: Process Water, Drinking Water and Water Waste Issues, New Orleans, LA, September 14-15, 1995.

[Simpson 2001] Simpson, G., "Control of Biofilm in Cooling Towers with Chlorine Dioxide," Fourth International Symposium on Chlorine Dioxide, Las Vegas, NV, February 15-16, 2001.

[Simpson and Miller 2000] Simpson, G., and Miller, J., "Control of Biofilm with ClO_2," Association of Water Technologies Annual Meeting, Honolulu, HA, October 28 – November 5, 2000.

[Simpson et al. 1993a] Simpson, G., Laxton, G., Miller, R., and Clements, W., "A Focus on Chlorine Dioxide: The 'Ideal' Biocide," Paper No. 472, Corrosion 93, New Orleans, LA, March 8-12, 1993.

[Simpson et al. 1993b] Simpson, G., Laxton, G., Miller, R., and Clements, W., "A Focus on Chlorine Dioxide: Biocide of Choice for 'Stressed' Cooling Water Systems," WaterTech 93, Houston, TX, 1993.

[Srinivasan et al. 1995] Srinivasan, R., Stewart, P., Grieve, T., Chen, C., and Xu, X., "Biofilm Parameters Influencing Biocide Efficacy," *Biotechnology and Bioengineering*, 46, 553(1995).

[Starkey 1945] Starkey, R., "Transformations of Iron by Bacteria in Water," *Journal of the American Water Works Association*, 37(10), 963(October, 1945).

[Stewart et al. 1994] Stewart, P., Griebe, T., Srinivasan, R., Chen, C., Yu, F., deBeer, D., and McFeters, G., "Comparison of Respiratory Activity and Culturability during Monochloramine

Disinfection of Binary Population Biofilms," *Applied and Environmental Microbiology*, 60(5), 1690(May1994).

[Sussman and Ward 1977] Sussman, S. and Ward, W., "Microbiological Control with Chlorine Dioxide Helps Save Energy," *Materials Performance*, 16(7), 24 (1977).

[Sussman and Ward 1979] Sussman, S. and Ward, W., "Chlorine Dioxide is an Attractive Treatment Alternative," *Water and Sewage Works*, 120 (1979).

[Sutherland 1990] Sutherland, I., Biotechnology of Microbial Exopolysaccharides, Cambridge University Press, 1990.

[Synan 1979] Synan, J. F., "Chlorine Dioxide - An Effective Biocide for Recycled or Reused Water Systems," Transactions of the Citrus Engineering Conference, 25, 66 (1979).

[Tatnall 1981] Tatnall, R., "Fundamentals of Bacteria Induced Corrosion," *Materials Performance*, 32(September 1981).

[Thompson 1992] Thompson, Skip, Galveston County Water Authority, Texas City, TX, private communication, March, 1992.

[Trulear and Wiatr 1988] Trulear, M. and Wiatr, C., "Recent Advances in Halogen Based Biocontrol," Technical Paper No. 19, Corrosion 88, March 21-25, St. Louis, MO, 1988.

[Updegraff 1955] Updegraff, D., "Microbiological Corrosion of Iron and Steel," University of Oklahoma Short Course on Corrosion, Norman, OK, April 7, 1955.

[Videla et al. 1991] Videla, H., Sautu, A., Gomez de Saravia, S., Guiamet, P., and de Mele, M., Gaylarde, C., and Beech, I., "Impact of Glutaraldehyde on Biofouling and MIC of Different Steels. A Laboratory Assessment," Paper No. 105, Corrosion 91, Cincinnati, OH, March 11-15, 1991.

[Wadsworth and Simpson 1997] Wadsworth, J. and Simpson, G., "Control of Biofilm in Alkaline Whitewater Systems with Chlorine Dioxide," TAPPI Papermakers Conference, Nashville, TN, October 6–9, 1997.

[Waite et al. 1978] Waite, T., Jorden, R., and Kawaratani, R., "Evaluation of Alternative Chemical Treatments for Biofouling Control in Electric Power Facilities," in Water Chlorination: Environmental Impact and Health Effects, (Jolley, R. L, et. al, eds.), 2, Ann Arbor Science Publishers, Ann Arbor, MI, 1978.

[Ward et al. 1978] Ward, W., Lee, J., and Freymark, S., "Advantages of Chlorine Dioxide as a Biocide," *Ammonia Plant Safety*, 20, 64 (1978).

[Watkins and Costerton 1987] Watkins, L. and Costerton, J., "Growth and Biocide Resistance of Bacterial Biofilms in Industrial Systems," *Chemical Times and Trends*, 1987.

[Welch and Folinazzo 1958] Welch, J. and Folinazzo, J., "Use of Chlorine Dioxide for Cannery Sanitation and Water Conservation," *Food Technology*, 179 (March 1958).

[Werdehoff and Singer 1987] Werdehoff, K. and Singer, P., "Chlorine Dioxide Effects on THMFP, TOXFP, and the Formation of Inorganic By Products," *Journal of the American Water Works Association*, 107(September 1987).

[White 1986] White, G., Handbook of Chlorination, Van Nostrand Reinhold Company, New York, 1986.

[Whittenburg and Dow 1977] Whittenburg, R. and Dow, C., *Bacteriological Reviews*, 41, 754 (1977).

[Xu et al. 1996] Xu, X., Stewart, P. and Chen, X., "Transport Limitation of Chlorine Disinfection of *Pseudomonas aeruginosa* Entrapped in Alginate Beads," *Biotechnology and Bioengineering*, 49, 93(1996).

[Yu and McFeters 1994] Yu, F. and McFeters, G., "Physiological Responses of Bacteria in Biofilms to Disinfection," *Applied and Environmental Microbiology*, 60(7), 2462(July 1994).

[Yu et al. 1993] Yu, F., Pyle, B., and McFeters, G., "A Direct Viable Count Method for the Enumeration of Attached Bacteria and Assement of Biofilm Disinfection," *Journal of Microbiological Methods*, 17, 167(1993).

Index

Index

www.ingramcontent.com/pod-product-compliance
Lightning Source LLC
LaVergne TN
LVHW062116200926
848484LV00007B/1684
* 9 7 8 0 9 7 7 1 9 8 5 1 1 *